Reduction, Modification and Valorisation of Sludge (REMOVALS)

EU report Series

Reduction, Modification and Valorisation of Sludge (REMOVALS)

Christophe Bengoa, Azael Fabregat,
Josep Font and Frank Stüber

Publishing
London • New York

Published by **IWA Publishing**
Alliance House
12 Caxton Street
London SW1H 0QS, UK
Telephone: +44 (0)20 7654 5500
Fax: +44 (0)20 7654 5555
Email: publications@iwap.co.uk
Web: www.iwapublishing.com

First published 2011

Typeset by MPS Limited, a Macmillan Company, Chennai, India.

British Library Cataloguing in Publication Data
A CIP catalogue record for this book is available from the British Library

Library of Congress Cataloging-in-Publication Data
A catalog record for this book is available from the Library of Congress

ISBN13: 9781843393450
ISBN: 184339345X

Contents

Preface........xi

Chapter 1
Reduction and stabilisation of the excess and toxicity of sludge in anaerobic conditions 1
Summary........1
Project objectives........1
Work performed........1
End results........2
Methodologies and approaches employed........2
Approach employed........2
Experimental methodology........3
Achievements related to the state of the art........6
Fundamentals of thermophilic sludge stabilisation........6
Effect of temperature on thermophilic anaerobic digestion (TAD)........8
Effect of hydraulic retention time on TAD........10
Impact of the pre-treatment on the raw sludge........11
Conclusions........25
Impact of the technology on its industry or research sector........25
Overall view........25
Mesophilic anaerobic digestion........28
Thermophilic anerobic digestion........30
Comparison between the two proposals........31
Thermal pre-treatment........31
Hydrogen peroxide addition........32
Conclusive remarks........34
References........34

Chapter 2
Sludge stabilisation in autothermal conditions 37
Summary 37
Project objectives 37
Work performed 37
End results 38
Methodologies and approaches employed 38
Approach employed 38
Experimental methodology 39
Achievements related to the state of the art 41
Fundamentals of autothermal thermophilic aerobic digestion 41
Baseline effect of HRT on ATAD at 55°C 42
Experimental characterisation of aeration and mixing: simulation studies 44
Optimisation of aeration and mixing 45
Assessment of better aeration and impact of pre-treatments 47
Numerical study on mixing and aeration 47
Concluding remarks 48
Impact of the technology on its industry or research sector 48
Overall view 48
Applicability of ATAD 49
References 51

Chapter 3
Production of hydrogen from sludge 53
Summary 53
Project objectives 53
Work performed 53
Methodologies and approaches employed 54
Fermentation apparatus 54
Evaluation of pre-treatment strategies 55
Stabilisation of hydrogen production 55
Two-phase fermentation of sewage biosolids 55
Optimisation of sparging process 55
Achievements related to the state of the art 56
Current knowledge regarding fermentative hydrogen production 56
Advancement from state of the art 57
Impact of the technology on its industry or research sector 60
References 61

Chapter 4
Utilisation of enzymes from sludge to optimise activated sludge digestion 63
Summary 63
Project objectives 63
Work performed 63
End results 64
Methodologies and approaches employed 64
Extraction of enzymes from activated sludge 64
Achievements related to the state of the art 65
Fundamentals about the extraction of enzymes from activated sludge 65
Definition of enzymatic activity used for each type of enzyme 68
Extraction of enzymes by stirring disintegration: evaluation of the disintegration time effect and the use of additives on the protease and lipase recovered 68
Extraction of enzymes by ultrasonication: evaluation of the disintegration time and the ultrasonic power on the protease and lipase recovered 70

Extraction of enzymes by ultrasonication: extension of the study to the detection of seven hydrolytic enzymes ... 71
Purification of the lipase fraction ... 74
Impact of the technology on its industry or research sector ... 75
Research impacts ... 75
Social and environmental impacts ... 75
Industrial and economical impacts ... 76
References ... 77

Chapter 5
Gasification of sewage sludge ... 79
Summary ... 79
Project objectives ... 79
Work performed ... 79
End results ... 79
Achievements related to the state of the art ... 80
Introduction ... 80
State of the art and research objectives ... 80
Methodologies and approaches employed ... 82
Characterization of research objects – effect of sewage sludge origin on pyrolysis behaviour ... 83
Influence of the pyrolysis extent on gasification process ... 85
Gasification reagent impact on the yield and kinetics ... 87
Kinetics of sewage sludge pyrolysis and gasification ... 87
Impact of the technology on its industry or research sector ... 90
References ... 91

Chapter 6
Production of organic substrate from sludge for enhancement of nutrient removal ... 93
Summary ... 93
Project objectives ... 93
Work performed ... 93
End results ... 94
Methodologies and approaches employed ... 94
Disintegration ... 94
Fermentation ... 95
Sequencing batch reactors ... 95
Achievements related to the state of the art ... 95
Costs of sludge disposal ... 96
Influent COD/N ratio and required N removal ... 97
Extent of filamentous bacteria presence in sludge ... 97
Energy requirements of disintegration method ... 97
Conclusions ... 97
Impact of the technology on its industry or research sector ... 98
References ... 99

Chapter 7
Advanced technology for the biological nitrogen removal of the reject water from the sludge dewatering systems ... 101
Summary ... 101
Project objectives ... 101
Work performed ... 101
End results ... 102
Methodologies and approaches employed ... 102
Partial nitrification pilot plants ... 102
Methodologies ... 103

Achievements related to the state of the art ... 103
Automatic control loops to achieve partial nitrification ... 103
Full and stable partial nitrification process in an activated sludge system ... 105
Full and stable partial nitrification process in a biofilm airlift reactor ... 106
Coupling the partial nitrification with a denitrification process ... 106
Mathematical model for describing the performance of the partial nitrification ... 107
Impact of the technology on its industry or research sector ... 110
Social, environmental and industrial benefits of the intensive treatment of the reject water ... 110
Economical viability of the partial nitrification technologies developed in project ... 111
Proposed alternatives ... 111
Economical evaluation of the alternatives ... 112
Investment costs ... 112
Operational costs ... 113
Economical comparison between alternatives ... 114
References ... 114

Chapter 8
Minimisation of sludge production by utilisation of biological potential in membrane bioreactors ... 117
Summary ... 117
Methodologies and approaches employed ... 117
Plant configuration ... 117
Sampling and analysis ... 118
Achievements related to the state of the art ... 118
Plant performance ... 118
Biomass accumulation ... 119
Sludge characteristics ... 119
Model development ... 122
Sludge production model ... 122
Aeration model for the biological process and fouling prevention ... 123
Sludge handling model ... 124
Sensitivity analysis ... 124
Cost considerations ... 124
Conclusion ... 126
Nomenclature ... 126
References ... 126

Chapter 9
Destruction of organic compounds in sewage sludge suspensions by ultrasound and catalytic wet air oxidation ... 129
Summary ... 129
Project objectives ... 129
Work performed ... 129
End results ... 129
Methodologies and approaches employed ... 130
Methodologies ... 130
Approaches ... 130
Achievements related to the state of the art ... 131
Preliminar ... 131
Influence of power ultrasound as pre-treatment process ... 131
Synthesis and characterization of catalysts ... 133
Operation in a continuous trickle bed reactor ... 138
Impact of the technology on its industry or research sector ... 139
References ... 140

Chapter 10
Production of activated carbon from sludge 143
Summary 143
Project objectives 143
Work performed 143
End results 144
Methodologies and approaches employed 145
Sludge preparation 145
Sludge carbonisation/preparation 145
Sample characterisation 146
Hardened carbon preparation 146
Achievements related to the state of the art 146
Production of high surface area SBAs 146
Development of a one-step production method for the production of SBAs 147
Utilisation of the response surface methodology method to evaluate the optimum steam activation conditions 147
Determination of the influence of the sludge type on the properties of the SBAs produced by physical activation 148
Production of high surface area SBAs by physical activation 148
Production of SBAs with a high EDC uptake 149
Development of post-activation techniques to mitigate the high leaching propensity of SBAs 149
Production of leaching resistant, hardened pellets 149
Appraisal of the effect of seasonal variation 150
Impact of the technology on industry or research sector 150
Research impacts 150
Social and environment impacts 151
Industrial and economic impacts 152
Conclusions 155
Acknowledgements 155
References 155

Chapter 11
Utilisation of activated carbon as catalyst in wet air oxidation 157
Summary 157
Project objectives 157
Work performed 157
End results 158
Methodologies and approaches employed 158
Integrated approach employed 158
Experimental methodology 158
CFD simulation of CWAO in trickle bed reactor 160
Achievements related to the state of the art 161
Introduction 161
Pioneering the use of sludge based carbons as a catalytic material in wet air oxidation of phenolic effluents 161
Catalytic wet air oxidation over sludge based active carbons in stirred batch reactor 161
Catalytic wet air oxidation over sludge based active carbons in continuous trickle bed reactor 164
CFD investigation of pilot scale trickle bed reactor at cold and reacting flow conditions 166
Impact of the technology on its industry or research sector 166
Research impacts 166
Social and environment impacts 166
Industrial and economic impacts 167
Conclusion 169
References 169

Chapter 12
Utilisation of activated carbon as catalyst in AD-OX process 171
Summary 171
Project objectives 171
Work performed 171
End results 171
Methodologies and approaches employed 172
Achievements related to the state of the art 172
Introduction: state of the art on activated carbon use and regeneration 172
Characterization of the sludge based carbonaceous materials and comparison with commercial activated carbons 174
Adsorption isotherms on SBCM, comparison with two commercial AC 175
Autoclave oxidation on SBCM, comparison with two commercial AC 176
AD-OX fixed bed experiments in a mini automated set-up 177
AD-OX fixed bed experiments in a semi industrial pilot reactor 179
AD-OX on real effluents 181
Fenton and other advanced oxidation processes for AC regeneration 181
Conclusion 182
Impact of the technology on its industry or research sector 183
References 184

Chapter 13
Utilisation of activated carbon in adsorption processes 187
Summary 187
Methodologies and approaches employed 187
Achievements related to the state of the art 189
Physico-chemical characterisation 189
Sorption of dyes 189
Sorption of metal ions and metalloid 190
Sorption of aromatic compounds 190
Fixed bed experiments 191
Conclusions and impact of the technology on its industry or research sector 192
References 193

Chapter 14
Utilisation of activated carbon in biofilters for elimination of industrial waste gases 195
Summary 195
Project objectives 195
Work performed 195
End results 195
Methodologies and approaches employed 196
Bioreactor setup 196
Methodologies 196
Approaches employed 196
Achievements related to the state of the art 197
Evaluation of biofilters' performance parameters 197
Evaluation of SBC as filter bed material in a biofilter 201
Model development, calibration and validation 202
Impact of the technology on its industry or research sector 203
Social, environmental and industrial benefits of the use of sludge-based carbon as filter bed in a biofilter 203
Economical benefits of the use of sludge-based carbon as filter bed in a biofilter 204
References 206

Preface

The overall objective of this project is to develop a technological policy that permits the reduction or the modification or the valorisation of excess sludge. This policy should be applicable and implemented by every municipal wastewater treatment plant. The project aims at developing strategies for the safe disposal and reuse of waste sludge. The scope envisages developing several processes for reducing both amount and toxicity of sludge, with simultaneous transformation into green energy vectors such as methane or hydrogen. Also, valuable materials will be obtained from sludge, such as activated carbons, which will serve to gas decontamination, water purification and in advanced oxidation processes, which could open new applications for that waste, thus valorising it.

Activated sludge is the most popular biochemical process for municipal and industrial wastewater treatment. However, a great amount of excess sludge (ES) resulting from activated sludge processes in wastewater plants is becoming a serious issue, especially in industrial countries. In addition, this huge amount is expected to yearly increase entailing a lack of final disposal lands and punishing the accomplishment of strict regulations on environmental protection. New processes with introduction of ES digestion by ozonation and thermophilic microbial digestion have been proved to be useful in reducing ES production in activated sludge by recycling the treated ES into the aeration tank. These methods aim at reducing the ES formation and have the main disadvantage of the potential accumulation of no-biodegradable components in the effluents. Considering the vast amount of ES and its variety, the development of other alternative ES-reuse technologies is still required as the unfavourable economic impact must be minimised.

The ES reduction can be achieved by two different ways. On one hand, modification/optimisation of the operation parameters for the classical sludge stabilisation processes can result in a better management and a reduction of the treated sludge with a higher associated production of methane or even hydrogen. Simultaneously, the toxicity can be reduced to meet the sanitary regulations before sludge disposal. On the other hand, new uses of sludge are foreseen, e.g. the production of activated carbons and the recovery of organic acids or valuable enzymes, which will be subsequently used in other applications. For instance, the activated carbon will be tested for wet oxidation with a trickle bed reactor using the activated carbon as oxidation promoter, biofilters for liquid and gaseous organic contaminant removal where the fixed bed is activated carbon, classical adsorption processes and finally enzymatic removal with a torus reactor. All these treatment modules could be used directly by industries as novel one-step processes or previous specifically targeted treatments before other classical techniques.

Co-ordinator details:
Universitat Rovira i Virgili
Dr. Azael Fabregat, Professor, project co-ordinator.
Dr. Christophe Bengoa, Associated Professor, executive co-ordinator.
Dr. Josep Font, Associated Professor.

Dr. Frank Stüber, Associated Professor.
CREPI, Department of Chemical Engineering, Universitat Rovira i Virgili, Avinguda Països Catalans, 26, Tarragona 43007, Spain.
Tel: +34 977559643, Fax: +34 977559621, E-mail: azael.fabregat@urv.cat.
Tel: +34 977558619, Fax: +34 977559621, E-mail: christophe.bengoa@urv.cat.

Contractor details:

Université de Nantes
Dr. Jack Legrand, Professor.
Laboratoire de Génie des Procédés Environnement Alimentaire (GEPEA), Université de Nantes, C.R.T.T., Boulevard de l'Université, BP 406, 44602 Saint Nazaire Cédex, France.
Tel: +33 (0)240172633, Fax: +33 (0)240172618, E-mail: jack.legrand@univ-nantes.fr.

Universitat Autònoma de Barcelona
Dr. Julián Carrera, Associated Professor.
Departament d'Enginyeria Química, Escola Tècnica Superior d'Enginyeria, Campus UAB, 01893 Bellaterra (Cerdanyola del Vallès), Spain.
Tel: +34 935812141, Fax: +34 935812013, E-mail: Julian.Carrera@uab.es.

University of Glamorgan
Dr. Alan Guwy, Professor.
Sustainable Environment Research Center, University of Glamorgan, Llantwit Road Treforest, CF37 1DL, Pontypridd Wales, United Kingdom.
Tel: +44 1443482239, Fax: +44 1443482285, E-mail: ajguwy@glam.ac.uk.

Institut National Polytechnique de Tolouse
Prof. Henri Delmas, Professor.
Laboratoire de Génie Chimique, UMR 5503, Institut National Polytechnique de Tolouse, B.P. 1301, Rue Paulin Talabot, 5, 31106 Tolouse Cédex 01, France.
Tel: +33 562885888, Fax: +33 562885600, E-mail: Henri.Delmas@ensiacet.fr.

Gestió Ambiental i Abastament, S.A.
Mr. Jaume Cabré Fontseré, Engineer.
EDAR de Reus, Partida Pórpora, finca 52, 43206 Reus, Spain.
Tel: +34 977753700, Fax: +34 977757984, E-mail: JCabre@gaiasa.es.

Tratamientos y Recuperaciones Industriales, S.A.,
Mr. Jaume Bresó i Coch, Engineer.
Tratamientos y Recuperaciones Industriales, S.A., Avinguda Europa, s/n, 43120 Constantí, Spain.
Tel: +34 977296522, Fax: +34 977522387, E-mail: jbresoc@fcc.es.

Institute of Chemical Technology of Praha
Dr. Pavel Jenicek, Professor.
Department of Water Technology and Environmental Engineering, Faculty of Environmental Technology, Institute of Chemical Technology, Technicka 5, 16628 Praha 6, Czech Republic.
Tel: +420 220443155, Fax: +420 224310770, E-mail: pavel.jenicek@vscht.cz.

Technical University of Lodz
Dr. Stanislaw Ledakowicz, Professor.
Faculty of Process and Environmental Engineering, Technical University of Lodz, Zeromskiego 116, 90924 Lodz, Poland.
Tel: +48 426313715, Fax: +48 426365663, E-mail: stanleda@p.lodz.pl.

Technische Universitaet Berlin
Dr. Matthias Kraume, Professor.
Department of Chemical Engineering, Technische Universitaet Berlin, Strasse des 17. Juni 135, 10623 Berlin, Germany.
Tel: +49 3031423701, Fax: +49 3031421134, E-mail: matthias.kraume@tu-berlin.de.

Universidade de Coimbra
Dr. Rosa Mª Quinta Ferreira, Professor.
Department of Chemical Engineering, Faculty of Sciences and Technology, Universidade de Coimbra, Colegio de S. Jeronimo, Largo D. Dinis, 3000-141 Coimbra, Portugal.
Tel: +351 239798723, Fax: +351 239798703, E-mail: rosa.quinta-ferreira@eq.uc.pt.

Institut de Recherche sur la Catalyse
Dr. Michèle Besson, Senior Researcher.
Institut de Recherche sur la Catalyse, Centre National de la Recherche Scientifique (CNRS), 2, Avenue Albert Einstein, B.P. 1335, 69609 Villeurbanne, France.
Tel: +33 (0)472445358, Fax: +33 (0)472445399, E-mail: michele.besson@ircelyon.univ-lyon1.fr.

Imperial College of Science, Technology and Medicine
Dr. Nigel Graham, Professor.
Department of Civil and Environmental Engineering, Imperial College of Science, Technology and Medicine, Exhibition Road, South Kensington Campus, SW7 2AZ London, United Kingdom.
Tel: +44 2075946121, Fax: +44 2075946124, E-mail: n.graham@imperial.ac.uk.

Salsnes Filter AS
Mr. Ivar Solvi, Engineer,
Salsnes Filter AS, Verftsgt.11, PO Box 279, 7801 Namsos, Norway.
Tel: +47 74274860, Fax: +47 74274859, E-mail: ivar.solvi@salsnes-filter.no.

Chemviron Carbon Limited
Mr. Neil Townend, Engineer.
Chemviron Carbon Limited, Edgar House, Locket Road, WN4 8DE Ashton in Makerfield, United Kingdom.
Tel: +44 (0)7740492685, Fax: +44 1942275600, E-mail: ntownend@calgoncarbon-eu.com.

Seen Technologie Ltd.
Mr. Pawel Matusiak, Engineer.
Design and Investment Realisation Office, Seen Technologie Ltd., Toszecka, 102, 44-117 Gliwice, Poland.
Tel: +48 322794613, Fax: +48 322794615, E-mail: pawel.matusiak@seen.pl.

K&H Kinetic A.S.
Dr. Pavel Fialka, Director.
K&H Kinetic A.S., Zlatnicka, 33, 339 01 Klatovy, Czech Republic.
Tel: +420 376356104, Fax: +420 376322771, E-mail: fialka@kh-kinetic.cz.

REMOVALS website: http://www.etseq.urv.es/REMOVALS/

Chapter 1

Reduction and stabilisation of the excess and toxicity of sludge in anaerobic conditions

Esther Torrens Serrahima, Frank Stüber, Azael Fabregat Llangostera, Agustí Fortuny Sanromà, Christophe Bengoa, Josep Font Capafons (Universitat Rovira i Virgili), Jaume Cabrè Fonseré, Iñaki Oriol, Raúl García (Gestió Ambiental i Abastament, S.A.), Jaume Bresó and Eva Rovira (Tratamientos y Recuperaciones Industriales S.A.)

SUMMARY

Project objectives

The study carried out was addressed in a three action line in order to finally improve the performance of thermophilic digestion to get the higher profit from the stabilised sludge before its further disposal. In all the cases, waste sludge was provided by Gestió Ambiental i Abastament, S.A., who also collaborated in the experimental task and mainly in the data interpretation and the economic potential of its implementation at high scale.

Taking as a baseline the more frequent mesophilic conditions, which do not assure sanitary conditions for agricultural use of the stabilised sludge, the main objective is establishing practical and economic operation conditions for thermophilic digestion of excess sludge coupled with sludge pre-treatments to improve its digestability. For such a goal, thermal pre-treatment and peroxidation using hydrogen peroxide were selected as good candidates. It was expected that the pre-treatment contributes to amend the pathogen content of the original sludge and also improve sludge condition for biogas production. The combination of the pre-treatment with the subsequent thermophilic digestion must assure sanitation and, as a secondary objective, improve biogas production and yield.

Work performed

The study was started with the construction and conditioning of a bench-scale sludge stabilisation system (5 L) to reproduce performing conditions of presently operating systems at industrial scale. Three lines were operated in this way, two at thermophilic conditions and one at mesophilic conditions for comparison. The thermophilic anaerobic digestion (TAD) was explored at several hydraulic retention times (HRT) and different temperatures within the range 40–60ºC. Total solids, volatile solids, COD reduction, biogas production and quality, and pathogen's presence were the parameters mainly assessed.

 ISBN: 9781843393450. Published by IWA Publishing, London, UK.

Then, on one side, thermal pre-treatment of the sludge was conducted up to 200°C at autogenous pressure at several contacting time. On the other side, peroxidation of the sludge was conducted by the addition of H_2O_2, within a range of 0–1.5 g H_2O_2/g COD at temperatures in the range 30–60°C for several reaction times, and the resulting sludge accordingly characterised. The pre-treated sludges were subsequently digested at thermophilic conditions in reactors operating batch wise to assess the impact of the pre-treatment conditions on their digestability, following biogas production and quality.

Finally, the coupling of selected pre-treatments with thermophilic digestion in the bench-scale stabilisation system was conducted to prove the feasibility of the approach at large scale and assessing the quality of the stabilised sludge.

End results

Results show that the thermophilic temperatures studied (50°C, 55°C and 60°C) are not yet enough for complete pathogenic destruction at the hydraulic times fixed. Although a high reduction of solids is found, the biogas production is lower than expected. In addition, thermophilic temperature at 60°C gives poor performance compared to the other two temperatures. The most favourable temperature is thus 55°C giving an acceptable biogas production and quality.

Thermal pre-treatment in the range 80–110°C for 1 h is recommended as temperatures beyond do not noticeably improve the solubilisation of organic matter. No significant variation in TS, VS and homogeneous COD is observed except for the most extreme thermal condition treatment at 200°C, where between 19–24% of TS is reduced. For instance, at just 80°C the TDS and the VDS is over four times higher than the initial content and absence of pathogens was found.

Although non detectable organic matter removal (mineralisation) was seen due to the addition of hydrogen peroxide at 30°C at a dose lower than 0.87 g H_2O_2/g COD, there is a solubilisation of the solid matter, which turns out in an increase of soluble COD and TDS. A concentration as low as 0.03 g H_2O_2/g COD is enough to solubilise more than 70% of the initial COD. Doses beyond 0.87 g H_2O_2/g COD drive to mineralisation of organic matter but non-acceptable costs.

Biogas production is better enhanced by thermal pre-treatment followed of TAD, so up to 80% more production can be obtained under controlled conditions. More modest enhancement was given by peroxide pre-treatment. In addition, most of the biogas production occurs during the first 300 h, which must allow cutting the typical HRT applied in direct TAD.

To test these findings at bench-scale, the effect of the thermal pre-treatment at 80°C for 1 h and of the pre-treatment with hydrogen peroxide at a dose of 0.2 g H_2O_2/g COD at 30°C in combination with the anaerobic digestion was followed both at 33°C (mesophilic range), and at thermophilic temperatures of 50°C and 55°C with an HRT of 20 days. Overall, there is an enhancement of homogeneous COD removal using thermal pre-treatment, especially at 55°C reaching 67% conversion. Sludge pre-treated with hydrogen peroxide showed more modest results when compared with thermal pre-treatment and sludge without pre-treatment. The reduction of pathogenic microorganisms is also enhanced by the use of pre-treatments, which is a major impact, which now could allow using of stabilised sludge for agricultural use.

Although a more long following period had been needed (remember that the WP2 was stopped), it is believed that, at least, 80% of digested sludge could be directed to agricultural use, this is, no landfill required. In those conditions, savings are estimated in 1 €/inhab per year, which assures a very short payback period for any re-engineering needed in current WWTP, mainly if cogeneration systems are already in operation.

METHODOLOGIES AND APPROACHES EMPLOYED

Approach employed

To draw an efficient combined treatment to give stabilised sludge simultaneously accomplishing sanitary conditions (Figure 1.1), several bench-scale digestion reactors at mesophilic and thermophilic conditions were constructed and operated to reproduce existing installations. The reactors were operated in a daily semi-continuous (daily draw-fill) mode at different Hydraulic Residence Time (HTR) and tested at one mesophilic temperature, 33°C, and two thermophilic temperatures, between 50°C and 60°C.

Pre-treatments were applied in order to improve the digestability and sanitary conditions. On one hand, thermal pre-treatment at several temperatures up to 200°C was applied over raw sludge and the pre-treated sludge characterised. On the other hand, peroxidation using hydrogen peroxide was performed at doses below 1.5 g H_2O_2/g COD and then characterised.

After investigating the effect of the thermal and peroxidation treatments over raw sludge and how its properties change, the next step was to combine them with the thermophilic anaerobic digestion (TAD), first in batch mode and carefully controlled conditions in order to observe the basic trends over biogas production and other sludge properties.

The last step was finally to study the overall effect of anaerobic digestion of selected pre-treated sludge in the bench-scale reactors, at mesophilic conditions, 33°C, and at thermophilic temperatures, then at 50 and 55°C. This allowed checking the enhancement in the anaerobic digestion. For the best conditions, a cost estimation of the savings was performed.

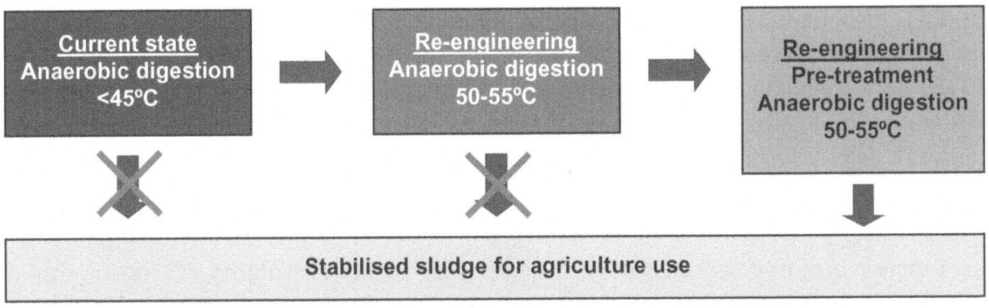

Figure 1.1 Proposed re-engineering of existing anaerobic digestion systems for excess sludge

Experimental methodology

Semi-continuous anaerobic digestion plant

A bench-scale reactor system with 5 litres effective reactors was constructed in order to study the three different operation conditions simultaneously (Figure 1.2). The reactor temperature is maintained with a thermostatic bath at the desired temperature. A water level control was installed in the bath for balancing water loss by evaporation. A magnetic stirrer provides continuous agitation into the reactor. A pump is used for sludge feeding and sampling. Biogas generated in the anaerobic digestion is measured by water displacement and then collected in an 1 litre gas sampling glass cylinder for gas analysis.

Figure 1.2 Bench-scale reactors for the anaerobic digestion of excess sludge

Raw excess sludge for feeding was always provided by the Reus Wastewater Treatment Plant. This was a 65/35 volume mixture of primary and secondary waste sludge, in order to keep the same conditions that Reus WWTP has in the full-scale mesophilic anaerobic reactor presently operating. The characteristics of the sludge used during this study are presented in Table 1.1.

Table 1.1 Main characteristics of raw sludge from Reus WWTP

Parameter	Maximum value	Minimum value	Average value
COD (g/l)	89.5	20	51
TS (% w/w)	7.6	1.6	4.25
VS (% w/w)	4.56	1.27	3.25

As stated, the sludge used for feeding digestion reactors comes directly from operative WWTP. This implies several fluctuations, sometimes strong as a result of seasonal population and climate, in its characteristics. COD easily ranges from 20,000 to 90,000 mg O_2/L being the average around 50,000 mg O_2/L, also solid content and volatile solids give similar fluctuations, an average of 3.3 and 2.5 has been measured respectively. Feed has an average value of pH around 6 and after anaerobic digestion sludge slightly increase being more alkaline, which is in accordance to the optimal pH range for maintaining the different population bacteria (Lay *et al.*, 1997).

For the start-up, 10% innoculum from the full-scale mesophilic (33°C) anaerobic reactor of the Reus WWTP or the full-scale thermophilic anaerobic reactor of the Reus WWTP operating at 43°C was add to the reactors and then left to acclimatise into the corresponding temperatures. One of the reactors was kept at 33°C (mesophilic conditions) and the others set at 50 and 60°C, respectively, although the latter changed to 55°C after a few weeks operating. In these conditions, three different HRT, 15, 20 and 25 days, were examined.

Sludge pre-treatments

For thermal pre-treatments, a closed jacketed glass reactor with an active volume of 500 ml was used to pre-treat 300 ml of raw sludge at temperatures below 80°C. The reactor temperature was maintained with a thermostatic bath at the desired set-point (Figure 1.3). Pre-treatments at higher temperature, up to 200°C, were performed in a 1000 ml autoclave reactor equipped with a magnetically driven stirrer. A stirring rate of 500 rpm was fixed and sludge volumes of 300 ml (Figure 1.4). Pressure was fixed with air. Treatment time was 1 hour once the sludge had reached the set point temperature (20–40 min) at free pH. After treatment the cooling lasted about 20–40 min.

Figure 1.3 Experimental equipment for low thermal pre-treatment over sludge

Figure 1.4 Experimental equipment for high temperatures pretreatment over sludge

For chemical pre-treatment a closed jacketed glass reactor (Figure 1.5) with a total volume of 500 ml was filled with 300 ml of raw sludge. Then, hydrogen peroxide was continuously added at the ratio of 0–1 g H_2O_2/g COD with a peristaltic pump at a constant flow range of 1–1.3 ml/min, without pH adjustment. Temperature reactor was kept constant by passing water at 30°C from a thermostatic bath through the reactor jacket. Time of reaction was set to be 1 hour. Hydrogen peroxide doses were in the range 0.02–1 g H_2O_2/g COD.

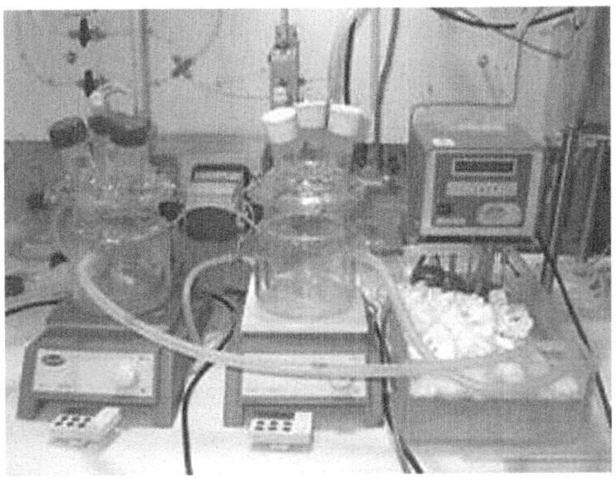

Figure 1.5 Experimental equipment for peroxidation pre-treatment over raw sludge

Batch tests to combine pre-treatments with thermophilic anaerobic digestion

Batch tests were carried out in 120 ml serum flasks with a useful volume of 80 ml and in triplicate. The bottles were incubated in a controlled temperature oven at the thermophilic temperature of 55°C. The flasks were carefully stirred before gas measurement. The start up was conducted by adding the substrate to the innoculum, then the flasks where closed with a septum and an aluminium crimp. Buffer and external nutrients were added in the reactors in order to achieve the optimal conditions for anaerobic microorganisms (Lu *et al.*, 2007). Acetate was used as control model substrate to indicate the quality of the innoculum, measuring the methanogenic activity. The innoculum was extracted from a lab-scale reactor operating at 55°C with an HRT of 20 days. For batch tests, the organic loading rate (OLR) was set at 0.3 g COD_{added}/g $VSS_{innoculum}$. Biogas production was measured from the innoculum without any substrate (endogenous) and from each of the pre-treated substrates, including the raw sludge. To keep anaerobic conditions, both the liquid contained in the flasks and the free headspace were purged with nitrogen. The biogas produced was volumetrically measured. Tests finished when there was no measurable biogas production for a reasonable period. Determination of COD (chemical oxygen demand), TS (total solids), TSS (total soluble solids), VS (volatile solids), and VSS (volatile soluble solids) were carried out as described in Section 'Analytical methods'. The quality of biogas and the volatile fatty acids were analysed by gas chromatography.

Analytical methods

In order to give an accurate picture of the thermophilic digestion efficiency and the quality of the end products, the next parameters were periodically monitored.

Total Solids: (%TS) Total Solids, TS, analysis is determined according to the Standard Method 2540B, 17th Ed (APHA, 1989). TS measures the solid part left in the vessel after sample evaporation.

Volatile Solids (%VS): Volatile Solids, VS, analysis is determined according to Standard Method 2540G, 17th Ed (APHA, 1989). It offers a rough approximation of the amount of organic matter in the solid fraction.

Alkalinity: Alkalinity analysis is determined according to the Standard Method 2320A (APHA, 1989). Alkalinity indicates the acid-neutralizing capacity of the water expressed as milligrams of $CaCO_3$ per litre.

Chemical Oxygen Demand: Chemical Oxygen Demand, COD, analysis is determined according to Standard Method 5220C, 17 Ed (APHA, 1989). COD is used to measure the content of organic matter of sludge.

Total Organic Carbon: Total Organic Carbon, TOC, is measured in a TC Multi Analyser 2100 N/C equipment from Analytic Jena with a non-diffractive IR detector.

Total Carbon: Total Carbon, TC, is measured in a TC Multi Analyser 2100 N/C equipment from Analytic Jena with a non-diffractive IR detector. TC is performed with no sample acidification.

Ammonia and organic nitrogen: Ammonia and organic nitrogen is determined according to UNE-EN 25663:1994 method. Nitrogen data evaluate the treatability of wastewater by biological processes.

Conductivity: Indicates the ability of an aqueous solution to carry an electric current, it depends on the presence of ions and their concentration. It is normally expressed as milisiemens per centimetre.

Total Phosphorous: Total Phosphorous is determined according to the UNE-EN 1189:1996 method.

Biogas analysis: Gas composition is determined by a gas chromatography 6890N Network GC System (Agilent Technologies) using a column packed with porapack Q 50/80 provided by Tecnokroma with a Thermal Conductivity Detector (TCD).

Volatile Fatty Acids: The measurement of organic acids is tested for controlling the anaerobic digestion, since they are intermediates during the anaerobic digestion. They are normally expressed as milligrams of acetic acid per litre.

Qualitative volatile fatty acids measurement is performed by HPLC 1100 (Agilent Technologies) liquid chromatograph with a phase reverse column C18 Hypersil ODS from Agilent at a UV wavelength of 210 nm. Acids measured are formic acid, acetic acid, butyric acid, isovaleric acid and valeric acid.

E-coli: The population of E-coli bacteria was determined in an external laboratory following the method ISO/FDIS 9308-3 (1998)/Cor 1:2000. Water Quality – Enumeration of Escherichia coli in Surface and Waste Water – Part 3: Miniaturized Method (Most Probable Number) by Inoculation in Liquid Medium. International Organization for Standardization, Geneva.

Salmonella: The detection of salmonella in sludge was also determined in an external laboratory following these methods:

1. APHA 1998. Standard methods for the examination of water and wastewater, 20 ed., New York: American Public Health Association.
2. ISO/DIS 19250:2003. Water quality. Detection and enumeration of Salmonella (ISO/TC 147/SC).
3. Goossens H., Wauters G., De Boeck M., Janssens M. and Butzler J. P. (1984). Semisolid selective motility medium for isolation of Salmonella from fecal specimens. *Journal of Clinical Microbiology*, **19**, 940–941.
4. Ewing W. H. and Ball M. M. (1996). The biochemical test of the genus Salmonella. National Center for Disease Control and Prevention, Atlanta, Georgia, USA.

Carbohydrates: Carbohydrates and soluble carbohydrates were measured by the phenol-sulphuric method (Dreywood *et al.*, 1946; Dubois *et al.*, 1956 and Bougrier *et al.*, 2007).

H_2O_2: Hydrogen peroxide concentration was measured in the supernatant following the Standard Iodometric Method 4500-Cl B (APHA, 1989).

Specific references for analysis methods

AWWA, 'Standard methods for the examination of water and wastewater', 17th ed. 1989, Washington: American Public Health Association (APHA), American Water Works Association, Water Pollution Control Federation (WPCF).

Bougrier C., Delgenes J. P. and Carrere H. (2007). Impacts of thermal pre-treatments on the semi-continuous anaerobic digestion of waste activated sludge. *Biochem. Eng. J.*, **34**(1), 20–27.

Dreywood R. (1946). Qualitative Test for Carbohydrate Material. *Industrial and Engineering Chemistry-Analytical Edition*, **18**(8), 499–499.

DuBois M., Gilles K. A., Hamilton J. K., Rebers P. A. and Smith F. (1956). Colorimetric Method for Determination of Sugars and Related Substances. *Analytical Chemistry*, **28**(3), 350–356.

Lu J., Gavala H. N., Skiadas I. V., Mladenovska Z. and Ahring B. K. (2007). Improving anaerobic sewage sludge digestion by implementation of a hyper-thermophilic prehydrolysis step. *J. Environm. Manag.*, **88**(4), 881–889.

ACHIEVEMENTS RELATED TO THE STATE OF THE ART

Fundamentals of thermophilic sludge stabilisation

Anaerobic digestion is the most widely sludge treatment process used in municipal Waste Water Treatment Plants (WWTP) to stabilise primary and secondary (waste activated sludge, WAS) sludge generated in the wastewater treatment (Metcalf, 2003). The most extended anaerobic process is at mesophilic temperatures due to its low energy requirements; however anaerobic digestion process at thermophilic ranges is characterised by accelerated biochemical reactions, upper biogas production and higher growth rate of microorganisms, leading to a possible reduction of the hydraulic retention time or space requirements. Moreover, thermophilic anaerobic digestion could drive to EPAs class A biosolids which are suitable for land application (Watanabe *et al.*, 1997) and, according to literature (Suh and Rousseaux, 2002), the combination of anaerobic digestion and agricultural land application is presently the most environmentally friendly alternative thanks to its lower emissions and lower energy consumption.

Anaerobic digestion is the microbiological oxidation of organic matter without the presence of air. The process involves a coordinated and interdependent set of reactions promoted by different bacteria population, which degrades organic matter into intermediary products in turn converted into methane. The digestion process can be divided in four sequential steps: hydrolysis, acidogenesis, acetogenesis and methanogenesis (Ros *et al.*, 2003).

During the hydrolysis, it takes place the conversion of particulate and soluble organic matter into soluble products such as amino acids, sugars and fatty acids by extracellular enzymes secreted by microorganisms. Hydrolysis can be the rate limiting process for anaerobic digestion, specially for sludge with high solid content, while methanogenesis is considered rate limiting in the fermentation of soluble substances (Pavlostathis *et al.*, 1991 and Puchadja *et al.*, 2006). In the acidogenesis step, acetate, hydrogen and carbon dioxide as well as volatile fatty acids (VFA) and alcohols are produced from the soluble organic matter by acidogenic fermentative bacteria. This step is often the fastest in the anaerobic digestion (Pavlostathis *et al.*, 1991). Acetogen bacteria are organisms that are able to reduce CO_2 to acetate via the acetyl coenzyme A (acetyl-CoA) or Wood-Ljungdahl metabolic pathway (Müller *et al.*, 2003), being responsible for acetogenesis step. Acetogen bacteria are strictly anaerobic bacteria, which can grow by the conversion of hexoses, C2 and C1 compounds, such as HCOOH and CO, into acetate. They can inhabit in a wide range of ecosystems due to their high resistance to extreme conditions such as pH, temperature and salinity; to date nineteen bacteria of this kind have been described (Gottschalk *et al.*, 1986).

During the fermentation, hexoses (equation 1) can exclusively be converted to acetate; this fermentation is referred to as homoacetogenesis (Müller *et al.*, 2003). Also other authors report that are capable of autotrophic growth on H_2/CO_2 (equations 2 and 3). However, homoacetogenesis (equation 4) in digested sludge may be maintained as a result of heterotrophic growth on sugars or other organic compounds (Ryan *et al.*, 2005). Acetogenesis requires very low hydrogen partial pressure to favour the thermodynamics of the reaction. In anaerobic digestion processes, this is achieved by the combination with hydrogenotrophic methanogenesis where hydrogen is consumed and homoacetogenesis where hydrogen is used to reduce carbon dioxide to acetate.

$$C_6H_{12}O_6 \rightarrow 3CH_3COOH \qquad \text{Eq. 1}$$

$$4H_2 + 2CO_2 \rightarrow CH_3 - COOH + 2H_2O \qquad \text{Eq. 2}$$

$$4CO + 2H_2O \rightarrow CH_3 - COOH + 2CO_2 \qquad \text{Eq. 3}$$

$$4CH_3OH + 2CO_2 \rightarrow 3CH_3 - COOH + 2H_2O \qquad \text{Eq. 4}$$

In methanogenesis, methanogens can be found in natural environments rich in organics but free from oxygen. Methanogens are strict anaerobes that obtain energy by converting carbon dioxide (CO_2), hydrogen, formate, methanol, acetate and other organic compounds into methane and carbon dioxide. Hydrogenotrophic methanogens generate methane from CO_2 and H_2 following equation 5, while acetoclastic methanogens produce methane from acetate (equation 6). About two thirds of the methane produced during the digestion directly comes from the degradation of acetate (Prescott *et al.*, 1999). Methanogens have slow growth rates and are usually considered rate limiting for the anaerobic process (Pavlostathis *et al.*, 1991).

$$CO_2 + 4H_2 \rightarrow CH_4 + 2H_2O \qquad \text{Eq. 5}$$

$$CH_3COOH \rightarrow CH_4 + CO_2 \qquad \text{Eq. 6}$$

Anaerobic digestion process is influenced by factors such as temperature, availability of nutrients, pH, alkalinity and toxics. Microorganisms require carbon, nitrogen and phosphorous sources for growth. The amount of carbon has to be in balance with the amount of nitrogen. The optimum C:N ratio for anaerobic digestion is suggested to be in to range of 20:1 to 30:1 (Wheatley *et al.*, 1980). Besides carbon and nitrogen, certain amounts of micronutrients such as sulphur, vitamins and trace of minerals (Fe, Ni, Mg, Ca, Ba, Mo, Se, Co) are needed to stimulate growth (Mata-Álvarez *et al.*, 2000). Also the sludge source influences its characteristics, for example primary sludge is solid organic material composed mainly by carbohydrates (app. 34%), proteins (app. 16%) and lipids (app. 17%), while secondary sludge consist mainly of bacterial cells, having a high protein content (app. 36%) but lower carbohydrates, app. 20% (Miron *et al.*, 2000). WAS is known to be more difficult to digest than primary sludge (Bougrier *et al.*, 2007).

Three main temperature intervals are suitable for anaerobic populations to degrade organic matter. They are termed psychrophilic, mesophilic and thermophilic intervals (see Table 1.2). The temperature is the most important parameter since it influences on the microorganism activity, equilibrium reactions, gas solubility, and the type of microorganisms present in the media. Temperature selects the predominant microorganisms in the reactor and also controls its growth rate.

Table 1.2 Temperature ranges for various anaerobic populations

Bacteria population	Temperature range
Psychrophilic	0–20°C
Mesophilic	15–45°C
Thermophilic	45–75°C

The pH in the digester is important since the microbial groups involved have optimal growth at different pH values. To maintain dynamic equilibrium in the anaerobic system pH between 6.0 and 8.3 are reported to be the best. However methanogenic bacteria are known to be more sensible to pH changes than hydrolytic and acidogenesis ones. pH is also important since certain compounds, e.g. NH_3, H_2S and VFA, can become toxic at specific pH. Thus an important parameter is the buffering capacity, usually measured as the alkalinity.

Thermophilic process enhances biochemical reactions (Zábranská *et al.*, 2000). Rate reaction increases with temperature in a factor of 2 or 3 in comparison to mesophilic conditions (Metcalf and Eddy, 1987) so that less time is needed to reduce the same amount of organic matter (Zábranská *et al.*, 2000) and the loading organic rate can be potentially elevated. Additionally, the same organic matter reduction can be obtained in shorter periods and consequently the reactor volume is smaller, reducing the investment cost (Romero *et al.*, 1990).

Therefore, changes in temperature operation to thermophilic ranges can result in a better exploitation of the existing infrastructures. There are some studies about the different effects in thermophilic anaerobic digestion. De la Rubia *et al.* (2006) report higher removal of volatile solids for thermophilic temperature ranges than mesophilic at the same hydraulic retention time (HRT). In contrast is reported that COD removal shows no significant differences when temperature changes (Garber *et al.*, 1975), however other studies indicate that there are higher COD reduction in thermophilic temperatures. Theoretically, as thermophilic bacteria give higher reaction rates, biogas production should be higher; this is not yet fully confirmed as some authors report lower values (Rimikus *et al.*, 1982) and other higher (Zábranská *et al.*, 2000).

Pathogenic destruction in anaerobic treatments should be clearly more efficient in thermophilic ranges (Wantanabe *et al.*, 1997) founding inactivation of pathogenic bacteria at 55°C and for HRT of 20 days, while at 35°C was not observed complete inactivation. Also, it is reported that E. coli is inactivated at temperatures above 60°C (Middelberg *et al.*, 1995).

Effect of temperature on thermophilic anaerobic digestion (TAD)

As mentioned in the experimental procedure one reactor was always kept into the mesophilic temperature of 33°C and the other two reactors at different thermophilic temperatures, at 50°C and 60°C, the latter then changed to 55°C due to the poor results obtained at 60°C.

Acclimatisation of the bacterial population into these three different thermophilic temperatures has shown to be a critical parameter and large periods can be needed. This is caused by two reasons; firstly the sludge used as innoculum for thermophilic temperatures came from the sludge anaerobic digestion plant of Reus WWTP (Tarragona, Spain) operating at 43°C and then it was needed to reach the new temperature through small temperature change steps followed with rest periods for allowing replacement of bacteria (De la Rubia *et al.*, 2005 and Boušková *et al.*, 2005). For example, Figures 1.6–1.8 show the acclimatisation periods for changing the conditions in the thermophilic reactor at 60°C into 55°C at a TRH of 20 days for the principal parameters in anaerobic process: COD, TS and VS.

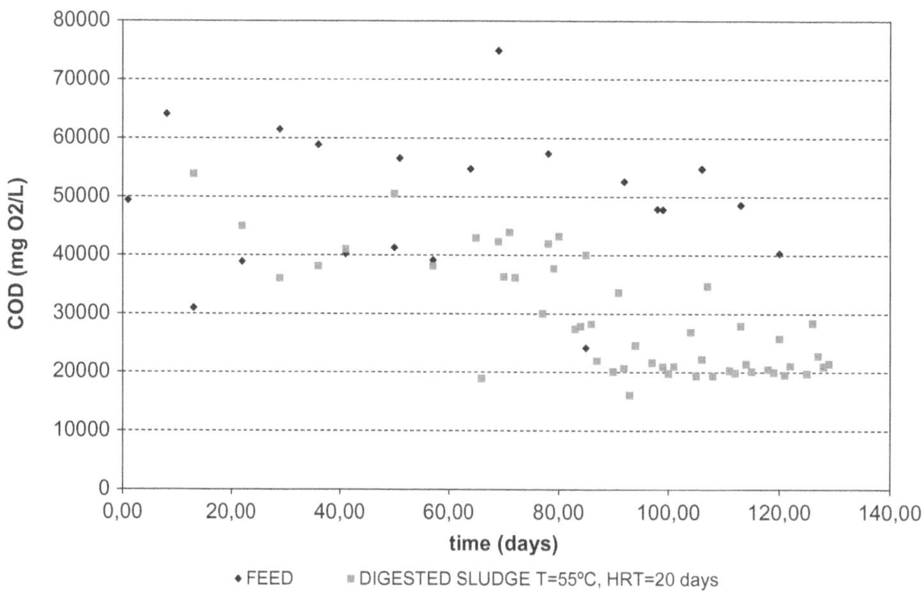

Figure 1.6 Evolution of COD during the acclimatisation period into the new temperature of 55°C at HRT = 20 days

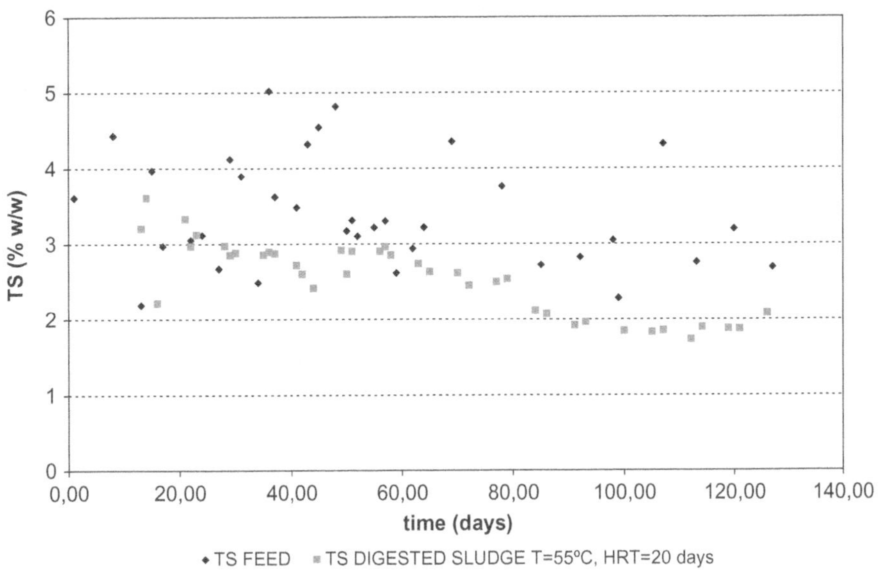

Figure 1.7 Evolution of TS during the acclimatisation period into the new temperature of 55ºC at HRT = 20 days

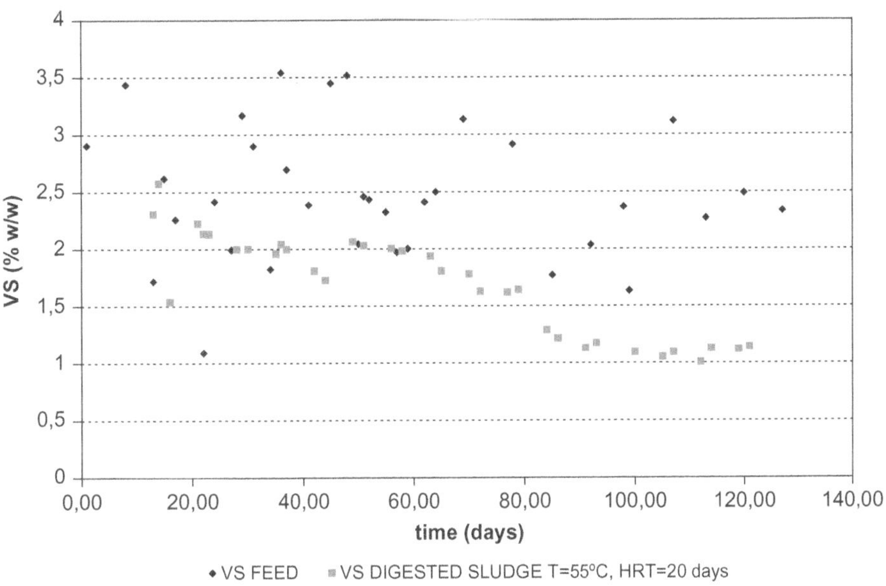

Figure 1.8 Evolution of VS during the acclimatisation period into the new temperature of 55ºC at HRT = 20 days

As seen, the time needed to reach stable conditions in the anaerobic digestion was around 100 days, 5 times the HRT. In contrast, smaller time is needed when changing HRT, as it can be seen, an acclimatisation period of 40 days was needed at 50ºC to change from a HRT of 20 days into 15 days. In turn, mesophilic reactor at 33ºC was innoculed with anaerobic sludge treatment at the same temperature, so no problems were derived. Temperature was carefully controlled as it is crucial for correct digestion, determining the predominant type of microorganisms present in the media and also having influence in rate reaction (Pavlostathis *et al.*, 1991).

The influence of temperature at 33ºC, 50ºC, 55ºC (60ºC) in anaerobic digestion was investigated in order to evaluate the expected improvement of thermophilic digestion over mesophilic digestion and also to check the temperature effect over the control parameters. Table 1.3 collects the values obtained for the different variables followed. The values corresponds to an average of approximately one month of stable operation. The reactors were considered stable approximately between two and three HRT, and when COD, TS and VS values were maintained constant.

Considering the elimination of organic matter present in the sludge, expressed either in terms of homogeneous COD removal or VS removed, it can be seen that there is not a significant improvement at thermophilic temperatures over mesophilic control reactor. However, biogas production is higher at 55ºC. Only some authors have also reported similar values, which is contrary to that expected, thermophilic digestion giving higher biogas production with no significant increase of VS reduction (Zábranská *et al.*, 2000).

Table 1.3 Effect of the temperature in anaerobic digestion at a HRT of 20 days

Parameter	Feed	Digester temperature (°C)			
		33	50	55	60
COD reduction (%)	–	59	58	52	33
$TS_{reduction}$ (%)	–	42	48	41	41
$VS_{reduction}$ (%)	–	57	59	54	49
pH	6.22	7.30	7.82	7.76	7.15
Alkalinity (mg/L)	1636	3475	3925	6083	4297
E. Coli (NMP/g)	601000	362	116	250	320
Salmonella (NMP/g)	Presence	Presence	Sometimes	Sometimes	Sometimes
$COD_{soluble}$ (mg/L)	3591	3496	6769	15876	10682
$BOD_{5\ soluble}$ (mg/L)	747	423	1387	4629	5667
$TOC_{soluble}$ (mg/L)	1714	435	1026	4442	4880
$TC_{soluble}$ (mg/L)	3410	1973	3035	5091	6703
Total $P_{soluble}$ (mg/L)	104	48	42	184	129
Ammoniacal and organic $N_{soluble}$ (mg/L)	234	1042	1377	1631	1649
Conductivity (μS/cm 20°C)	2755	6820	8171	8803	13860
Volumetric ratio CH_4/CO_2	–	3.13	2.77	3.81	2.57
Biogas production (L/kg VS)	–	771	215#	1021	39

#*This value is erroneous. A gas leak affected the measure.*

In view of the pH value, it was observed that, in general, thermophilic digestion is slightly more alkali than that mesophilic, except for 60°C. The larger acidity and the lower biogas production at this higher temperature indicate that methanogenic bacteria seem to be inhibited. Accordingly, the digestion tank undergoes an accumulation of VFA, because of the incapability of being transformed into methane and, consequently, the expected decrease in pH.

Regarding to sludge analysis of the supernatant liquid, it can be stated that the higher the temperature of the digestion, the larger the soluble organics in the media mesuared as $COD_{soluble}$ and $TOC_{soluble}$. It seems that hydrolytic bacteria are more active at high temperatures, but the organic-consumer microorganisms (methanogenic) activity is not accordingly increased. Conductivity also increases with temperature, probably due to release of ionic components and the organic acids produced during the degradation of the organic matter, hydrolysis and acidogenesis steps.

TAD only surpasses MAD, in terms of biogas production per kg of VS eliminated, at the thermophilic temperature of 55°C. 60°C seems to be too high temperature for methanogenic bacteria, and at 50°C, probably the population of thermophiles is not large enough. Therefore, it is confirmed that thermophilic temperature of 55°C is more suitable for thermophilic conditions and leads to higher biogas production and quality (methane content) than mesophilic conditions.

In despite of the high pathogen destruction, i.e. *E-coli* and Salmonela, which achieves a reduction of up to 3 $\log_{10}$ (three orders of magnitude), the removal is not yet enough to accomplish sanitary conditions for use to agricultural land.

Effect of hydraulic retention time on TAD

Tables 1.4 and 1.5 summarise the results from mesophilic temperature of 33°C and thermophilic temperature of 50°C, respectively at different HRTs. In this range of HRT (15, 20 and 25 days), COD removal did not show significant differences. Apparently, 20 days is the more suitable HRT as for the other two HRTs tested even a slightly decrease in TS and VS reduction can be noticed.

Biogas production increased with the decrease of HRT since more organic loading rate is fed into the reactor and a larger amount of VS is removed (De la Rubia *et al.*, 2006), however at HRT of 20 days there is a discrepancy with respect to this trend, which is related to fluctuations in the raw sludge characteristics, although also a gas leak in the system must be reported, which unfortunately was not detected. In contrast quality of the biogas in terms of methane content is higher at the highest HRT.

Pathogen destruction was expected to be better at thermophilic conditions as HRT increases since bacteria are subjected at high temperature during more time and less amount of microorganisms are daily fed, however values obtained at the HRT of 25 do not show this tendency, suggesting that some remaining bacteria are more reluctant to heating than expected and survive to larger periods.

Although some literature seems to suggest that thermophilic range should be enough to the full hygienisation of the sludge (Carrington *et al.*, 1982 and Wagner *et al.*, 2008), it has been found that some especially reluctant bacteria withstands the typical thermophilic temperature of 55°C. It is well known that pathogens are inactivated during exposure to heat, which to be effective and relatively rapid must be above their optimum growth temperature. The period of exposure is dependent on the temperature and on the species of the organism (Feacham *et al.*, 1983). However, depending on the bacteria population, the kind of pathogen and even operating conditions, sometimes they need more drastic conditions to be effectively destroyed. Therefore, heating a higher temperature is needed or exposure to oxidation agents with the objective to eliminate these remaining bacteria, which also helps to improve the digestability of the raw sludge as degradation of organic matter also achieved improves the hydrolysis step.

Table 1.4 Effect of the HRT in anaerobic digestion at the mesophilic temperature of 33°C

Parameter	Hydraulic retention time (day)		
	15	20	25
$COD_{reduction}$ (%)	58	59	55
$TS_{reduction}$ (%)	33	42	37
$VS_{reduction}$ (%)	44	57	53
pH	7.44	7.30	7.52
Alkalinity (mg/L)	4128	3475	3948
E. Coli (NMP/g)	631	362	1800
Salmonella (NMP/g)	Presence	Sometimes	Sometimes
$COD_{soluble}$ (mg/L)	1999	3496	4112
$BOD_{5\ soluble}$ (mg/L)	315	423	375
$TOC_{soluble}$ (mg/L)	583	435	361
$TC_{soluble}$ (mg/L)	1453	1973	1036
Total $P_{soluble}$ (mg/L)	68	48	32
Ammoniacal and organic $N_{soluble}$ (mg/L)	1026	1042	899
Conductivity (μS/cm 20°C)	6958	6820	6141
Volumetric relation CH_4/CO_2	3.26	3.13	4.77
Biogas production (L/kg VS)	2159	771	1750

Table 1.5 Effect of the HRT in anaerobic digestion at termophilic temperature of 50°C

Parameter	Hydraulic retention time		
	15 days	20 days	25 days
$COD_{reduction}$ (%)	59	58	56
$TS_{reduction}$ (%)	33	48	35
$VS_{reduction}$ (%)	44	59	52
pH	7.92	7.82	7.97
Alkalinity (mg/L)	5164	3925	5234
E. Coli (NMP/g)	281	116	741
Salmonella (NMP/g)	Sometimes	Presence	Sometimes
$COD_{soluble}$ (mg/L)	5269	6769	4241
$BOD_{5\ soluble}$ (mg/L)	1260	1387	1446
$TOC_{soluble}$ (mg/L)	1354	1026	1122
$TC_{soluble}$ (mg/L)	2257	3035	1986
Total $Phosphorous_{soluble}$ (mg/L)	59	42	33
Ammonical and organic $N_{2\ soluble}$ (mg/L)	1349	1377	1371
Conductivity (μS/cm 20°C)	7881	8171	8228
Volumetric relation CH_4/CO_2	3.13	2.77	4.90
Biogas production (L/kg VS)	1598	215#	1185

 # This value is erroneous. A gas leak affected the measure.

Impact of the pre-treatment on the raw sludge

As previously described, anaerobic digestion is the microbiological oxidation of organic matter without the presence of air. The process involves a coordinated and interdependent set of reactions promoted by different bacteria population, which degrades organic matter into intermediary products in turn converted into methane. The digestion process can be divided in four steps: hydrolysis, acidogenesis, acetogenesis and methanogenesis (Ros and Zupancic, 2003 and Gavala *et al.*, 2003). Hydrolysis can be the rate limiting process for anaerobic digestion, specially for sludge with high solid content, while methanogenesis is considered rate limiting in the fermentation of soluble substances (Ros and Zupancic, 2003; Gavala *et al.*, 2003; Puchajda and Oleszkiewicz, 2006; Wang *et al.*, 1997 and Pavlostathis and Giraldo-Gomez, 1991). A number of pre-treatment processes: thermal, chemical, biological and mechanical pre-treatments and their combination (Bougrier *et al.*, 2007; Lu *et al.*, 2008; Chen *et al.*, 2007; Elliott and Mahmood, 2007; Wong *et al.*, 2006; Cacho-Rivero and

Suidan, 2006a,b; Vlyssides and Karlis, 2004 and Valo *et al.*, 2004) have been explored in order to improve hydrolysis, by solubilisation of sludge solids and disintegration or lysis of the cell walls releasing the intracellular content and letting them more accessible for the anaerobe microorganisms (Kim *et al.*, 2003 and Muller, 2001).

Thermal pre-treatment has been mostly studied in WAS with COD content about 20 g O_2/L at low temperature, from 60°C to 120°C, and it was found that hydrolysis was promoted by thermal pre-treatment in anaerobic digestion, resulting in an increase of 52.1% in methane generation and 26.6% of organic matter destruction using WAS pre-treated at 60°C and HRT in mesophilic conditions of 8 days (Wang *et al.*, 1997). However at higher temperatures, problems of toxicity were found in WAS thermal pre-treatment. Stuckey and McCarty (1984) studied a temperature range within 150–270°C and observed an increase of methane production as increasing pre-treatment temperature up to an optimum at 175°C whereas at higher temperature sludge biodegradability declined, which was attributed again to the formation of toxic compounds. Also, Valo *et al.* (2004) found that the degree of WAS solubilisation was mostly influenced by the temperature selection rather than time, at temperature ranges of 130–170°C. Only 15 min was enough for most of the solubilisation during thermal hydrolysis being the optimum at 170°C, where COD solubilisation reached 83% at pH 12. Common treatment time values were in the range 30–60 min.

Most studies focused on the investigation of the pre-treatment duration and temperature but only using one type of sludge, in addition with low solids concentration, while in most of the WWTP primary and secondary sludge are generated and then they are combined before anaerobic digestion. Gavala *et al.* (2003) found that pre-treatment step at 70°C of primary sludge followed by thermophilic anaerobic digestion was more efficient in terms of methane yield and methane production rate than mesophilic processes, whereas pre-treated WAS has only a positive effect on methane production potential when mesophilic digestion followed. They concluded that the selection of the pre-treatment duration as well as anaerobic digestion temperature should depend on the ratio of primary and secondary sludge.

The addition of chemicals as a pre-treatment has also been reported. It was found that sludge hydrolysis is improved by pre-treatments in the following order: alkaline > acidic > neutral and blank test (Chen *et al.*, 2007). Most of the investigations have been focused in alkali addition and their combination with other sludge disintegration techniques, such as thermal, to improve hydrolysis (Vlyssides and Karlis, 2004; Valo *et al.*, 2004; Kim *et al.*, 2003 and Tanaka and Kamiyama, 2002). Wang *et al.* (1997) studied chemical addition combined with microwave of WAS within 60–120°C temperature ranges and 0–3 wt% of H_2O_2 addition. They obtained a solubilisation of 100% of COD at 80°C and 3 wt%. Cacho Rivero *et al.* (2006a,b), applied several H_2O_2 doses at 90°C and 65°C as a co-treatment with mesophilic anaerobic digestion; they found that the major effect of the thermo-oxidative treatment was the solubilisation of the particulate organic matter as increase of soluble COD was observed, although no increase of organic matter removal was seen.

Despite the evaluation of several pre-treatments, there are a few studies that combine them with thermophilic digestion. Ferrer *et al.* (2007) studied thermal pre-treatment at 70°C for 9h and found a positive effect on gas production, a 50% of increase of biogas production, however at higher temperatures it was not found any effect. Also Bougrier *et al.* (2007) pre-treated sludge and found an increase in methane production of 25% at 190°C. Sanitary conditions are not reported, neither combination between peroxidation combined with anaerobic digestion.

The hypothesis contrasted is the application of thermal and peroxidation pre-treatment over a real sludge from the municipal Reus WWTP, which is a mixture of primary and secondary sludge, where the former predominates, giving high COD content sludge, over 50 g O_2/L, followed by thermophilic anaerobic digestion. As thermophilic conditions have demonstrated that are not enough to guarantee the sanitation conditions imposed in a close future for agricultural land use of the stabilised sludge. Therefore, a combination of thermophilic digestion using a thermally or chemically pre-treated sludge seems to be a viable option for those WWTP with mesophilic digestion implemented as they only would require a minimum re-engineering (Appels *et al.*, 2008).

Effect of temperature on thermal pre-treatment

After some preliminary tests for evaluating the effect of the pre-treatment time, it was decided to establish the duration of the thermal pre-treatment at 1 h at the desired temperature, as further increase of time, especially at temperatures lower than 110°C, does not improve solubilisation of organic matter. The following Figures 1.9 to 1.14 graphically show the evolution of several sludge characteristics when the different thermal pre-treatment was applied over the same initial raw sludge.

No significant variation in TS, VS and homogeneous COD was observed except for the most extreme thermal condition at 200°C, where between 19 and 24% of TS is reduced. What is more noticeable is the progressive increase of the dissolved solids as the temperature raises. For instance, at just 80°C the TDS and the VDS is over four times higher than the initial content. Of course, the increase is more impressive, for instance, at 170°C, reaching seven folds the original values. Measures of COD and TOC correspondingly correlate with the above results. The increase in soluble matter strongly suggests that digestability should have been improved.

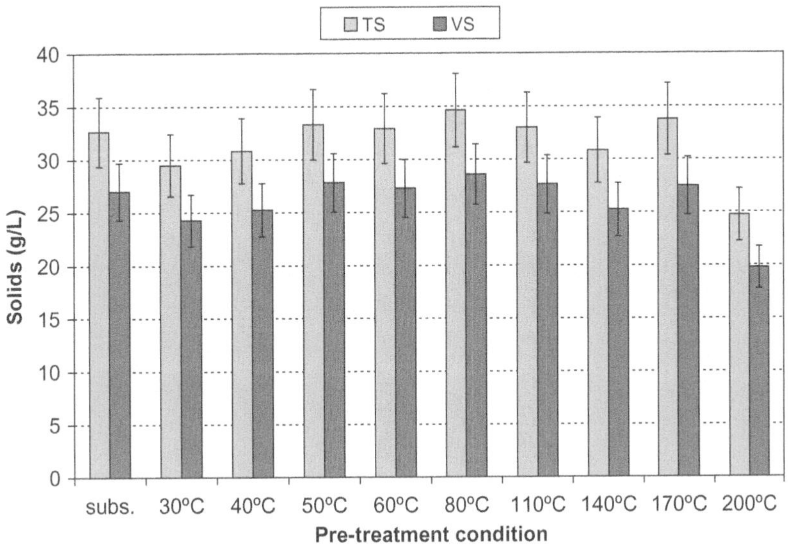

Figure 1.9 TS and VS at different thermal pre-treatment temperatures for 1 h

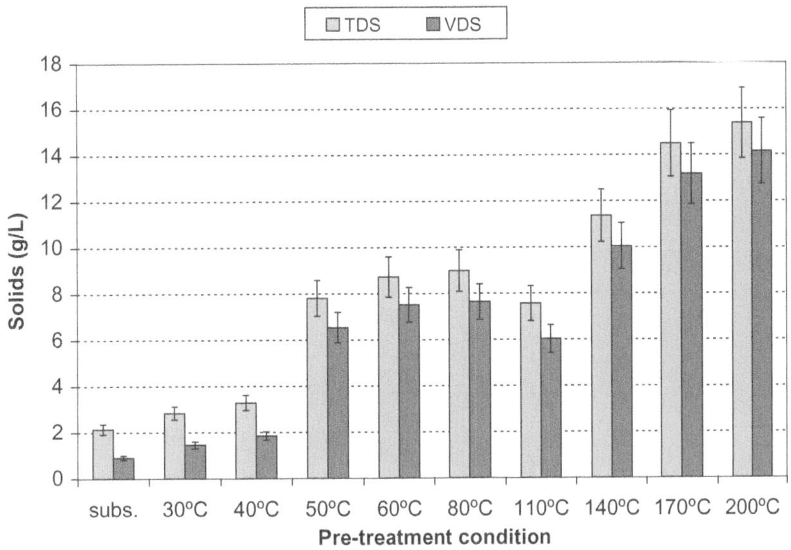

Figure 1.10 TDS and VDS at different thermal pre-treatment temperatures for 1 h

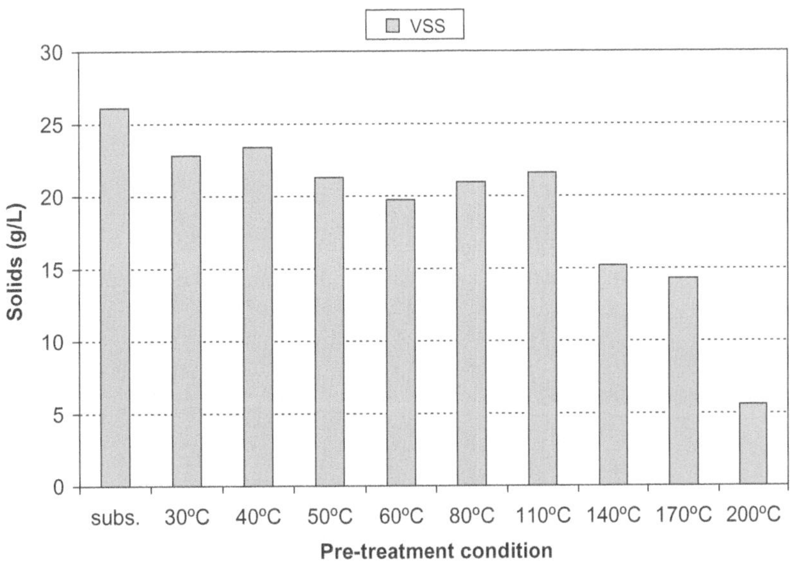

Figure 1.11 VSS at different thermal pre-treatment temperatures for 1 h

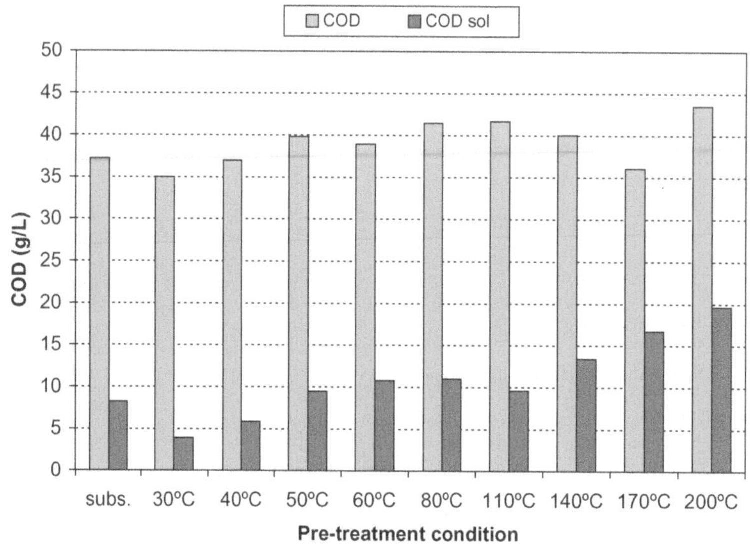

Figure 1.12 COD and soluble COD at different thermal pre-treatment temperatures for 1 h

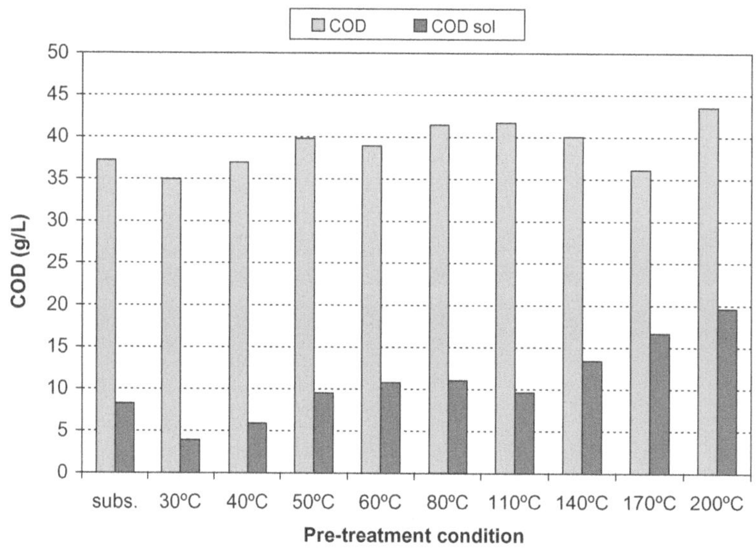

Figure 1.13 Soluble TC and soluble TOC at different thermal pre-treatment temperatures for 1 h

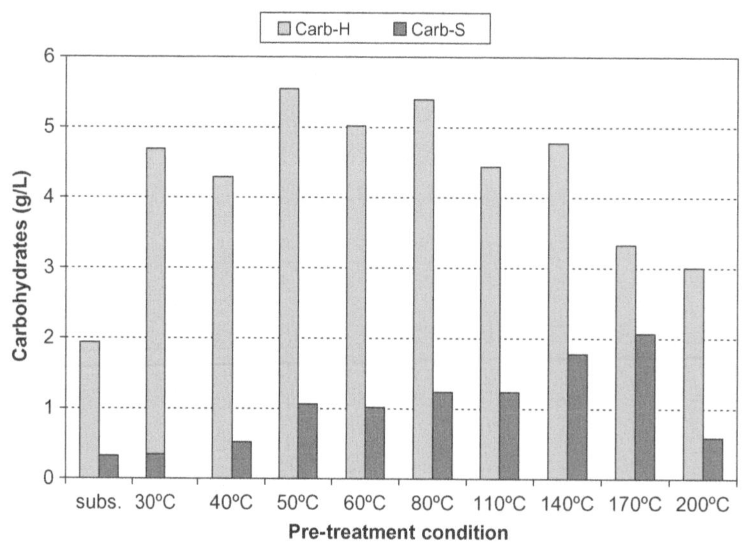

Figure 1.14 Homogeneous and soluble carbohydrates at different thermal pre-treatment temperatures for 1 h

Effect of H_2O_2 dose on peroxidation pre-treatment

Chemical pre-treatment was performed by the addition of H_2O_2, within a range of 0–1.5 g H_2O_2/g COD at a constant temperature of 30 or 60°C. Tables 1.6 and 1.7 summarise the conditions studied.

Table 1.6 Experimental conditions of H_2O_2 pre-treatment over sludge at 30°C

$[H_2O_2]/[COD]$ (w/w)	$[H_2O_2]_0$ (g/L)	% volume added	Addition time (min)	$[H_2O_2]_f$ (g/L)
0.00	0.00	0.00	0	n.d.
0.03	2.30	0.80	2	n.d.
0.07	5.80	2.00	5	n.d.
0.14	11.4	3.90	11	n.d.
0.29	21.9	7.90	19	n.d.
0.58	40.8	15.7	43	n.d.
0.87	57.3	23.6	65	n.d.
1.16	71.8	31.5	74	1.4
1.45	84.7	39.3	93	3.1

Table 1.7 Experimental conditions of H_2O_2 pre-treatment over sludge at 60°C

$[H_2O_2]/[COD]$ (w/w)	$[H_2O_2]_0$ (g/L)	% volume added	Addition time (min)	$[H_2O_2]_{30\,min}$ (g/L)	$[H_2O_2]_{18\,h}$ (g/L)
0.00	0.00	0.00	0.00	–	–
0.03	1.30	0.44	1.21	n.d.	–
0.07	3.30	1.10	2.60	1.7	–
0.11	8.40	2.27	6.77	–	–
0.22	16.3	5.74	13.6	4.73	–
0.55	24.3	8.81	24.2	18.7	1.73
0.69	29.8	11.0	30.3	32.3	3.74
0.83	35.0	13.2	31.2	31.5	2.72
1.11	44.9	17.6	41.6	38.5	5.09
1.38	54.1	22.0	52.0	51.9	7.65

In the further discussion, the nomenclature – I indicates the initial value of the (raw) sludge property before the addition of the hydrogen peroxide but taking into account the dilution effect. Similarly, – C indicates the value of the parameter in the control reactor (at the desired temperature for each concentration of H_2O_2), again just due to the dilution effect caused by the addition of the hydrogen peroxide volume. And finally – H_2O_2 is the value of the parameter measured in the reactor where hydrogen peroxide was added.

As illustrated in Figure 1.15, there is no evident effect over COD content for the different hydrogen peroxide doses tested at 30°C. However, at the latter two concentrations, hydrogen peroxide was still present in the solution at the end of the tests (Table 1.6), which can disturb the measurement of COD (Kang *et al.*, 1999 and Talinli and Anderson, 1992; U.S. Peroxide), overestimating the calculated value.

COD results from sludge pre-treated with hydrogen peroxide at 60°C will not be presented. As Table 1.7 indicates, residual H_2O_2 was found at the end of the tests in almost all the cases. Therefore, due to its interference in the COD analysis, the results are not confident.

For a better discussion of these parameters, Figures 1.16–1.19 show TS and VS expressed as total mass. Figures 1.16 and 1.17 present the results at 30°C, where it is observed the evident decrease in TS and VS at the higher hydrogen peroxide doses. For a ratio of hydrogen peroxide added of 1.16 and 1.45 g H_2O_2/g COD, 22.5 and 14% of TS elimination and 35.5 and 31.5% of VS reduction are respectively obtained. In Figures 1.18 and 1.19, the results of TS and VS at 60°C are drawn, which depict the elimination of solid and organic matter also collected in Table 1.8.

As it can be seen, there are not significant differences in terms of COD, TS and VS between the initial and control sludge, which agrees with the results obtained in the previous section where was concluded that the thermal pre-treatment at 30 and 60°C was not enough to eliminate organic matter. However it seems that the addition of hydrogen peroxide enhances the TS and VS removal, especially at 60°C at for the higher H_2O_2 doses, where it seems that the temperature increase improves the activity of the hydrogen peroxide for matter elimination.

Although non remarkable organic matter removal was observed after the addition of hydrogen peroxide at concentrations below 0.87 g H_2O_2/g COD at 30°C, it seems to be a solubilisation of the solid fraction, which results in an increase of the soluble COD and TDS, as Figure 1.20 suggest. A ratio of 0.03 g H_2O_2/g COD is already enough to solubilise over 70% of the initial COD. No considerable changes were observed in terms of VDS.

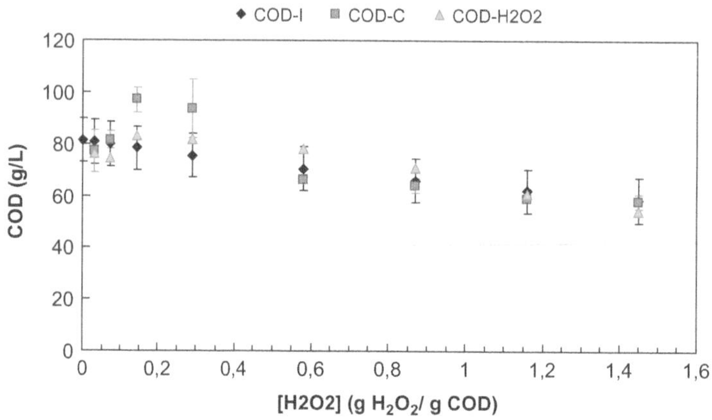

Figure 1.15 Final COD at different H_2O_2 doses at 30°C and 30 min

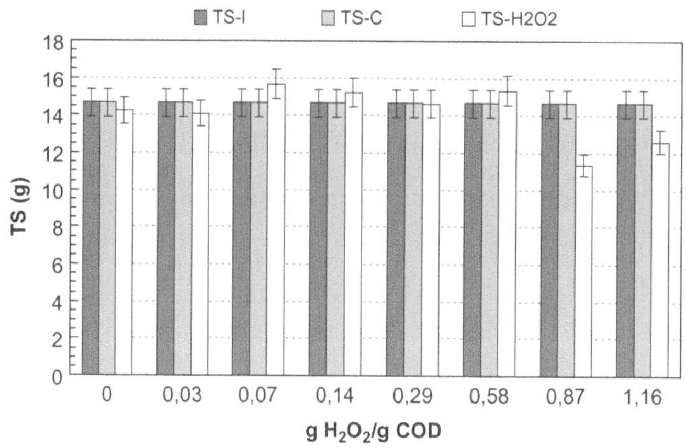

Figure 1.16 TS mass at different H_2O_2 doses at 30°C and 30 min

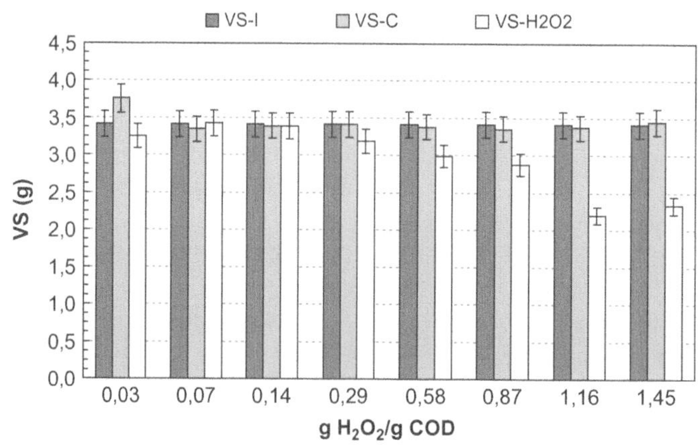

Figure 1.17 VS mass at different H_2O_2 doses at 30°C and 30 min

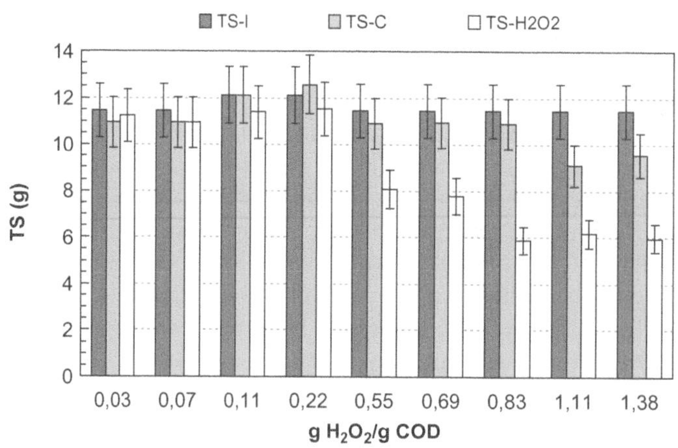

Figure 1.18 TS mass at different H_2O_2 doses at 60°C and 30 min

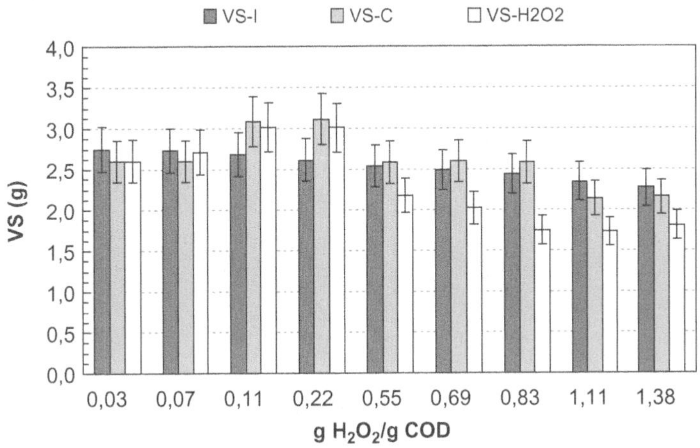

Figure 1.19 VS mass at different H_2O_2 doses at 60°C and 30 min

Table 1.8 Elimination of TS and VS by the addition of H_2O_2 at 60°C

H_2O_2/COD (w/w)	TS removal (%)	VS removal (%)
0.55	26.0	14.3
0.69	28.8	18.7
0.83	46.1	28.3
1.11	32.0	26.4
1.38	37.5	20.0

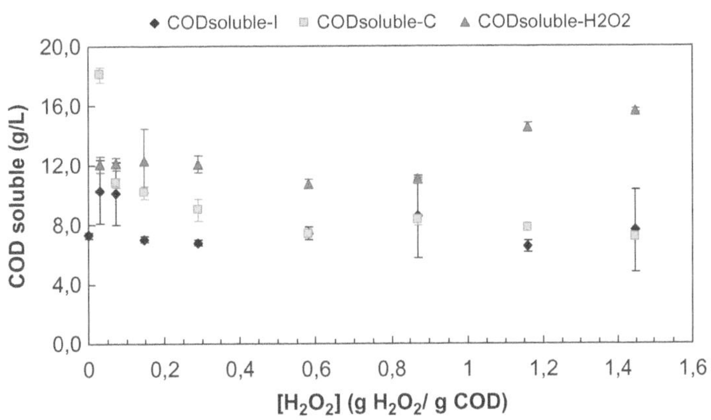

Figure 1.20 Soluble COD at different H_2O_2 doses at 30°C and 30 min

Similarly, in Figure 1.21 and Figure 1.22, where solids in terms of mass are presented, it is more clearly seen the increase of the dissolved material.

In contrast, experiments at 60°C, presented in Figure 1.23 and 1.24, demonstrate that, at doses below 0.22 g H_2O_2/g COD, the TDS are higher in the hydrogen peroxide reactor, while, at ratios above 0.55 g H_2O_2/g COD, the TDS are lower than in the control reactor.

This indicates that, at 60°C, there is significant solubilisation of the solid matter compared to the initial value, as above concluded, but at ratios beyond 0.55 g H_2O_2/g COD, elimination of suspended material also occurs, mainly mineral matter, so that no clear changes in VDS are observed.

Figures 1.25 and 1.26 present the organic matter of the sludge expressed as VSS after the peroxide pre-treatment at 30 and 60°C, respectively. As observed in these figures, at 30°C only at the two higher doses applied, 1.16 and 1.45 g H_2O_2/g COD, there is a reduction of VSS, 41 and 34.5% respectively. In contrast, at 60°C, it is already observed elimination of VSS at doses higher than 0.55 g H_2O_2/g COD, as observed in Table 1.9 that shows that the higher the hydrogen peroxide dose, the larger the reduction of VSS, except for the highest peroxide dose tested.

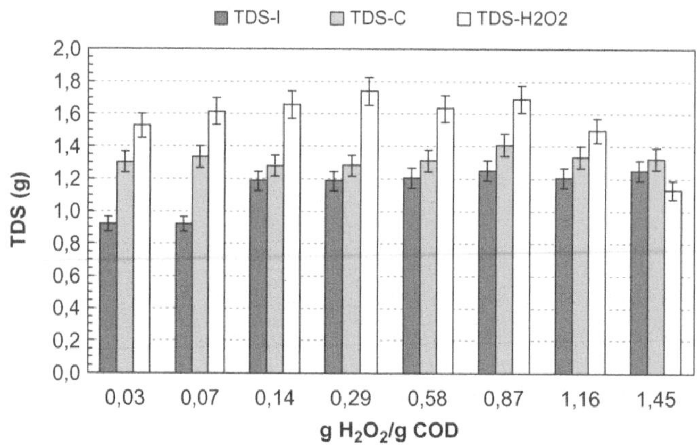

Figure 1.21 TDS mass at different H_2O_2 doses at 30°C and 30 min

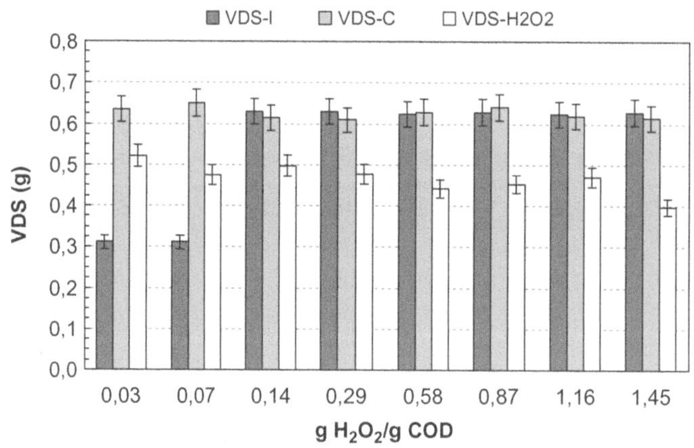

Figure 1.22 VDS mass at different H_2O_2 doses at 30°C and 30 min

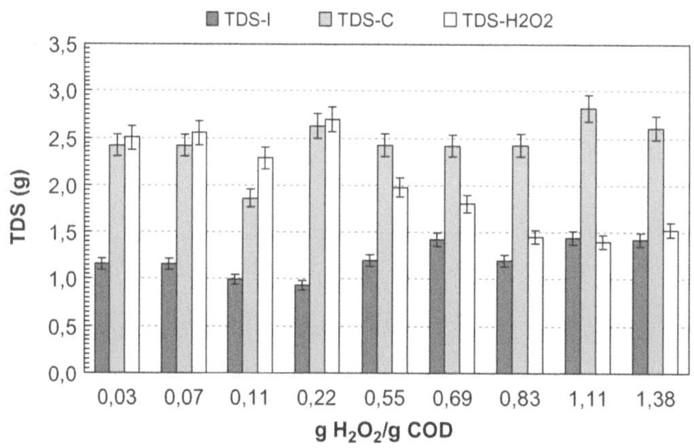

Figure 1.23 TDS mass at different H_2O_2 doses at 60°C and 30 min

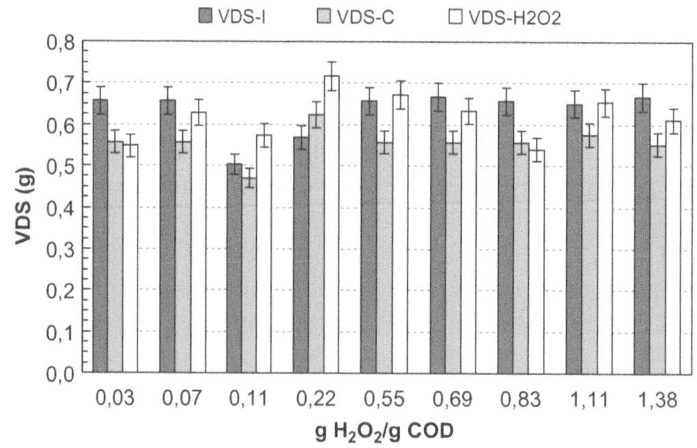

Figure 1.24 VDS mass at different H_2O_2 doses at 60°C and 30 min

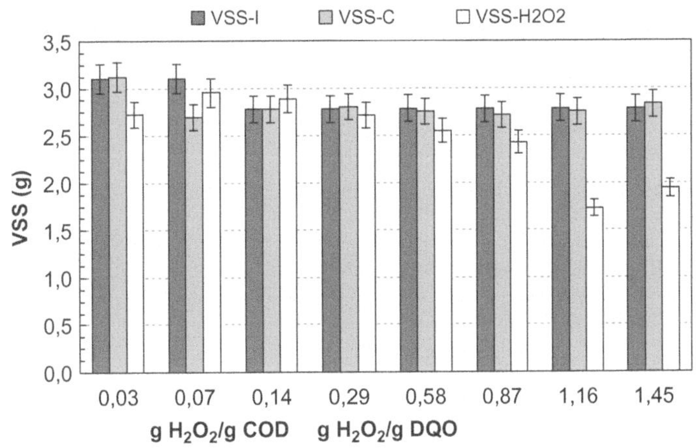

Figure 1.25 VSS mass at different H_2O_2 doses at 30°C and 30 min

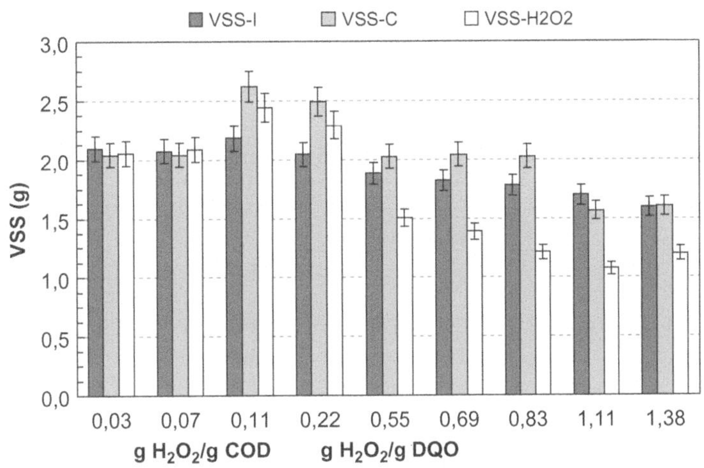

Figure 1.26 VSS mass at different H_2O_2 doses at 60°C and 30 min

Table 1.9 VSS elimination after chemical peroxide pre-treatment at 60°C and 30 min

H_2O_2/COD (w/w)	VSS removal (%)
0.55	20.1
0.69	23.7
0.83	32.1
1.11	37.7
1.38	24.9

It was also inspected the importance of the time exposed to H_2O_2, once this was added into the sludge. The reason is that, as previously commented, residual hydrogen peroxide was found at short times of reaction, especially at 60°C. This study was done at 0.55, 0.69, 0.83, 1.11 and 1.38 g H_2O_2/g COD, analysing the samples 18 h after the addition of the H_2O_2 maintaining the reaction at the temperature selected.

There is no significant variation on the TS against time, neither in the control reactor nor in the hydrogen peroxide reactor, irrespective of the dose. However, VS content decreases along the time. Around 20% of the VS is additionally eliminated when the exposure to the hydrogen peroxide is extended from 30 min to 18 h. Regarding to the total dissolved solids, TDS, at reaction times of 18 h, the dissolved solids considerable increase, but only part of them are volatile, specially at the dose of 0.83 g H_2O_2/g COD, leading to a 240% raise in terms of TDS if compared with the results obtained at 30 min.

Combination of pre-treatment and thermophilic digestion in batch tests

Batch tests of thermophilic digestion at 55°C were conducted over the set of pre-treated sludge in order to assess the impact over biogas production and quality and the other referenced variables. The characteristics of the sludge used in the experiments are shown in Tables 1.10 and 1.11 for thermal and hydrogen peroxide pre-treatment respectively. Tables 1.12 and 1.13 summarise biogas production parameters taken at the end of the test for the thermophilic anaerobic digestion at the different thermal and peroxidation pre-treatments, respectively.

Table 1.10 Initial conditions of the sludge in thermal pre-treatments

Sludge type	COD_{homog} (g/L)	COD_{sol} (g/L)	$OLR_{measured}$ (g_{COD}/g_{VSS})	TS (g/L)	VS (g/L)	TSS (g/L)	VSS (g/L)
Innoculum	37.3	9.8	–	22.1	14.4	18.5	11.4
Raw sludge	68.5	7.6	0.24	42.0	35.3	38.0	33.4
30°C	78.7	11.5	0.28	48.7	38.8	43.8	36.0
40°C	76.0	11.6	0.27	46.1	36.4	40.8	33.3
50°C	67.2	12.8	0.24	39.4	33.2	32.9	28.3
60°C	63.0	13.7	0.22	38.5	31.5	31.1	25.5
80°C	69.2	14.4	0.24	42.0	33.2	33.2	26.2
110°C	55.7	n.a.	0.20	31.4	29.0	23.5	23.1
140°C	58.3	14.5	0.20	37.5	34.0	26.8	25.2
170°C	59.9	14.4	0.21	34.9	27.2	19.1	13.8

Table 1.11 Initial conditions of the sludge in hydrogen peroxide pre-treatments

Sludge type	COD_{homog} (g/L)	COD_{sol} (g/L)	$OLR_{measured}$ (g_{COD}/g_{VSS})	TS (g/L)	VS (g/L)	TSS (g/L)	VSS (g/L)
Innoculum	41.3	8.9	–	24.8	15.2	21.0	12.5
Raw sludge	62.4	5.4	0.30	34.3	27.3	30.0	25.1
P-0.02	65.4	10.7	0.31	33.0	25.9	26.8	21.9
P-0.05	62.6	11.2	0.30	33.6	26.5	27.5	22.3
P-0.1	59.1	12.8	0.28	32.7	25.8	25.9	20.9
P-0.2	60.2	12.4	0.29	31.7	24.6	25.0	19.7
P-0.4	57.4	11.2	0.28	32.8	26.0	25.9	20.8
P-0.6	69.7	2.5	0.33	32.7	26.2	26.0	21.2
P-0.8	54.0	3.0	0.26	32.6	26.2	25.9	21.1
P-1.0	75.8	3.1	0.36	29.6	23.5	23.0	18.4

Table 1.12 Characteristics of biogas in thermophilic anaerobic digestion over thermally pre-treated sludge#

Parameter	Endogen.	Acet.	Raw slud.	P-30	P-40	P-50	P-60	P-80	P-110	P-140	P-170
Total production of biogas (ml)	178	190	214	234	243	252	249	254	251	257	264
Concentration of methane (%)	71.7	71.5	66.2	71.4	70.7	62.9	70.3	69.7	71.5	72.4	72.2
Total methane production (ml)	127	136	142	167	172	159	175	177	179	186	191
Biogas yield (l/g) (BP. $COD_{reactor}$)	0.33	0.39	0.45	0.49	0.51	0.53	0.52	0.53	0.52	0.54	0.55
Biogas yield (l/g) (BP. $VS_{reactor}$)	0.13	0.16*	0.18	0.19	0.20	0.21	0.20	0.21	0.20	0.21	0.21
Substrate biogas production (ml)	–	12.4	36.7	56.0	65.3	74.7	71.0	76.4	73.2	79.2	86.4
Substrate methane production (ml)	–	8.44	14.6	39.4	44.2	31.2	47.4	49.6	52.0	58.5	63.2
Biogas yield (l/g) (BP. VS_{added})	–	0.11*	0.35	0.48	0.60	0.75	0.75	0.77	0.84	0.78	1.06
Methane yield (l/g) (BP. VS_{added})	–	0.07*	0.14	0.34	0.41	0.31	0.50	0.50	0.60	0.57	0.77
COD-h remaining (g/l)	7.7	5.5	7.6	9.5	10.2	10.9	10.7	10.4	6.0	5.9	11.0
VS-h remaining (l/g)	6.6	7.5	7.7	7.7	7.6	7.4	7.4	7.5	7.3	7.2	7.8

#All measurements of biogas were performed at P = 1atm, T = 17°C.
**1.42g COD/g VSS.*

Although the OLR of all experiments was expected to be the same, measurements of homogeneous COD over pre-treated sludge at the initial time of the tests differs from the value measured to prepare the test. As reported before, homogeneous COD measurements have a significant error entailed due to the non homogeneity of the sample and hydrogen peroxide interferes COD analysis. This fact derives some bias to the results, although the general trends are as stated.

Table 1.13 Characteristics of biogas in thermophilic anaerobic digestion over H_2O_2 pre-treated sludge#

Parameter	Endogen.	Acet.	Raw slud.	P-0.02	P-0.05	P-0.1	P-0.2	P-0.4	P-0.6	P-0.8	P-1.0
Total production of biogas (ml)	231	277	266	267	257	296	294	277	282	289	269
Concentration of methane (%)	65.9	72.7	57.6	65.0	64.7	65.8	63.6	64.4	60.3	64.3	64.9
Total methane production (ml)	152	201	154	174	166	194	187	178	170	186	175
Biogas yield (l/g) (BP. $COD_{reactor}$)	0.48	0.62	0.60	0.60	0.58	0.66	0.66	0.62	0.63	0.65	0.60
Biogas yield (l/g) (BP. $VS_{reactor}$)	0.18	0.25*	0.22	0.22	0.22	0.25	0.25	0.23	0.24	0.24	0.22
Substrate biogas production (ml)	–	46.0	35.3	36.2	25.5	64.6	63.3	45.7	51.1	58.3	38.4
Substrate methane production (ml)	–	49.0	1.17	21.4	13.6	42.0	34.7	25.9	17.7	33.7	22.5
Biogas yield (l/g) (BP. VS_{added})	–	0.35*	0.43	0.47	0.32	0.83	0.86	0.59	0.65	0.74	0.54
Methane yield (l/g) (BP. VS_{added})	–	0.37*	0.01	0.28	0.17	0.54	0.47	0.33	0.22	0.43	0.32
COD-h remaining (g/l)	13.9	24.5	13.9	11.3	16.5	16.1	29.2	16.9	19.6	23.0	3.9
VS-h remaining (l/g)	6.6	6.5	7.3	7.0	7.2	7.1	7.1	7.1	7.5	7.2	6.7

#All measurements of biogas were performed at P = 1atm, T = 17ºC.
**1.42g COD/g VSS.*

Thermal pre-treatment enhances biogas production in all reactors tested, as Figure 1.27 shows. All the pre-treated sludge leads to higher biogas generation than raw sludge. It must be noted that endogenous biogas production represents around 70% of the total biogas production reported, whilst the rest, 30%, owns to the substrate, i.e., 'pre-treated raw sludge'. Higher biogas production was observed as the temperature of the pre-treatment rises. Temperatures higher than 80ºC doubles the biogas production in comparison to the biogas formed by substrate without pre-treatment (raw sludge) if the endogeneous production is not considered (17.1% considering it). Also, biogas yield increases as the pre-treatment temperature does, up to a maximum of 2.08 L biogas/g VS_{added} (0.95 L biogas/g COD_{added}) at 170ºC.

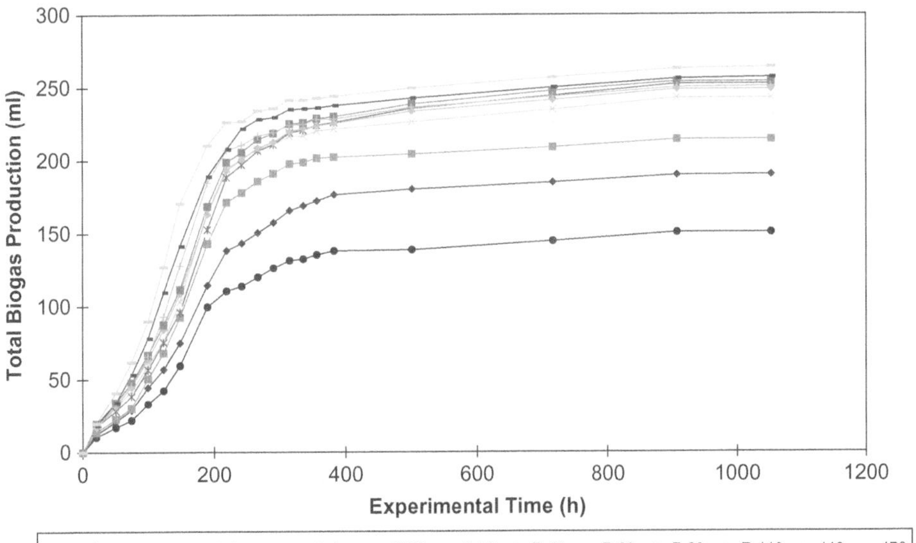

Figure 1.27 Total biogas production of thermally pre-treated sludge in thermophilic digestion batch tests

In tests concerning chemical pre-treatment (see Table 1.11 for initial sludge characteristics) endogenous biogas production represents 80% of the total biogas produced by raw sludge. Biogas production is also improved using sludge pre-treated with hydrogen peroxide as Figure 1.28 illustrates. Biogas production is increasingly enhanced by the dose of hydrogen peroxide added up to 0.1 to 0.2 g H_2O_2/g COD, with an increase around 70% when compared to non-treated raw sludge if the endogeneous production is not considered (10.5% considering it). However, higher doses do not provide significant further improvement even a considerable decrease of biogas production, after peaking, until only 26% additional biogas. Similarly, biogas yield increases with at 0.2 g H_2O_2/g COD dose up to 1.283 L/g VS_{added}, which is two fold that given by raw sludge, but lower than that obtained after thermal treatment. Addition of further hydrogen peroxide again does not provide further biogas yield enhancement.

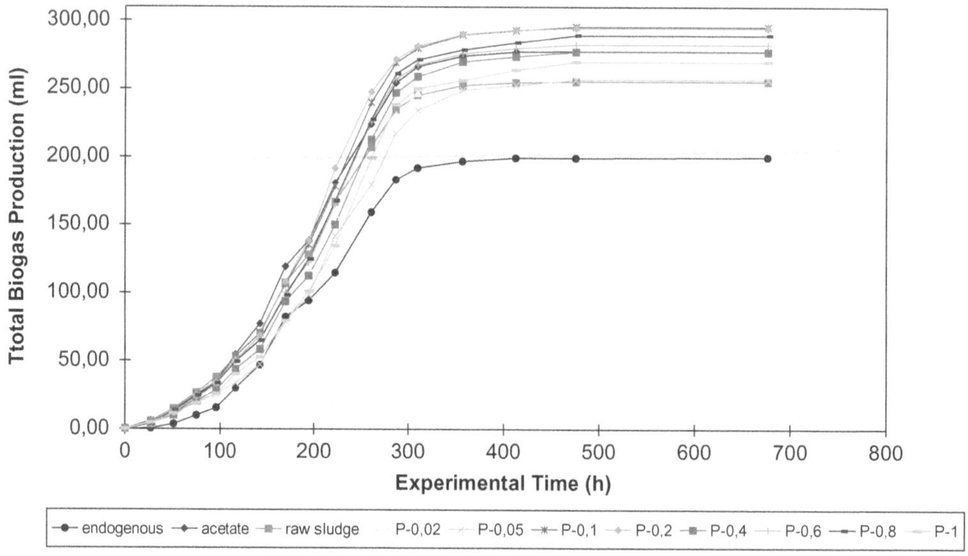

Figure 1.28 Total biogas production of sludge pre-treated with hydrogen peroxide in thermophilic digestion batch tests

Most of the biogas production in both thermal and chemical pre-treatments occurs in the first 300 h of the test, as the specific biogas production rate clearly indicates (Figures 1.29 and 1.30), however thermal pre-treatment beyond 110°C leads to much higher specific biogas rates than the rest of conditions tested. This fact could allow to almost half the actual HRT applied for thermophilic digestion without any pre-treatment.

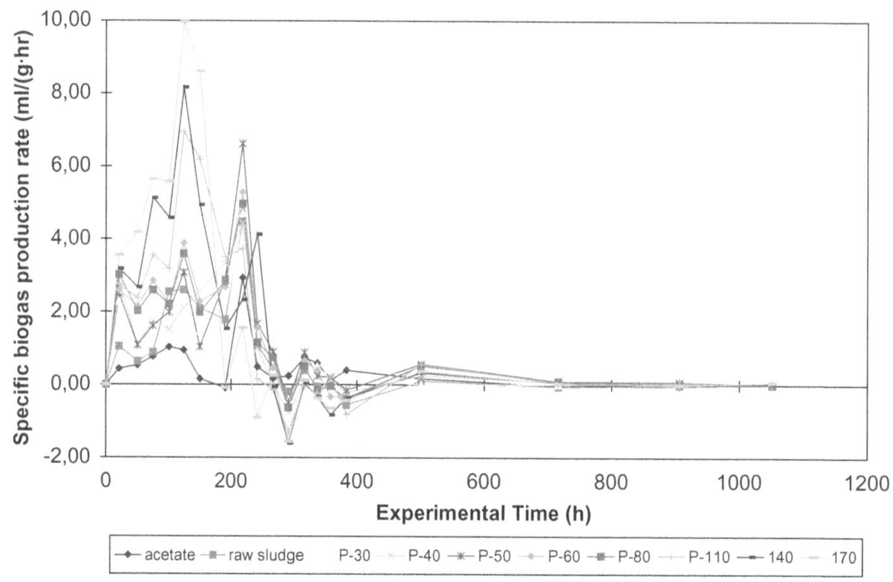

Figure 1.29 Specific biogas production rate of thermally pre-treated sludge in thermophilic digestion batch tests

The quality of the biogas expressed in different terms can be found in Tables 1.12 and 1.13 for thermal and chemical pre-treatment, respectively. It can be seen that both thermal or peroxidation pre-treatment enhance not only the biogas production but also the methane fraction obtained during the subsequent thermophilic anaerobic digestion. For instance, it must be noted that the methane fraction is up to 6% richer. When dealing with peroxidated sludge, the methane fraction still is somewhat better 7%. In any case, the impact of the pre-treatments on the biogas quality is rather marginal in comparison to the net biogas production improvement.

Combination of pre-treatment and thermophilic digestion in bench-scale reactors

In the next section, the results obtained for the anaerobic digestion in the semi-continuous bench-scale reactor are shown. It was studied the effect of the thermal pre-treatment at 80°C for 1 h and also the effect of the pre-treatment with hydrogen peroxide at a dose of 0.2 g H_2O_2/g COD at 30°C in combination with the anaerobic digestion at the three different temperatures tested, 33°C (mesophilic range), and the two thermophilic temperatures 50°C and 55°C with an hydraulic retention time of 20 days. Figure 1.31 depicts the evolution of the homogeneous COD of the sludge at the conditions studied.

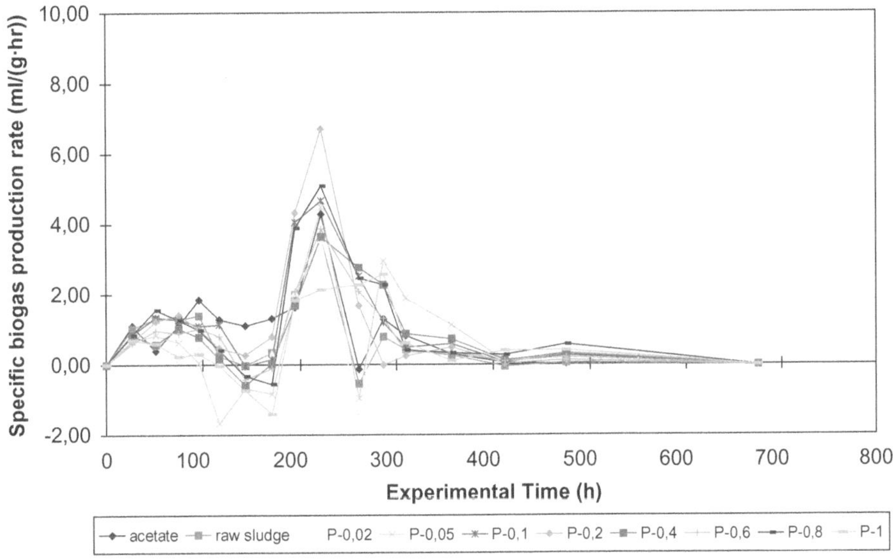

Figure 1.30 Specific biogas production rate of sludge pre-treated with hydrogen peroxide in thermophilic digestion tests

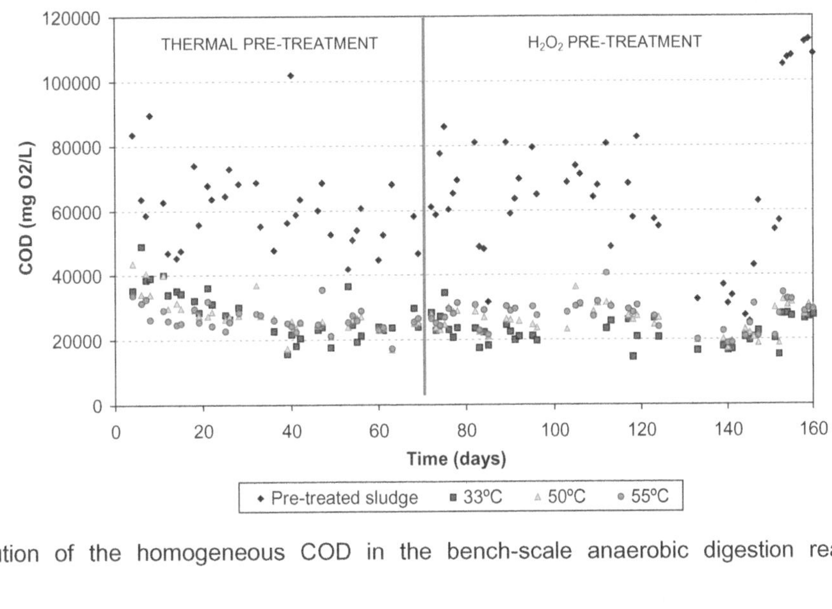

Figure 1.31 Evolution of the homogeneous COD in the bench-scale anaerobic digestion reactors after feeding pre-treated sludge

As it can be seen concerning to thermal pre-treatment, 40 days were necessary to acclimate the digestors at the new feed conditions, from raw sludge (without pre-treatment) into thermal pre-treated sludge. It can be noted a significant decrease of COD in the exited sludge after feeding thermally pre-treated sludge because of the change in the composition of the dissolved fraction of the sludge, as then it had more organic matter solubilised than the sludge without pre-treatment. After the further change from thermal to peroxide pre-treated sludge, the reactors achieved more easily the new steady operation because there was no drastic change in the characteristics of the influent in this case.

Figure 1.32 shows the evolution of total solids and volatile solids over the different reactors during the same period, which follow similar tendencies than COD removal, although they are more sensible towards fluctuations of the influent characteristics.

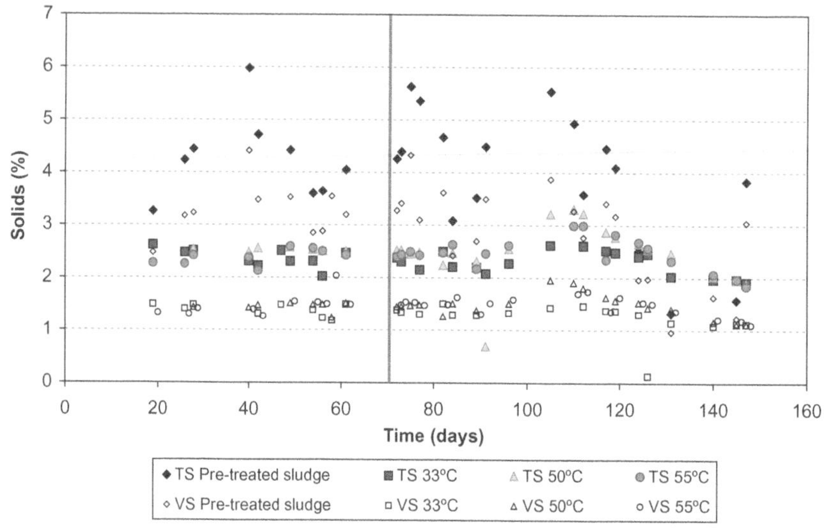

Figure 1.32 Evolution of the TS and VS in the semi-continuous anaerobic digestion of pre-treated sludge

Table 1.14 summarises average results of the main parameters in the anaerobic digestion for the three types of feed compared: sludge without pre-treatment, sludge thermally pre-treated and sludge pre-treated with hydrogen peroxide. It must be pointed out that a significant increase in homogeneous COD removal was found for TAD at 55°C using thermally pre-treated sludge, which increased up to 67% (52% with non-treated sludge), even the influent containing about twice soluble COD. This indicates a much higher destruction of organic matter as a result of the pre-treatment. In turn, neither TS nor VS removal, measured as conversion, are significantly modified by the use of thermally pre-treated sludge, while the biogas quality was found to be somewhat lower as the CH_4/CO_2 volumetric ratio goes from 2.81 with non-treated sludge to 2.61 using thermally pre-treated sludge. Unfortunately, the biogas production was not measured by failure of the apparatus and was not possible reproduce the test because the decision of the European Commission of stopping this work package. However, given the higher COD removal, some additional biogas production should be expected. For peroxidised sludge, a meagrer increase of COD conversion was obtained at 55°C, which passes from 52 to 59%. On the other hand, significantly lower values of TS and VS removal were observed. This probably is related to the decrease of both TS and VS resulting from the pre-treatment with hydrogen peroxide, which leaves less amenable solid matter for the subsequent TAD. In general terms, the improvement during the subsequent TAD is less apparent when hydrogen peroxide was used than for thermal pre-treatment, except for the pathogen reduction below commented.

Table 1.14 Data for the anaerobic digestion in semi-continuous reactors for the three types of feed studied: sludge without pre-treatment, and sludge thermally or chemically pre-treated

Parameter	Sludge without pre-treatment				Sludge thermally pre-treated$^{\#}$				Sludge pre-treated with hydrogen peroxide$^{\#\#}$			
	Influ-ent	33°C	50°C	55°C	Influ-ent	33°C	50°C	55°C	Influ-ent	33°C	50°C	55°C
COD removal (%)	–	59	58	52	–	61	59	67	–	65	60	59
TS removal (%)	–	42	48	41	–	48	43	45	–	26	26	24
VS removal (%)	–	57	59	54	–	60	57	55	–	47	46	44
pH	6.2	7.3	7.8	7.7	6.1	7.3	7.6	7.8	6.2	7.5	7.7	7.7
COD soluble (mg/L)	3591	3496	6769	15876	7862	2822	3684	4257	6731	732	3137	3812
TOC soluble (mg/L)	1714	435	1026	4442	2387	881	718	1124	2457	247	739	1074
Vol. ratio CH_4/CO_2	–	2.13	2.77	2.81	–	2.28	2.55	2.61	–	2.25	2.79	2.47
Biogas (L/kg VS)	–	771	215	1021	–	n.a.	n.a.	n.a.	–	1171	1188	1099
Salmonella (NMP/g)	P	P	P	S	–	P	S	S	–	S	S	0
E.Coli (NMP/g)	$6{\cdot}10^5$	362	116	250	–	n.a.	n.a.	n.a.	–	n.a.	n.a.	n.a.

$^{\#}$*At 80°C for 1 h.*
$^{\#\#}$*0.2 g H_2O_2/g COD at 30°C.*

The reduction of pathogenic microorganisms was always enhanced by the use of pre-treatments, which obviously is a major impact. However, only for one case (hydrogen peroxide pre-treatment and TAD at 55°C) pathogens were never found in the systematic analyses. In any case, a considerable additional pathogen's removal was observed in the other cases, for which acceptable effluent was obtained with a frequency nearly 80% of cases in contrast with the more or less 50% given when non pre-treated sludge was used. As the pre-treated influents were found free of detectable pathogens, it is believed that the contamination comes from the microorganisms already present in the digestion tank before moving to pre-treated influents. Probably, a simple sterilisation and re-starting of the digestors will serve to accomplish with pathogen content requirements in 100% of cases. Unfortunately, the work package stopping decision taken by the European Commission prevented this already planned action from being carried out.

Conclusions

Bench-scale anaerobic digestion reactors can confidently reproduce operation at large scale. Therefore, a rapid (at a biological scale) assessment of the impact due to designed re-engineering for existing sludge digestion plants can be carried out for failure prevention.

Thermophilic anaerobic digestion, at 55°C for HRT 15–20 days is the most favourable conditions in terms of COD, TS and VS removal, biogas production and methane content. However, it is not enough to assure reduction of pathogens to safe levels at any time. Therefore, pre-treatments are needed to achieve the objective of pathogen reduction.

As cost-effective pre-treatments, heating and peroxidation were inspected in a wide range of conditions. Thermal pre-treatment at temperature below 110°C allows solubilising solids, without removal of organic matter.

Thermal pre-treatments highly enhance biogas production in the subsequent TAD (almost two fold at temperature beyond 80°C in controlled tests if endogeneous production is not considered).

Peroxide pre-treatment is much more aggressive, giving significant removal of TS (up to nearly 40%) and VS (up to nearly 30%) at moderate to high peroxide doses. Only doses below 0.2 g H_2O_2/g COD are recommended as VS are increased without loss of TS, which would have a detrimental effect on biogas production.

Peroxide pre-treatments also enhance biogas production in the subsequent TAD (70% increase at a dose of 0.2 g H_2O_2/g COD if endogeneous production is not considered). Doses higher than 0.2 g H_2O_2/g COD do not provide additional benefits.

No pathogens (at least in detectable amounts) were found after pre-treatments beyond 80°C and for any hydrogen peroxide dose tested.

In bench-scale tests, thermal pre-treatment gives an increased COD removal, 67%, in front of non pre-treated sludge, 52%, maintaining biogas production and quality. Peroxide pre-treatment gives less impressing improvements, which make it less interesting.

In bench-scale tests, both thermal and peroxide pre-treatment give significant higher frequency of stabilised sludge accomplishing sanitary conditions (around 80%). Although, sanitised sludge was not found in 100% cases, it is believed to be due to the pathogens already present in the digestion tank before starting to feed with pre-treated sludge, which recommends digestion tanks to be previously sterilised.

A cost effective (assessed in the next section) and feasible re-engineering option for the existing mesophilic (or thermophilic at low temperature) sludge digestion units is the migration to thermophilic conditions at 55°C coupled with a unit to pre-treat the raw sludge at temperature ranging 110°C (conservative option) for one hour. This should assure enough sanitation conditions for subsequent use of the stabilised sludge for agriculture land.

IMPACT OF THE TECHNOLOGY ON ITS INDUSTRY OR RESEARCH SECTOR

Overall view

One of the tools being developed for environmental, social and economical evaluation is the Life Cycle Assessment (LCA), which main objective is to promote sustainable development related to a process/product/service. There are some authors that have evaluated LCA over wastewater treatment plants including the sludge management and have developed indicators that combine environmental, social and economic sustainability in order to evaluate different wastewater treatment technologies and also for sludge handling (Muga *et al.*, 2008; Balkema *et al.*, 2002 and Palme *et al.*, 2005).

Some environmental indicators are energy use (resource utilisation), and performance of the technology (pathogens, COD, N, P). Economic indicators include capital, operational and management and user costs. Social indicators are cultural acceptance of the technology through public participation and measurement of whether there is an improvement in the community from the specific technology through increased job opportunities, better education or local environment. This study (Muga *et al.*, 2008) demonstrated that there are different degrees of sustainability in the way a particular treatment technology is selected and then operated. It is very difficult to find the 'best overall option' as it is not easy to find the technology that satisfies a degree of social acceptance like odourless, visual impact, and employment in the community, space, etc, minimizing costs and simultaneously being the best environmental option. However, in most of these studies there is one step that is subjective to the author and that implies weighting some of the parameters under study, which depending on the scientist are decisive to the final conclusion of sustainability assessment. Nevertheless, there are several indicators in this type of analysis that are decisive in the evaluation, such as methane recovery, and composting but with previous hygienisation requirements.

Concerning sludge studies, there are several authors that include not only sludge management but the overall WWTP process and evaluate how to improve the operational of a municipal wastewater treatment plant using this type of studies LCA as a decision tool for environmental improvement (Lundin *et al.*, 2002; Suh and Rousseaux, 2002; Palme *et al.*, 2005; Pasqualino *et al.*, 2009; Peregrina *et al.*, 2006; Hong *et al.*, 2009; Gallego *et al.*, 2008 and Houillon *et al.*, 2005).

Several possible scenarios are shown in Figures 1.33 and 1.34. For example, Lundin *et al.* (2002) evaluated from an environmental point of view a LCA for urban water systems, which included sludge treatment and final disposal in landfill, agricultural land, and incineration (including transportation). This paper found some environmental sustainable indicators important for this assessment, among others some related to sludge are: quantity of sludge to landfill, transportation of sludge for disposal, energy recovered from biogas, recycling of N and P to agricultural land, etc. Suh and Rousseaux (2002) studied the environmental impact over different sewage sludge treatments in the European context: incineration, agricultural land application or landfill and sludge stabilisation processes such as lime stabilisation, composting or anaerobic digestion. The project showed that the combination of anaerobic digestion and agricultural land application was the most environmentally friendly process. About 48% of destruction of VS is considered to generate biogas during the digestion, with a methane content of 66%. Palme *et al.* (2005) evaluated four options of sludge handling: 1. spreading of pasteurised sludge on agricultural land; 2. Co-incineration with household waste; 3. Separate incineration followed by phosphorus recovery by the bio-Con process; 4. Fractionation by acid hydrolysis for recovery of phosphorous with the Cambi-Krepro process. Potential reductions in atmospheric emissions of CO_2, NO_x, and sulphur oxides caused by the reduction of fossil resources due to sludge incineration favour the three incineration options (co-incineration, bio-con and cambio-krepro) where energy recovery replaces alternative energy source. Agricultural uses was the lest favorable environmental option as it can hypothetically transfer potential contaminants of sewage sludge. Hospido *et al.* (2005) made a LCA over anaerobic digestion versus pyrolysis and incineration of urban wastewater sludge. Again anaerobic digestion followed by land application leaded both higher energy and nutrient recovery, and seems an acceptable option, however more efforts are needed to know the negative impact of sludge spreading over the land.

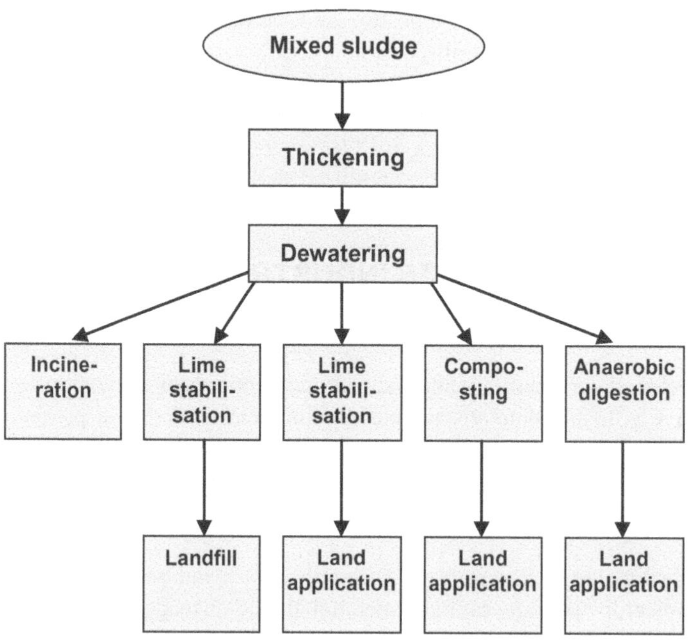

Figure 1.33 Different scenarios of sludge treatment by Suh and Rousseaux (2002)

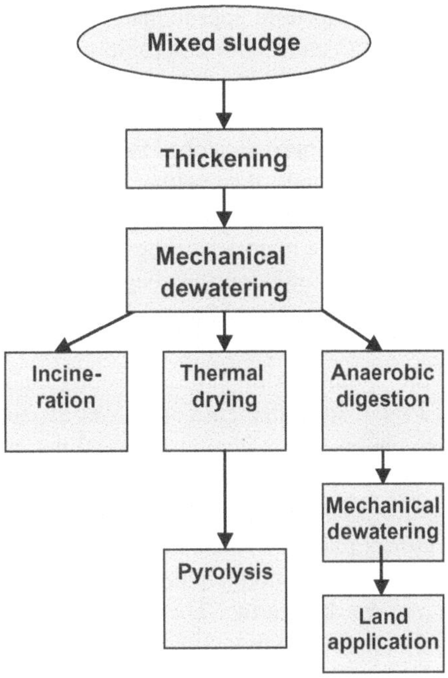

Figure 1.34 Different scenarios of sludge treatment by Hospido *et al.* (2005)

Our approach is not as ambitious as an LCA study of sludge handling, which requires a dedicated study not included among the objectives of this project and that is beyond the scope of this work package. The starting point of our analysis is the fact that the most widespread option for sludge stabilisation is mesophilic anaerobic digestion, although a number of digestion units at low thermophilic temperature are also in operation, followed by application of the stabilised sludge in agricultural land. A cost-effective re-engineering of such units needs to be provided to meet the requested sanitary requirements of the stabilised sludge without putting in compromise the economics of the WWTP.

Of course, there are a number of industrial applications related to sludge, which are offered at commercial scale. For instance, the Cambi Process is designed for enhancing the digestion by accelerating the hydrolysis through previous thermal treatment. The process consists in first of all, pre-dewatering the sewage sludge to 15–20% of dry solid content. Then, the sludge is pre-heated in a first reactor, while the sludge is continuous fed to the reactor. The second reactor, where steam at 12 bar is added to achieve temperatures of 150–160°C and pressures of about 8 bar, is batch operated. Before discharging from the second reactor, the pressure is decreased to approximately 2 bar. The hydrolysed sludge is then cooled down further to 35°C, diluted and subsequently fed to the digester. Typical retention times in the digester range from 10–12 days. The hydrolysis process dissolves about of 27% of the COD in the sludge, which followed by anaerobic digestion converts up to 62% of the VS into biogas, in comparison to 30–50% in the conventional digestion process. It is estimated that 50% of the digester volume is saved, and accomplishes the sanitary conditions required, so it can subsequently be used in agriculture. For example, this technology has been applied in the WWTP of Hamar (Norway).

Similarly, BioThelys is a process developed by Veolia Environnement. The thickened sludge is heated at temperatures between 150–180°C and 8–10 bar of pressure with steam injection in a tank with a retention time of 30–60 min. This patented process has been implemented for instance in three WWTP – Saumur in 2006, Chateau Gontier in 2007 and S.I.V.O. in 2008, in France. Reported performances of the process are that it removes between 38–44% of volatile material, and it can produce a biogas enhancement of 30–50% if compared with conventional mesophilic anaerobic digestion. Also, it is claimed that accomplishes sanitary conditions, and thus it is useful for fertilising uses (www.veoliawaterst.com).

The MicroSludge proces is a patented chemical and pressure treatment that liquefies waste activated sludge (WAS) in order to improve its degradation by subsequent anaerobic digestion. It has been developed by Paradigm Environmental Technologies, in Vancouver (Canada). The process uses caustic pre-treatment to weaken the activated sludge flocs and then uses a high-pressure cell to provide a large and sudden pressure drop that fractures the sludge flocs and solubilises a large fraction of the floc COD.

Alternatively, Kemira Kemwater AB is actively developing the KREPRO technology. This process is a post-treatment for digested sludge in order to enhance the phosphorous removal. The digested sludge is first dewatered in a centrifruge to a 20% dry solid content. After that, the sludge is hydrolysed with sulphuric acid and pre-heated from ambient temperature to 100ºC and then it is heated again to hold approximately 140ºC for 30–40 minutes. Then, the sludge is subsequently cooled down to approximately 50ºC and fed to a decanter-centrifugue and where the sludge is dewatered up to a dry solid content of 45%. The supernatant is fed to the phosphorus precipitation stage, where alkali is added as a primary base for pH adjustment, followed by a ferric salt addition for precipitation of the phosphorus. This ferric phosphate is dewatered by a decanter centrifuge to approximately 35% dry solids by weight. As this fraction of the ferric phosphate contains a low concentration of heavy metals, it can either be reprocessed to different phosphorus products or be used directly in agriculture as a slow release fertiliser. A recovery of about 75% of the phosphorus is projected. The liquid phase from the centrifuge, which contains the majority of the recovered coagulant and soluble organic matter, is recycled back to the wastewater treatment process, utilising the coagulant for phosporus precipitation and as a carbon source for increased denitrification.

Even more drastic options are offered from the 50's, like the well-known Zimpro technology (now owned by Siemens Water Technologies) based in the Zimmermann process, which basically is a wet oxidation conducted at severe pressure, temperature conditions (www.water.siemens.com). Wet air oxidation is the oxidation of soluble or suspended components in water using oxygen as the oxidizing agent. When air is used as the source of oxygen the process is referred to as wet air oxidation. The oxidation reactions occur at temperatures of 150 to 320ºC and pressures from 10 to 220 bar. The wet oxidation process can pre-treat difficult wastewater streams, making them amenable for discharge to a conventional biological treatment plant for polishing. Although now it is rather offered to deal with highly refractory wastewaters, it was historically designed for municipal wastewater sludge applications. At lower temperatures and pressures, sludge is conditioned to improve dewatering. This is referred to as Low Pressure Oxidation. At higher temperatures and pressures, biological sludge can be destroyed, as an alternative to incineration.

The main drawback of the above options is that they are proprietary technologies, some patented, which usually means an extra cost in the application of these solutions, requires major re-engineering of the existing installations, specialised workers for operating the new equipment in safe manner and severe operation conditions with additional measures for safe operation. Instead, the objective of this work package was to demonstrate that the application of freely available technology with minor re-engineering allows easy equipment design and coupling with the existing installations.

The purpose of this section cannot be to give full economical and environmental assessment of the implementation of the technologies studied to WWTP industry throughout Europe. The European Community covers a range of climatic conditions and latitude, and an example of this diversity is shown in Table 1.15. Consequently, because of the wide range, it is difficult to make recommendations that cover the entire Community.

Table 1.15 Diversity of climate and latitude within the EU (Carrington, 2001)

	Latitude	Temperature	Precipitation
Finland, Helsinki	60N	−9 to +16°C	200 to 1100 mm
Greece, Athens	38N	+10 to +28°C	38 to 1270 mm

The estimation of the operation costs was conducted for the municipal WWTP from Reus, which is an average size town (Catalonia, Spain), with a depuration capacity for a population of 60000 equivalent inhabitants and assesses the transformation of the current mesophilic digestion units into thermophilic conditions and the implementation of pre-treatments. Similar analyses can be done for other facilities or the results directly extrapolated.

It must be noted that a main hypothesis of the approach is that the WWTP is equipped with a cogeneration system to recover energy from biogas. The estimation has been done taking into account only the changes in volatile matter and the derived biogas production, but no the probable higher yield of biogas, thus taking a conservative option. The costs related to construction of the facilities are not considered, because the impacts are negligible in front of those from the long operation period.

Mesophilic anaerobic digestion

The anaerobic mesophilic digestion is considered the classical stabilisation process and will be used only for comparison purposes.

Calculation hypothesis

The calculation basis for the feed sludge volume to the digester is taken as 100 m^3/day. In Table 1.16, the input volatile solid and total solid data is shown. The output values are taken from experimental data or present operation costs.

Table 1.16 Data used in the calculations

Parameter	Input	Output
Sludge volume (m^3/day)	100	100
% DM	3.70	2.11
% VM	77.0	59.6
Dehydrated sludge dryness, %	24.5	–
Polyelectrolyte dose, %	18	–

Mass balance

Taking into account the data shown in Table 1.16 and the hypotheses from previous section, the total solids and volatile solids are calculated at the output of the digester. The results are shown in Table 1.17.

Table 1.17 Mass balance summary

Parameter	Input	Output
Sludge volume (m^3/day)	100	100
DM (kg/day)	3700	2110
VM (kg/day)	2850	1257

From Table 1.16 and Table 1.17, the necessary amount of polyelectrolyte for dehydrating the sludge and the amount of dehydrated sludge is calculated. This data are shown in Table 1.18.

Table 1.18 Dehydrated sludge data

Parameter	Value
Dilute polymer (kg/day)	18000
Dehydrated sludge production (kg/day)	8612

The biogas production can be expected to be 0.9 Nm^3/kg VM destroyed (XX Curso, 2002). Table 1.19 collects the biogas production and the electricity generation from this gas, 2.15 kW/Nm^3 (Sugranyes Palet, 1999).

Table 1.19 Biogas production

Parameter	Value
VM destroyed (kg/day)	1593
Biogas production (Nm^3/day)	1434
Electricity production (MWh/day)	3.08

Operation costs

The main operation costs are the dehydrated sludge management and transport and the polyelectrolyte consumption. Experimental data obtained show that 97% of the sludge produced in mesophilic conditions does not accomplish the 86/278/CEE directive. Then, landfill management and transport costs are taken into account. The main data are shown in Table 1.20.

Table 1.20 Main operation costs

Parameter	Value
Polymer cost (€/kg)	2.94
Dilute polymer cost, 0.12% (€/t)	3.53
Agriculture sludge management (€/t)	9.40
Agriculture sludge transport (€/t)	11.2
Landfill sludge management (€/t)	50.3
Landfill sludge transport (€/t)	13.3

From Table 1.20, the overall operation costs are calculated and shown in Table 1.21 in form of ratio to the dry matter amount entering to the digester.

Table 1.21 Overall operation costs

Parameter	Cost (€/day)	Ratio (€/t DM)
Sludge management	537	145.1
Polymer	63.5	17.2
Energetic profit	176	47.5
Total	425	114.8

Thermophilic anerobic digestion

The anaerobic thermophilic digestion is accepted as the alternative stabilisation process. Two different temperatures were considered, 50 and 55°C.

Calculation hypothesis

Again, the calculation basis for the feed sludge volume to the digester is taken as 100 m^3/day. In Table 1.22 the input volatile solid and total solid data are shown. The output values are taken from experimental data.

Table 1.22 Data used in the calculations

Parameter	Input	Output 50°C	Output 55°C
Sludge volume (m^3/day)	100	100	100
% DM	3.70	2.06	2.24
% VM	77.0	59.2	61.6
Dehydrated sludge dryness, %	24.5	–	–
Polyelectrolyte dose, %	18	–	–

Mass balance

Taking into account the data shown in Table 1.22 and the hypotheses listed in Section 'Calculation hypothesis', the total solids and volatile solids in the digesters output are determined. The results are shown in Table 1.23.

Table 1.23 Mass balance summary

Parameter	Input	Output 50°C	Output 55°C
Sludge volume (m^3/day)	100	100	100
DM (kg/day)	3700	2060	2240
VM (kg/day)	2850	1219	1381

Taking into account these data and that shown in Section 'Calculation hypothesis', the necessary amount of polyelectrolyte for dehydrating the sludge and the amount of dehydrated sludge produced can be calculated. These data are shown in Table 1.24.

Table 1.24 Dehydrated sludge data

Parameter	Value 50°C	Value 55°C
Dilute polymer (kg/day)	18000	18000
Dehydrated sludge production (kg/day)	8408	9143

The biogas production can be taken as 0.9 Nm^3/kg VM destroyed (XX Curso, 2002). Later, the conversion of biogas to electricity is 2.15 kW/ Nm^3 (Sugranyes Palet, 1999). Both are collected in Table 1.25.

Table 1.25 Biogas production

Parameter	Value 50°C	Value 55°C
VM destroyed (kg/day)	1630	1469
Biogas production (Nm^3/day)	1467	1322
Electricity production (MWh/day)	3.15	2.84

Operation costs

The main operation costs are again the dehydrated sludge management and transport and the polyelectrolyte consumption. Also, it is estimated the profit of transforming biogas to electricity, through cogeneration, and subsequent sale to the power network. The main data per unit matter was already shown in Table 1.23.

With the data from Table 1.23 and from experimental data, an average of 14% thermophilic at 50°C sludge and 43% thermophilic at 55°C sludge must be managed for landfill (by accomplishment of 86/278/CEE directive). The operation costs are calculated and shown in Table 1.26. Also, the cost per dry matter unit that enters to the digesters is shown.

Table 1.26 Overall operation costs

Parameter	50°C		55°C	
	Cost (€/day)	Ratio (€/t DM)	Cost (€/day)	Ratio (€/t DM)
Sludge management	485	130.9	413	111.5
Polymer	63.5	17.2	63.5	17.2
Energetic profit	180	48.6	162	43.8
Total	368	99.5	314	84.9

Comparison between the two proposals

In order to highlight the economic savings that the anaerobic digestion in thermophilic conditions implies, a comparison is given in Table 1.27.

Table 1.27 Summary of economic data

Parameter	Mesophilic	Thermophilic at 50°C		Thermophilic at 55°C	
		Cost	Increase/ decrease	Cost	Increase/ decrease
Sludge management cost, (€/t DM)	145	131	−9.8%	112	−23.2%
Polymer cost, (€/tDM)	17.2	17.2	0%	17.2	0%
Electricity exportation, (€/tDM)	47.5	48.6	+2.3%	43.8	−7.8%
Total (€/tDM)	115	99.5	−13.3%	84.9	−26.0%

The thermophilic at 55°C sludge management cost decreases about 23% due to the greater level of fulfilment with agriculture regulations. The electricity exportation, in case of thermophilic at 50°C sludge, is also greater because there is a higher volatile matter destruction and thus a higher biogas production. In conclusion, the thermophilic treatment alone is already favourable in economic terms because it is able to produce acceptable sanitary sludge in many cases. It must be noted that most of the hypotheses have been conservatives.

Thermal pre-treatment

Calculation hypothesis

According to the results shown section 2.3, a process temperature below 110°C could be enough to improve the digestability (probably 90°C is an acceptable option) of the sludge with a contact time around 1 h. These conditions have also been considered for other authors using low COD content sludge (Paul *et al.*, 2006).

Mass balance

In the estimation, only the experimental data obtained about total solids and volatile solids will be considered. The results for the three alternatives are shown in Table 1.28.

Table 1.28 Mass balance summary for the thermal pre-treatments

Parameter	Mesophilic + Heating	Thermophilic 50°C + Heating	Thermophilic 55°C + Heating
DM Pre-treatment input (kg/day)	4290	4290	4290
VM Pre-treatment input (kg/day)	3508	3508	3508
DM digester input (kg/day)	4400	4400	4400
VM digester input (kg/day)	3410	3410	3410
DM digester output (kg/day)	2310	2530	2430
VM digester output (kg/day)	1350	1450	1540
Dehydrated sludge (kg/day)	9429	10327	9918

As stated in previous sections the biogas production can be taken as 0.9 Nm^3/kg VM destroyed, although it is likely that pre-treated sludge yields higher specific biogas production. The conversion of biogas to electricity is taken as 2.15 kW/Nm^3. Both are collected in Table 1.29.

Table 1.29 Biogas production

Parameter	Mesophilic + Heating	Thermophilic 50°C + Heating	Thermophilic 55°C + Heating
VM destroyed (kg/day)	2060	1960	1870
Biogas production (Nm^3/day)	1854	1764	1683
Electricity production (MWh/day)	3.98	3.79	3.61

Operation costs

Taking into account the agriculture and landfill sludge management cost and from experimental data, an average of 80% of the sludge could be managed to agriculture use – although probably the frequency of acceptable cases is actually higher-; then the sludge treatment cost is calculated. The results are shown in Table 1.30.

Table 1.30 Operation costs

Parameter	(Current) Mesophilic		Mesophilic + Thermal		Thermophilic 50°C + Thermal		Thermophilic 55°C + Thermal	
	Cost (€/day)	Ratio (€/tDM)	Cost (€/day)	Ratio (€/tDM)	Cost (€/day)	Ratio (€/tDM)	Cost (€/day)	Ratio (€/tDM)
Sludge management	537	145.1	275	64.2	302	70.3	290	67.5
Polymer	63.5	17.2	63.5	14.8	63.50	14.8	63.5	14.8
Energetic profit	176	47.5	227	52.9	216	50.4	206	48.1
Total	425	114.8	112	26.0	149	34.7	147	34.3

Hydrogen peroxide addition

According to the results, previously shown a dose of 0.2 g H_2O_2/g COD is convenient for a peroxidation pre-treatment of the raw sludge.

Calculation hypothesis

The more outstanding hypotheses are shown in Table 1.31.

Table 1.31 Calculation hypotheses

Parameter	Value	Parameter	Value
H_2O_2 dose (g/kg DM)	25	H_2O_2 cost (€/kg)[a]	0.8
$FeSO_4$ dose (g/kg DS)	1.67	Polyelectrolyte cost (€/kg)[b]	2.30
Polyelectrolyte dose (g/kg DM)	6.7	$FeSO_4$ cost (€/t)[c]	4.73
H_2SO_4 dose (g/kg DM)	145	H_2SO_4 cost (€/kg)[c]	0.100
$Ca(OH)_2$ dose (g/kg DM)	55.6	$Ca(OH)_2$ cost (€/kg)[c]	0.095
Polyelectrolyte solution (%)	0.1		

[a] *Paul* et al., *2006.*
[b] *Stockhaussen Degusa, SL.*
[c] *Kemira Iberica, SA.*

Mass balance

The necessary reactive amounts for applying the pre-treatment are also calculated. The results of the mass balance for the different processes are shown in Table 1.32.

Table 1.32 Mass balance summary

Parameter	Mesophilic + Peroxide	Thermophilic 50°C + Peroxide	Thermophilic 55°C + Peroxide
DM Pre-treatment input (kg/day)	2900	2900	2900
VM Pre-treatment input (kg/day)	2343	2343	2343
DM digester input (kg/day)	2580	2580	2580
VM digester input (kg/day)	2270	2270	2270
DM digester output (kg/day)	2200	2250	2330
VM digester output (kg/day)	1030	1280	1330
Dehydrated sludge (kg/day)	8980	9184	9510

Operation costs

In this case, the main operation costs are the dehydrated sludge management and transport, the polyelectrolyte consumption and the reactive consumption for the peroxidation. From experimental data, 20% sludge management to landfill is taken into account. In Table 1.33, the operation costs calculated are shown. Moreover, a ratio with the dry matter quantity that enters to the pre-treatment is also shown.

Table 1.33 Operation costs for peroxidised sludge stabilisation

Parameter	Mesophilic		Mesophilic + Peroxide		Thermophilic 50°C + Peroxide		Thermophilic 55°C + Peroxide	
	Cost (€/day)	Ratio (€/tDM)	Cost (€/day)	Ratio (€/tDM)	Cost (€/day)	Ratio (€/tDM)	Cost (€/day)	Ratio (€/tDM)
Sludge management	537	145	262	90.4	268	92.5	278	95.8
Polymer	63	17	63.5	21.9	63.5	21.9	63.5	21.9
Reactives	–	–	116	39.9	116	39.9	116	39.9
Energetic profit	176	48	137	47.1	109	37.6	104	35.7
Total	425	115	305	105.1	338	116.6	353.2	121.8

In conclusion, the combination of thermophilic digestion with thermal pre-treatment is an attractive alternative from the economical and environmental point of view, because it permits to reduce the dehydrated sludge management costs because a higher sludge amount fulfils with the regulations allowing its application to the agriculture. Instead, the benefits given by hydrogen peroxide pre-treatment are mostly offset by the additional cost of the reagents.

Conclusive remarks

After the estimation of the expected savings conducted for the two recommended pre-treatments, in the case of chemical pre-treatment with hydrogen peroxide, the benefits of the higher amount of stabilised sludge accomplishing sanitary conditions and the larger power production from additional biogas are mostly offset by the high cost of the reagents, for which hydrogen peroxide is the main contributor. Only in case that hydrogen peroxide could be obtained at a much lower cost, this option would be interesting in practice, from the point of view of existing installations that set aside the stabilised sludge to agricultural land.

On the contrary, thermal pre-treatment at temperature below 110°C for 1 h gives unitary costs of the stabilised sludge three times lower that without using any pre-treatment, because most of the stabilised sludge is now estimated that can be dedicated to agricultural use. The savings are calculated in approximately 1 € per equivalent inhabitant and year in comparison to the present cost for landfilling. As the re-engineering needed for existing plants in this case is minimum, these savings assure a very short payback period (2–4 years), even in case of overestimation of the benefits.

Therefore, the transformation of mesophilic units into thermophilic units adding a thermal pre-treatment (Figure 1.35) at non severe conditions (<110°C, 1 h) is economically viable as the higher operating costs are widely compensated by the increase of biogas production and the extended use of the stabilised sludge for agricultural land use. This option requires minimum re-engineering of the plants and is especially interesting for plants with power cogeneration units implemented. As for those, the savings can be estimated in 1 € per equivalent inhabitant and year, which must cover any investment needed with a payback period of less than four years, even when a cogeneration plant must be build up.

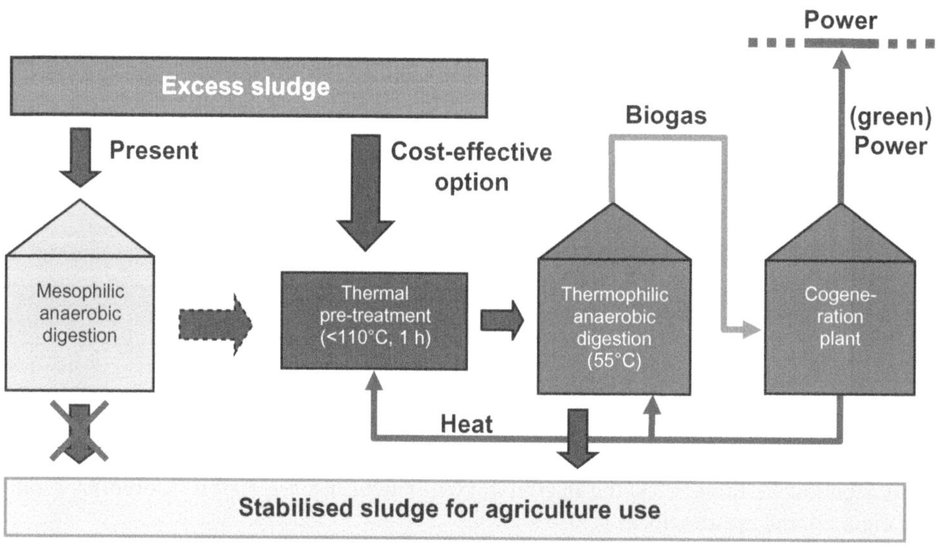

Figure 1.35 Re-engineering scheme proposed for the existing sludge mesophilic anaerobic digestion systems in WWTP

REFERENCES

Appels L., Baeyens J., Degrève J. and Dewil R. (2008). Principles and potential of the anaerobic digestion of waste-activated sludge. *Prog. Ener. Comb. Sci.*, **34**, 755–781.

Balkema A. J. (2002). Indicators for the sustainability assessment of wastewater treatment systems. *Urban Water*, **4**, 153–161.

Bougrier C., Delgenes J. P. and Carrère H. (2007). Impacts of thermal pre-treatments on the semi-continuous anaerobic digestion of waste activated sludge. *Biochem. Eng. J.*, **34**, 20–27.

Boušková A., Dohányos M., Schmidt J. E. and Angelidaki I. (2005). Strategies for changing temperature from mesophilic to thermophilic conditions in anaerobic CSTR reactors treating sewage sludge. *Wat. Res.*, **39**, 1481–1488.

Cacho Rivero J. A. and Suidan M. T. (2006a). Effect of H_2O_2 dose on the thermo-oxidative co-treatment with anaerobic digestion of excess municipal sludge. *Wat. Sci. Technol.*, **54**, 253–259.

Cacho Rivero J. A. and Suidan M. T. (2006b). Anaerobic Digestion of Thermo-Oxidized Excess Municipal Sludge. *Wat. Environm. Res.*, **78**, 510–521.

Carrington E. G. (2001). 'Evaluation of sludge treatments for pathogen reduction', Final Report, Study Contract No B4-3040/2001/322179/MAR/A2 for the European Commission Directorate-General Environment.

Carrington E. G., Harman S. A. and Pik E. B. (1982). Inactivation of Salmonella during anaerobic digestion of sewage sludge. *J. Appl. Bacteriol.*, **53**, 331–334.

Chen Y., Jiang S., Yuan H., Zhou Q. and Gu G. (2007). Hydrolysis and acidification of waste activated sludge at different pHs. *Wat. Res.*, **41**, 683–689.

De la Rubia M. A., Romero L. I., Sales D. and Pérez M. (2005). Temperature conversion (mesophilic to thermophilic) of municipal sludge digestion. *AICHE J.*, **51**, 2581–2586.

De la Rubia M. A., Perez M., Romero L. I. and Sales D. (2006). Effect of solids retention time (SRT) on pilot scale anaerobic thermophilic sludge digestion. *Process Biochemistry*, **41**, 79–86.

Elliott A. and Mahmood T. (2007). Pretreatment technologies for advancing anaerobic digestion of pulp and paper biotreatment residues. *Wat. Res.*, 41, 4273–4286.

Feacham R. G., Bradley D. J., Garelick H. and Mara D. D. (1983). Health aspects of excreta and wastewater management. John Wiley and Sons, Chichester.

Ferrer I., Climent M., Baeza M. M., Artola A., Vazquez F. and Font X. (2007). Effects on thermal and mechanical pretreatments of secondary sludge on biogas production under thermophilic conditions. *Chem. Eng. J.*, **133**, 335–342.

Gallego A., Hospido A., Moreira M. T. and Feijoo G. (2008). Environmental performance of wastewater treatment plants for small populations. *Resour. Conserv. Recycling*, **52**, 931–940.

Garber W., Ohara G., Colbauch J. and Raskit S. (1975). Thermophilic digestion at the Hyperion Treatment Plant. *J. Wat. Pollut. Control Fed.*, **47**(5), 950–961.

Gavala H. G., Y. U., Skiadas I. V., Wastermann P. and Ahring B. (2003). Mesophilic and thermophilic anaerobic digestion of primary and secondary sludge. Effect of pre-treatment at elevated temperature. *Wat. Res.*, **37**, 4561–4572.

Gottschalk G. (1986). Bacterial Metabolism 2nd edition, Springer-Verlag, New York.

Hong J., Hong J., Otaki M. and Jolliet O. (2009). Environmental and economic life cycle assessment for sewage sludge treatment processes in Japan. *Waste Manag.*, **29**, 696–703.

Hospido A., Moreira M. T., Martin M., Rigola M. and Feijoo G. (2005). Environmental evaluation of different treatment processes for sludge from urban wastewater treatments: Anaerobic Digestion versus Thermal Processes. *Intern. J. of LCA*, **10**, 336–345.

Houillon G. and Jolliet O. (2005). Life cycle assessment of processes for the treatment of wastewater urban sludg: energy and global warming analysis. *J. Clean. Produc.*, **13**, 287–299.

Kang Y. W., Cho J.-J. and Hwang K.-Y. (1999). Correction of hydrogen peroxide interference on standard chemical oxygen demand test. *Wat. Res.*, **33**, 1247–1251.

Kim J., Park C., Kim T.-H., Lee M., Kim S., Kim S.-W. and Lee J. (2003). Effect of various pretreatments for enhanced anaerobic digestion with waste activated sludge. *J. Biosci. Bioeng.*, **95**, 271–275.

Lay J. J., Li Y. Y. and Noike T. (1997). Influences of pH and moisture content on the methane production in high-solid sludge digestion. *Wat. Res.*, **31**(10), 1518–1524.

Lu J., Gavala H. N., Skiadas I. V., Mladenovska Z. and Ahring B. K. (2008). Improving anaerobic sewage sludge digestion by implementation of a hyper-thermophilic prehydrolysis step. *J. Environm. Manag.*, **88**, 881–889.

Lundin M. (2002). A Life cycle assessment based procedure for development of environmental sustainability indicators for urban wastewater systems. *Urban Water*, **4**, 145–152.

Mata-Álvarez J., Macé S. and Llabrés P. (2000). Anaerobic of organic solid waste. An overview of research achievements and perspectives. *Biores. Techn.*, **74**, 3–16.

Metcalf E. (2003). *Wastewater Engineering: Treatment, disposal, and reuse*. 4th ed., McGraw-Hill, Boston.

Middelberg A. P. J. (1995). Process-scale disruption of microorganisms. *Biotechnol. Adv.*, **13**, 491–551.

Miron Y., Zeeman G., Van Lier J. B. and Lettinga G. (2000). The role of sludge retention time in the hydrolysis and acidification of lipids, carbohydrates and proteins during digestion of primary sludge in CSTR systems. *Wat. Res.*, **34**, 1705–1713.

Muga H. E. and Mihelcic J. R. (2008). Sustainability of wastewater treatment technologies. *J. Environm. Manag.*, **88**, 437–447.

Müller, J. A. (2001). Prospects and problems of sludge pre-treatment process. *Wat. Sci. Technol.*, **44**, 121–128.

Pasqualino J. C., Meneses M., Abella M. and Castells F. (2009). LCA as a decision support tool for the environmental improvement of the operation of a municipal wastewater treatment plant. *Env. Sci. Tech.*, **43**, 3300–3307.

Paul E., Camacho P., Sperandio M. and Ginestet P. (2006). Technical and economical evaluation of a thermal, and two oxidative techniques for the reduction of excess sludge production. *Proc. Saf. Env. Protec.*, **84**(B4), 247–252.

Pavlostathis S. G. and Giraldo-Gomez E. (1991). Kinetics of anaerobic treatment a critical review. *Critical Reviews in Environm. Control*, **21**, 411–489.

Peregrina C. A., Lecomte D., Arlabosse P. and Rudolph V. (2006). Life cycle assessment (LCA) applied to the design of an innovative drying process for sewage sludge. *Proc. Saf. Env. Protec.*, **84**, 270–279.

Prescott L. M., Harley J. P. and Klein D. A. (1999). Microbiology, 4th ed. WCB/McGrawHill, New York.

Puchajda B. and Oleszkiewicz J. (2006). Thermophilic anaerobic acid digestion of biosolids: hydrolysis, acidification, and optimization of retention time of acid digestion. *J. Environ. Eng. Sci.*, **5**, 187–195.

Rimikus R., Ryan J. and Cook E. (1982). Full scale thermophilic digestion at the west-southwest sewage treatment works, Chicago, Illinois. *J. Wat. Control Fed.*, **54**, 1447–1457.

Romero Ll., Sales D. and Martinez E. (1990). Comparison of three practical processes for purifying wine distillery wastewater. *Proc. Biochem.*, **25**, 93–96.

Ros M. and Zupančič G. D. (2003). Thermophilic anaerobic digestion of waste activated sludge. *Acta Chim. Slov.*, **50**, 359–374.

Ryan P., McHugh S., Golden T. and Colleran E. (2005). Investigation of homoactogenic bacterial activity during mesophilic, thermophilic and phychrophilic anaerobic digestion, *Proceedings of ESAI Environ.*, 61–64.

Stuckey D. C. and McCarty P. L. (1984). The effect of thermal pretreatment on the anaerobic biodegradability and toxicity of waste activated sludge. *Wat. Res.*, **18**, 1343–1353.

Sugranyes Palet S. (1999). Design Project of the Cogeneration Plant from Biogas at the WWTP of Reus; Junta de Sanejament; Departament de Medi Ambient; Generalitat de Catalunya.

Suh Y.-J. and Rousseaux P. (2002). An LCA of alternative wastewater sludge treatment scenarios. *Resour. Conserv. Recycling*, **35**, 191–200.

Talinli I. and Anderson G. K. (1992). Interference of hydrogen peroxide on the standard COD test. *Wat. Res.*, **26**, 107–110.

Tanaka S. and Kamiyama K. (2002). Thermochemical pretreatment in the anaerobic digestion of waste activated sludge. *Wat. Sci. Technol.*, **46**, 173–179.

U.S. Peroxide, Product information: interferences with analytical methods (http://www.h2o2.com/intro/interferences.html).

Valo A., Carrere H. and Delgenès J. P. (2004). Thermal, chemical and thermo-chemical pre-treatment of waste activated sludge for anaerobic digestion. *J. Chem. Technol. Biotech.*, **79**, 1197–1203.

Vlyssides A. G. and Karlis P. K. (2004). Thermal-alkaline solubilization of waste activated sludge as a pre-treatment stage for anaerobic digestion. *Bioresource Technol.*, **91**, 201–206.

Wagner A. O., Gstraunthaler G. and Illmer P. (2008). Survival of bacterial pathogens during the thermophilic anaerobic digestion of biowaste: Laboratory experiments and in situ validation. *Anaerobe*, **14**, 181–183.

Wang Q., Noguchi C., Hara Y., Sharon C., Kakimoto K. and Kato, Y. (1997). Studies on anaerobic digestion mechanism: influence of pretreatment temperature on biodegradation of waste activated sludge. *Environm. Technol.*, **18**, 999–1008.

Watanabe H., Kitamura T., Ochi S. and Ozaki M. (1997). Inactivation of pathogenic bacteria under mesophilic and thermophilic conditions. *Wat. Sci. Technol.*, **36**, 25–32.

Wheatley B. I. and Hughes D. (1980). Factors affecting net energy production from mesophilic anaerobic digestion. In Anaerobic Digestion, eds Stratford, 131–150.

Wong W. T., Chan W. I. and Liao P. H. (2006). A hydrogen peroxide/microwave advanced oxidation process for sewage sludge treatment. *J. Env. Sci. Health Part A*, **41**, 2623–2633.

www.veoliawaterst.com

www.water.siemens.com

XX Curso. (2002). Tratamiento de Aguas Residuales y Explotación de Estaciones Depuradoras; Tomo II; Centro de Estudios y Experimentación de Obras Públicas (CEDEX); Ministerio de Medio Ambiente/Ministerio de Fomento; Madrid, España.

Zábranská J., Štěpová J., Wachtl R., Jeníček P. and Dohányos M. (2000). The activity of anaerobic biomass in thermophilic and mesophilic digesters at different loading rates. *Wat. Sci. Technol.*, **42**, 49–55.

Chapter 2

Sludge stabilisation in autothermal conditions

Esther Torrens Serrahima, Frank Stüber, Azael Fabregat Llangostera, Agustí Fortuny Sanromà, Christophe Bengoa, Josep Font Capafons (Universitat Rovira i Virgili), Jack Legrand, Karine Loubière (Laboratoire GEPEA, Université de Nantes), Jaume Cabrè Fonseré, Iñaki Oriol, Raúl García (Gestió Ambiental i Abastament, S.A.), Jaume Bresó and Eva Rovira (Tratamientos y Recuperaciones Industriales S.A.)

SUMMARY

Project objectives

The study carried out was exclusively addressed to explore the application of Autothermal Thermophilic Aerobic Digestion (ATAD) using pre-treated sludge in an optimised mixing environment giving enough aeration. Overall, an underlying objective was to experimentally demonstrate the feasibility of ATAD equipment design and operation using freely available technology and engineering tools avoiding the dependence on patented designs. In all the cases, waste sludge was provided by Gestió Ambiental i Abastament, S.A., who will also collaborate in the experimental task and mainly in the data interpretation and the economic potential of its implementation at high scale. Universitat Rovira i Virgili was responsible of the operation of the bench-scale ATAD reactor whilst Laboratoire GEPEA of Université de Nantes performed the theoretical and experimental investigation of mixing and aeration. Economic aspects of its application have also been considered.

Work performed

An ATAD bench-scale plant was designed and built-up using non-dedicated equipment, therefore non-optimised. The plant was working since month 18 in order to study the ATAD performance at thermophilic temperatures over different substrates (raw sludge and pre-treated sludge) and organic loading rates (HRT). After the start-up and acclimatising period, study of aerobic digestion in thermophilic conditions was performed at different HRT (10 and 8 days). Meanwhile, optimisation of the mixing and aeration system was performed.

 ISBN: 9781843393450. Published by IWA Publishing, London, UK.

Through experimental and numerical analysis tools, the mixing and aeration efficiency inside the bioreactor devoted to the ATAD process was firstly characterised and later an optimised design was proposed, built up and installed. The bioreactor chosen as reference is the one implemented by Universitat Rovira i Virgili, however a new one was designed and built-up by Laboratoire GEPEA of Université de Nantes, especially for hydrodynamics experiments (transparent walls, adapted instrumentation).

Thus, as not available in the literature, the agitation system initially implemented was fully characterised (power number curves without/with gas, gas-liquid flow maps and gas-liquid mass transfer coefficients in water). As the aeration capacities were insufficient (even in water) for satisfying the potentially maximum biological requirements, a new system was designed, consisting in introducing four baffles, an efficient impeller for gas dispersion (a home-made CD6 Chemineer® impeller) and a ring air sparger. Significant improvements were obtained in water in terms of Relative Power Demands and k_La. Some rheological measurements on sludge were performed and demonstrated that sludge behaves like a non-Newtonian fluid (shear-thinning, no viscoelasticity, with a small yield stress). Then, model fluids (aqueous solutions of CMC and xanthan gum) were able to be selected for specific experiments aiming at better understanding the impact of the non-newtonian rheology on aeration.

In parallel, a numerical study (Fluent®) was run. Recognising that the ATAD reactor is very complex from a CFD point of view (two-phase, mechanical agitation, biological reaction, non-newtonian fluid, heat and mass transfer), it is nowadays fully unrealistic to perform entire numerical simulation. Therefore, only partial simulations (without mechanical agitation) were performed and the information was used as a tool to investigate the coupling between transport phenomena and operating conditions. At present, the numerical strategy was validated using a benchmark reactor (an internal-loop airlift), its transposition to the ATAD reactor was begun. These numerical investigations are not yet finished and will be continued beyond the frame of this project.

End results

Operation of the ATAD plant demonstrated that, even in a non-optimised environment affected by oxygen mass transfer limitations, a high COD reduction was found, at a mean value of 62%, although important fluctuations were observed. In turn, TS and VS content decreased up to 34% and 40%, respectively. Pathogen content decreased from the original 2800000 NMP/g to just 23 NMP/g of E-coli, and from 5 NMP/g to 0 NMP/g of Salmonella. Although disappearance of organic matter does exist, autothermal conditions were not achieved as extra-heating was added to sustain the thermophilic temperature. This was caused by the small size of the bench-scale reactor, which leads to excessive heat loss. However, the COD removal achieved assures (by energy balance) that higher scale plants would work under auto-sustained operation.

Despite non self-maintenance of the reactor temperature, the feasibility of thermophilic aerobic digestion was demonstrated using non-dedicated equipment, no requiring owned or patented technologies. Freely available mixing systems are able to give enough mixing to reach acceptable level of COD conversion. However, the operation was still limited by the oxygen mass transfer rate as the oxygen concentration was very low in the reactor. The protocol for oxygen mass transfer measurement in the reactor developed by Laboratoire GEPEA of Université de Nantes, allowed proposing improvements in the agitation system in order to substantially increase the oxygen transfer (two to ten times). These modifications were implemented into the bench-scale plant, but the improvement in the operation was not possible to be substantiated because work was stopped in month 32.

Regarding to the theoretical and experimental investigation of mixing and aeration, the main achievements are (i) the optimisation of the agitation and aeration system, (ii) the implementation of experimental tools (measurements of the power dissipated by the impeller, the aeration capacity and homogeneity and the gas-liquid hydrodynamics); (iii) the characterisation of sludge rheology (shear-thinning, yield stress, no viscoelasticity), (iv) the first steps towards the numerical simulation of the ATAD reactor without mechanical agitation. One year more (end of the ongoing Ph.D. thesis) is still necessary to complete this last part. Beyond of this project, this research work will give elements at disposal for establishing a relationship between the aeration performances and the biological degradation (efficiency, thermophilic temperature, autothermal conditions) in such systems, applicable to a wide diversity of situations.

Therefore, ATAD is a very attractive option for small size municipal WWTP, for which high investment in anaerobic systems for biogas production is not justified.

METHODOLOGIES AND APPROACHES EMPLOYED

Approach employed

Demonstration of the feasibility of ATAD equipment design and operation was planned through collaboration basically between Universitat Rovira i Virgili and Laboratoire GEPEA of Université de Nantes. Autothermal Thermophilic

Aerobic Digestion (ATAD) was envisaged as an alternative suitable when digestors for biogas production are not economical. The starting point was to apply freely available technology and engineering tools for such a design, avoiding the dependence on patented designs and thus on specialised suppliers.

Universitat Rovira i Virgili was dedicated to the operation of ATAD using a non-optimised home-made reactor at bench-scale. ATAD was firstly applied to stabilise raw sludge at thermophilic temperature ad two different HRT (8 and 10 days). After implementation of the optimised agitation and aeration system, drawn by Laboratoire GEPEA of Université de Nantes, benefits of a highly effective oxygen mass transfer should be examined and compared. In addition, as raw sludge is not biodegradable enough as to be directly digested by the aerobic microorganisms, use of pre-treated sludge (as described in Chapter 1 'Reduction and stabilisation of the excess and toxicity of sludge in anaerobic conditions') was planned to be investigated in order to substitute the typical anaerobic digestion, previously or simultaneously occurring for providing more biodegradable matter. In all the cases, waste sludge was provided by Gestió Ambiental i Abastament, S.A. (GAIASA), who will also collaborate in the experimental task and mainly in the data interpretation and the economic potential of its implementation at high scale.

Parallelly, Laboratoire GEPEA of Université de Nantes performed the theoretical and experimental investigation of mixing and aeration using experimentally obtained data through rheology characterisation of sludge and in-situ measurement of oxygen availability, with selection of proper mixing and aeration system experimentally tested, for subsequent installation in the ATAD reactor from Universitat Rovira i Virgili. Economic aspects of its application have also been considered. Overall, an underlying objective was to experimentally demonstrate the feasibility of ATAD equipment design and operation using freely available technology and engineering tools avoiding the dependence of patented designs. Through the gained experience a simulation model, accounting for biological degradation later coupled to mechanical agitation, using CFD tools was conceived to aid the design of new equipment or the adaptation of existing one.

Finally, economic aspects of its application have also been considered. Cost estimation of a given plant was performed and thus the feasibility of its potential application in WWTP was positively assessed.

Experimental methodology

ATAD operation

A pilot plant consisting of a 5 L cylindrical reactor of stainless steel was constructed with an active effective reaction volume of 4 L. Figure 2.1 displays a picture of the pilot plant, where the reactor is supported with a metallic structure giving a compact design and mobility to the equipment. A thermocouple monitors the reactor temperature, two probes measure the pH and the dissolved oxygen concentration in the reactor. Another probe monitors the pH.

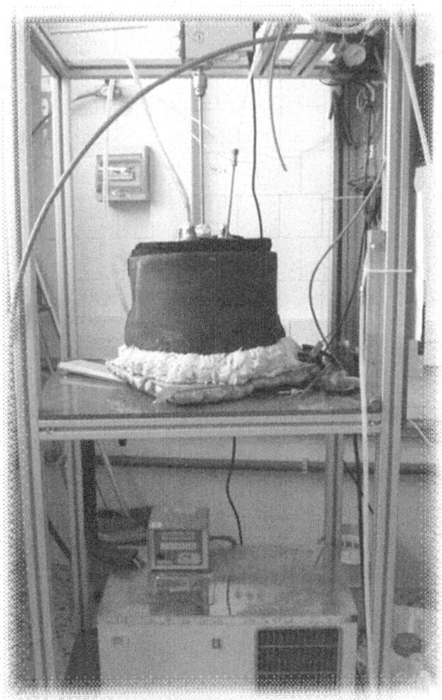

Figure 2.1 ATAD bench-scale plant

The possibility of foaming was taken into account for reactor design. Foaming layer seems to improve oxygen utilisation with the consequent enhancement of biological activity. Therefore, approximately one fifth of the reactor was kept free, and the reactor shape in height/diameter ratio of active volume was designed to be one.

The objective of achieving thermophilic ranges depends on taking advantage of the heat released for the oxidation reactions during the aerobic degradation of the organic matter. The design key points most affecting the operation temperature achieved include the efficiency of the aeration system, the proper reactor insulation, the foam management, and the sludge pre-thickening. Besides the self-heating advantage of autothermal digestion, it is also recommendable to use a double system that can be applied either as an auxiliary heating or as a heat recovery device. The acquisition of this auxiliary equipment contributes to have a flexible pilot plant where parameters such as temperature digestion can be accurately controlled and then investigated.

The insulation system should provide nearly adiabatic conditions in the reactor and minimise heat losses caused by air convection. Felt wool and polyurethane were chosen as the insulation material in the reactor due to its low thermal conductivity coefficient, 0.071 W/m K and 0.03 W/m K respectively, and easy availability.

After operating the pilot plant, it was seen that supplementary heat was needed to maintain thermophilic temperature conditions, even though the considerable destruction of organic matter was (by energy balance) enough to attain thermophilic ranges. After careful inspection of the possible reasons, it is believed that the surface/volume ratio of the reactor was not low enough to give suitable heat loss. It is known that industrial ATAD reactors shows a much lower ratio so heat loss is significantly reduced. The operation temperature was fixed at 55ºC.

Due to the high temperature process operation is also important to prevent water evaporation through the exhausted gas outflow. A glass water-cooler system was thus installed to condensate the evaporated water and avoid as many as possible changes in liquid volume. The designed reactor also incorporates a level indicator for controlling the actual reactor volume. The level indicator is made of transparent material for visual control. The influent air was passed through a humidification system to limit subsequent saturation in the reactor. Air-input flow rate was set in the range 16–20 L/h, above the recommended value of 4 $m^3/h/m^3$ of active reactor volume.

One of the most important factors for ATAD is aeration and mixing systems. Proper aeration and reactor homogenisation are decisive for meeting up the extremely high oxygen demand required in the aerobic process. Availability of oxygen in the liquid phase can be controlled by means of a solenoid valve that control the air inflow into the reactor in order to maintain a constant dissolved oxygen concentration, if desired. An aeration device was located at the reactor bottom, equipped with a membrane system that provides small air-bubbles in order to improve the contact area and facilitate oxygen transfer into the sludge. Mixing will be provided with an electrical stick mixer equipped with a paddle stirrer located at the end of the stick; the position of the mixer can be regulated. The agitation speed was set at 400 rpm.

Start-up of the ATAD plant was conducted using aerobic inoculum provided by Gestió Ambiental i Abastament, S.A. from its municipal WWTP. However the inoculum was taken at the mesophillic temperature of 30ºC. Therefore, it was necessary to slowly acclimate these microorganisms into thermophilic range. The temperature was increased at intervals of 5ºC, every two weeks. Gestió Ambiental i Abastament, S.A. also provided the fed raw sludge; its characteristics can be found in Chapter 1 'Reduction and stabilisation of the excess and toxicity of sludge in anaerobic conditions'. The feed was made intermittently, once a day.

In order to give an accurate picture of the autothermal digestion efficiency the following analytical parameters were chosen to assess process evolution and quality of end product: TS, VS, alkalinity, COD, BOD_5, TOC, TC, ammoniacal and organic nitrogen, conductivity, total phosphorous, E-coli, Salmonella, and dissolved oxygen. Details about analytical protocols can be found in Chapter 1 'Reduction and stabilisation of the excess and toxicity of sludge in anaerobic conditions'.

Optimisation and modelling

From a chemical engineering point of view, an ATAD reactor is a very complex process to model, as all the following phenomena interact with others: multiphase system (air bubbles dispersed into a biological non-newtonian liquid phase containing solid particles), presence of biological reactions and coupling between heat and mass transfer.

With regard to this complexity, tailored investigations are required for optimising mixing and aeration, but also for better understanding the basic phenomena involved and their coupling. At last, the main bottleneck will be to successfully establish the relationship between the gas-liquid mass transfer performances and the biological degradation (efficiency, thermophilic temperature, autothermal conditions). To carry on this research, it was decided, as a first step, to work with mock-ups, namely by using transparent fluids which properties (rheology, surface tension, density) are the closest to the ones of sludge. A methodology based on three key points, was adopted:

1. Choice of model fluids
   - Rheological measurements on sludge: characterisation of the non-newtonian properties (shear-thinning, viscoelasticity, time-dependency, yield stress).

- Selection of the model fluids: type (aqueous solutions of CMC or xanthan gum), concentration, stability in time (adjuvant introduction).

2. Experimental approach
   - Construction of a lab-scale reactor identical to the one of Universitat Rovira i Virgili (in terms of dimensions, impeller and aerators) and enabling gas-liquid hydrodynamics and mass transfer measurements to be performed;
   - Experimental characterisation of: (i) the power consumed (torque measurements), (ii) gas-liquid hydrodynamics, in terms of gas-liquid regime (visualisation technique), residence time distribution (tracer experiments), bubble size distribution and gas retention (double optical probe), (iii) mass transfer k_La (dissolved oxygen probes);
   - Experimental study and modelling of the influence on the mixing and on the aeration of air flow rate, agitation rate, mixer height and type, aerator design, rheological behaviour and temperature.
3. Numerical approach
   - Full validation of the numerical strategy for simulating gas-liquid hydrodynamics and mass transfer in a benchmark reactor (internal-loop airlift);
   - Application of the same methodology to the lab-scale reactor when no mechanical agitation is induced and when water is the liquid phase;
   - Numerical study for understanding and modelling the influence of the liquid phase rheology and the temperature on both hydrodynamics and mass transfer;
   - Introduction of the first tools for scaling-up (coupling with biological results, formulation of a process model based on numerical results).

ACHIEVEMENTS RELATED TO THE STATE OF THE ART

Fundamentals of autothermal thermophilic aerobic digestion

Today, according to Layden (2007), the Autothermal Thermophilic Aerobic Digestion (ATAD) is actually classified as an established process, when compared to the limiting operating experience of innovative or the still bench tested embryonic process. This technology is nowadays successfully performed to face with the regulations limits, as it produces class A bio-solids and offers several other benefits: high pathogen destruction, low space and volume, reduced energy requirements (self-heating operation), high sludge treatment rate. Such process enables the overall cost of the sewage treatment to be reduced, when compared to the thermophilic conditions obtained classically with additional energy inputs.

From a technological point of view, it consists of a process that usually makes use of a series of closed and insulated tanks. The tanks are aerated and mixed, similar to conventional aerobic digestion. When waste sludge is added to the tank, the biological organisms proceed to oxidise the volatile organic compounds, thus making that the temperature easily reaches the thermophilic range of 50–60ºC without any external heat input. This high temperature, along with significant volatile oxidation, provides for pathogenic organism destruction. In parallel, the biosolids volume that must be dewatered and/or removed by over 50%, is reduced.

There are two main operational type of industrial autothermal reactors, one patented by Fuchs (www.fuchs-germany.com) which consists in two reactors operated in a daily batch, the usual hydraulic retention time of the reactors are 5–6 days, the first reactor operates between 35–50ºC, and the second at higher temperatures 50–60ºC. The aeration system flow is around $4\,m^3/h\ m^3$. Other patented reactors were developed by NILSA (www.nilsa.com). The main difference with respect to Fuchs technology is that it works only with a single reactor, in the range 55–65ºC an HRT of 7 days and it requires only $2\,m^3/h\ m^3$ of air input. Both technologies reported about the same organic removal, VS reduction of 30–60% (Layden *et al.*, 2007a,b). Curiously, even the first commercial size ATAD built by Fuchs was placed in Bavaria (Germany) in the late 1977, today just some dozens of full-scale ATAD plants are in operation, which does not compare very well with the hundreds of existing WWTP.

The recent improvements carried out on the ATAD operation (known as the second generation) have focused on designing equipment that works at larger hydraulic residence time, HRT, (i.e. at lower influent replacement ratio), and developing high efficiency systems to the aeration of the sludge (prevention of local oxygen deficit, oxygen monitoring) (Staton *et al.*, 2001).

While there is an agreement on the key operating criteria of the process, there is still considerable misunderstanding of the process, frequent claims of special knowledge demand and many instances of inappropriate design (Kelly and Mavinic, 2003). Even if the first study focused on ATAD was done in the late 1960's (Kambhu and Andrews, 1969), only sporadic studies can be encountered afterwards. They are rarely based on investigations aiming at a full comprehensive characterisation of the ATAD process, but rather reporting continued operation of existing plants in operation (Layden, 2007).

For succeeding in autothermal digestion some requirements are reported in a few previous wastewater and sludge treatment studies (EPA, 1990).

- High strength waste is needed; approximately organic feed should ensure a minimum concentration of COD 40000–50000 mg/L and a minimum feed volatile solid content of 2.5% w/w.
- Total solids should be below a maximum concentration 4–6% w/w, to allow adequate mixing.
- Air-input value is recommended at 4 m^3/h m^3 of active reactor volume, in order to fulfill the high oxygen demand for treating organic matter producing enough energy for self-heating.

In this study, sludge provided by the Reus WWTP (owned by Gestió Ambiental i Abastament, S.A.) was used as substrate for ATAD. The feed consists of a mixture of primary and secondary waste sludge in order to keep the same conditions studied in Chapter 1, where thermophilic anaerobic digestion was explored. The mixture composition corresponds to the sludge rate produced in the physico-chemical and secondary wastewater treatments of the Reus municipal WWTP, which is typically encountered in small-medium WWTP (65% primary sludge, 35% secondary sludge). Its solid and volatile matter contents accomplish the main requirements for ATAD.

The main goal of ATAD is to decompose, in aerobic conditions, a fraction of waste organic solids present in excess sludge. As a result of these oxidation reactions, heat is released, which allows to increase the reactor temperature up to thermophilic conditions (above 40ºC) without supplementary heat. Because of the expected higher reaction rates in comparison to anaerobic processes, digester volume is lower than conventional anaerobic digestion and the HRT is typically between 5 and 15 days.

In the course of the project, a rather complete review of ATAD experience was published by Layden *et al.* (2007a,b) reporting some experiences in Europe, which complements a previous one authored by Kelly and Mavinic (2003) focused on examples in North-America. A close inspection of the reported data confirms that the objectives drawn in this research work are still valid and not too much progress has been done in the later years. Literally copying a sentence taken from Layden *et al.* (2007b) 'Very little published information or operational data are available for full-scale ATAD in Europe', anyway some conclusions arising from it are:

- ATAD technology (aeration and mixing) is mostly protected under patents, so potential users have limited available information about ATAD performance.
- Reported information mostly come from operating large-scale plants, with little chance to be extrapolated to different cases.
- Most of the ATAD plants gives VS reduction in the range 30–60%.
- Most of the ATAD plants reports deficient oxygen mass transfer rates, which results in poor performance and odours.

As also stated by Layden *et al.* (2007b): 'There is an urgent need for more up-to date, comprehensive studies on process performance, costs, and comparison with different sludge treatment options. Also, results from research, and information relating to operating experience need to be disseminated better to facilitate a more objective assessment of process performance, potential applicability and more widespread use'.

Baseline effect of HRT on ATAD at 55ºC

The experimentation of the ATAD performance was conducted in the bench-scale described previously. Reactor was maintained at the thermophilic temperature of 55ºC by means of a thermostatic bath. Air input was kept at the recommended range 16–20 L/h. In order to minimise water loss by evaporation, air input was saturated with water at 55ºC before entering the reactor.

Aerobic inoculum for the star-up of the ATAD reactort was provided by the Reus WWTP. However, the inoculum was taken at the mesophillic temperature of 30ºC. Therefore, it was necessary to slowly acclimate these microorganisms into thermophilic range. The strategy followed to acclimate the bacteria was to increase the temperature at intervals of 5ºC, starting from 30ºC, every two weeks. The feed was made intermittently, once a day. After some fluctuations in the operation, the ATAD reactor spent about three months to reach an acceptable steady operation.

Operation at two selected HRT (8 and 10 days) was then followed. Figures 2.2 and 2.3 show the evolution of final COD against time at the different HRT studied. High reduction of COD is observed, at a mean value of 62% of COD, although large fluctuations were observed with a minimum of COD removal of 6% and a maximum of almost 88% being found. Correlation with COD feed is clearly seen and this can be assigned as the main reason. As noted, feed sludge is weekly obtained directly from a real municipal WWTP and its conditions may differ depending on the population and mainly weather conditions. Therefore, weekly changes in COD and solid content require a close control of air flow rate with the consequent fluctuations. Whenever the percent COD removal is considered, the fluctuations are much less evident as they are efficiently regulated by the digestor with a hampering effect. Only in a few cases, oxygen transfer limitations resulted in very low COD removals with some eventual events related to the death of the aerobic microorganisms.

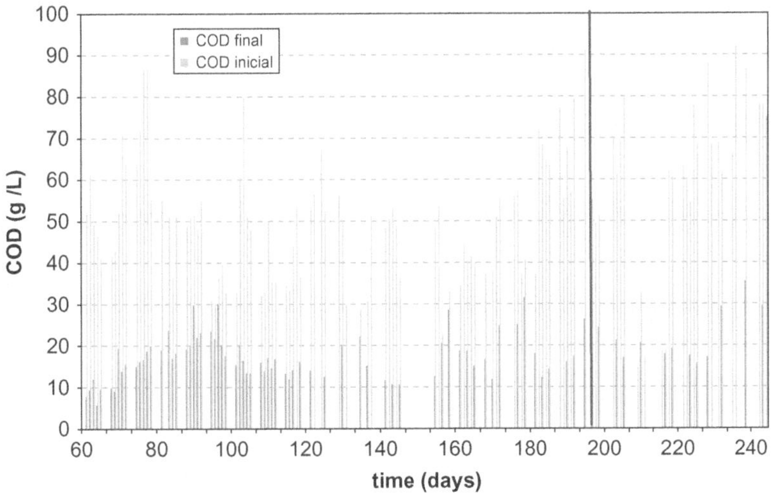

Figure 2.2 Evolution of the initial and final COD in the ATAD bench-scale plant

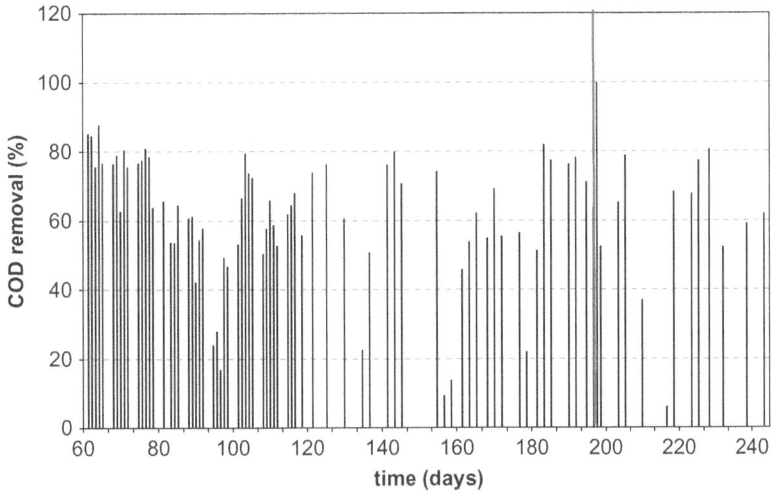

Figure 2.3 Evolution of the COD removal in the ATAD bench-scale plant

A similar behaviour than that shown by COD can be noted regarding to the solids (Figure 2.4) where TS and VS contents are displayed. TS and VS decrease up to 34% and 40% (average values), respectively. It must be pointed out that non significant differences are observed between 10 and 8 days of HRT in terms of COD or solid content removal. These values are comparable to those obtained in anaerobic digestion at higher hydraulic retention times (or lower organic loading rates). For instance, the more similar type of feed found in literature is reported by de la Rubia *et al.* (2006), where thermophilic anaerobic digestion at 55°C of sludge feed with pre-thickened primary and secondary sludge was studied. In this anaerobic system, around 30% of COD removal with an HRT of 20 days and 35% at HRT of 15 days were found. Therefore, ATAD provides much better elimination of COD with shorter HRT as expected. Also, results reported in Chapter 1 'Reduction and stabilisation of the excess and toxicity of sludge in anaerobic conditions' dealing with anaerobic digestion at the thermophilic temperature of 55°C and HRT of 20 days show a 52% of COD removal.

It must be outlined that this performance was obtained with a non-optimised design of the ATAD bench-scale plant. Only open technology was used without the use of any patented equipment. In such an environment, the results compare very well with the performance reported for many of the ATAD plants operating (usually based on Fuchs technology). This suggests that small size municipal WWTP can consider home-made designs fulfilling the pursued needs, without the participation of highly specialised companies and the subsequent savings in the investment needed, if open literature stating the performance of usual aerators and mixers is available.

Thermophilic aerobic treatment reached almost complete hygienisation as pathogen content decreases from the original 2,800,000 NMP/g to just 23 NMP/g of E-coli, and from 5 NMP/g to 0 NMP/g of Salmonella populations.

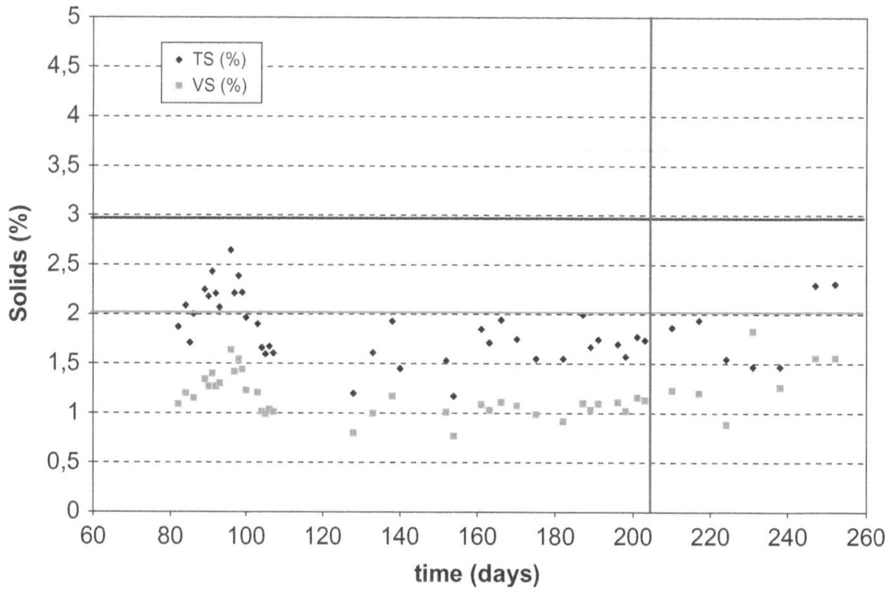

Figure 2.4 Evolution of the total and volatile solids in the ATAD bench-scale plant

Experimental characterisation of aeration and mixing: simulation studies

From a chemical engineering point of view, an ATAD reactor is a very complex process to model, as all the following phenomena interact with others: multiphasic system (air bubbles dispersed into a biological liquid phase containing solid particles), biological reaction, strong coupling between heat and mass transfers. Moreover, the physico-chemical and rheological properties of sludge depend on many parameters (total solid content, biological flocs, temperature, particle size) and change in time (origin, pre-treatment, operating conditions). All these characteristics strongly influence hydrodynamics (mixing), heat transfer (operating temperature), gas-liquid mass transfer (oxygen supply for microorganisms), and thus biological performances (organic matter degradation, pathogen destruction).

With regard to this complexity, tailored investigations are required for optimising mixing and aeration and for better understanding the phenomena involved and their coupling. A special attention should be paid on the relationship between the gas-liquid mass transfer performances and the biological degradation. For that, this research work proposed a combination of investigations carried out on: (i) the biological performance of degradation (rate, efficiency, oxygen requirement), (ii) and on mixing and aeration. This original integrative approach is the key to access to the better understanding of the ATAD process.

For the second part of the study, as a first step, it was decided to work with mock-ups, namely with using transparent fluids which properties (rheology, surface tension, density) are the closest than the ones of sludge. The methodology adopted is based on the keywords: rheology of sludge, choice of model fluids, experimental and numerical investigations, aiming at formulating the influence on aeration of sludge rheology and temperature. To conduct this research work, a Ph.D. thesis was begun in October 2007 at the laboratories of Laboratoire GEPEA of Université de Nantes.

The main achievements deal with: (i) the optimisation of the agitation and aeration system in the lab-scale ATAD reactor of Universitat Rovira i Virgili, (ii) the implementation of specific experimental tools (measurements of the power dissipated by the impeller, the aeration capacity and homogeneity, the local gas-liquid hydrodynamics properties); (iii) the characterisation of sludge rheology (issued from Gestió Ambiental i Abastament, S.A. and used by Universitat Rovira i Virgili), (iv) the first steps towards the numerical simulation of the ATAD reactor without mechanical agitation.

Usual investigations on power consumption and aeration capacity were carried out to, firstly characterise the performances of the system implemented by Universitat Rovira i Virgili in the lab-scale ATAD reactor, and secondly, to optimise them. The initial agitation system was based on an anchor-type paddle and an air sparger consisting in four porous ceramic nozzles distributed at the tank bottom. As not available in the literature (Nagata, 1975 and Paul *et al.*, 2003), it was first necessary to experimentally establish the power number curve of this agitation system under ungassed and gassed conditions (torque measurements), as well as the associated gas-liquid flow maps and aeration performances.

When compared to the usual impellers, this device appeared not able to supply enough oxygen with respect to the potential maximum biological needs of the ATAD operation. To improve these performances, this initial agitation system was changed: based on a literature review (Paul *et al.*, 2003), a six-concave-blades disk turbine (one of the most efficient impeller for dispersing gas) and a ring air sparger were chosen, and four baffles were introduced. An identical methodology of characterisation was used and demonstrated that a significant improvement was achieved in water when compared to the old system. The further experimental works will be devoted to the validation of the latter data in non-Newtonian model fluids (and at last in sludge), and to the establishment of a relationship between rheological properties and gas-liquid mass transfer performances. The results expected will be original, as:

- such methodology has never been applied to an ATAD process,
- the literature involving non-newtonian fluids in stirred aerated fluids is quite poor, except for fermentation application (Vlaev *et al.*, 2002; Espinosa-Solares *et al.*, 2002 and Cabaret *et al.*, 2008). The literature's overview indicates that the characterisation of both local gas-liquid hydrodynamics and mass transfer is uncommonly integrated in these kinds of systems, which constitutes yet a real limitation for understanding the coupling between the different phenomena and thus for optimising the process.

Concerning the sludge rheology, a state-of-art is presented by Seyssiecq *et al.* (2003). These authors explained the different methods used (advantages, limits, problems interfering with measurements) and also summarised the main characteristics encountered, in particular depending on the type of sludge and on their composition. They outlined the sludge variability and the need to carry out specific measurements for each sample. The results revealed that sludge in presence in the ATAD process is a non-newtonian fluid, with a shear-thinning behavior, a small yield stress, no viscoelasticity and no time-dependency. Moreover, a variability of these properties was observed as a function of the moment when sludge is sampled. The information collected on sludge is in agreement with the literature, and has been used in the present work to choice the 'best' model fluids (mock-up experiments), the ones behaving as close as possible like sludge (various aqueous solutions of CMC and Xanthan gum) for further investigation.

Optimisation of aeration and mixing

The new agitation system was designed so as to be respectful for the standard criteria used in stirred aerated tank (H = T, C = T/4). Figure 2.5a shows an elevation view of the tank modified. As well known for its efficiency to disperse gas by conserving high RPD under gassed conditions (Paul *et al.*, 2003), the six-concave-blades disk turbine was retained as new impeller. Even if home-made, the dimensions are identical to those of the commercial impeller CD6 Chemineer® (Figure 2.5b): 8.5 cm (D = T/3) in impeller diameter, 6.4 cm (3/4D) in disk diameter, six blades with 2.1 cm in length (D/4), 1.7 cm in vertical height (D/5), 0.8 cm in horizontal width (D/10). The four air nozzles are replaced by a ring sparger (Figure 2.5c) with a diameter equal to the impeller diameter (T) (Xuereb *et al.*, 2006), and perforated by 20 holes of 0.5 mm in diameter. Four baffles (in stainless steel) are added to the tank to prevent the vortex formation at the surface, enabling thus high rotation speeds to be applied. Their dimensions are: 30 cm in a total height, 2.1 cm (T/10) in width 1.0 cm in thickness; they are located at 4 mm (T/50) from the tank walls.

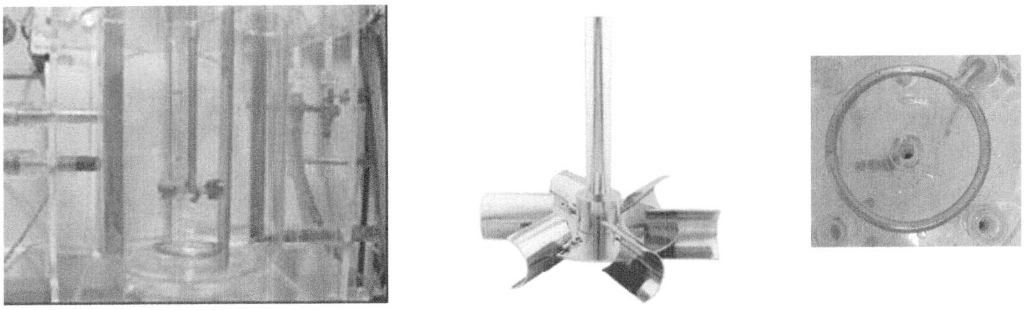

(a) (b) (c)

Figure 2.5 Lab-scale reactor equipped with the new agitation system: (a) Elevation view of the reactor, (b) Six-concave-blades disk turbine, (c) Ring sparger

Then, power consumption, gas-liquid regime and gas liquid mass transfer of the old and new systems were characterised. Only, the results corresponding to the latter are shown in order to assess the major improvements achieved.

To illustrate the effect of agitation, the variations of the mean volumetric mass transfer coefficients (averaged over the oxygen probes at the top and bottom) with the air flow rates are shown for both systems in water, without agitation (Figure 2.6) and at various rotation speeds (Figure 2.7). As expected, k_La increases with increasing gas flow rates and rotation speeds, whatever the agitation systems. This is mainly because, as the impeller speed increases, gas dispersion and bubble break-up are favoured, enhancing thus the surface area available for mass transfer. More important in the comparison of both agitation systems; it is observed that: (i) from Figure 2.6, the k_La values are higher when using a ring sparger instead of four nozzles, (ii) the new impeller significantly improves (from two to ten times higher) the aeration capacity (Figure 2.7).

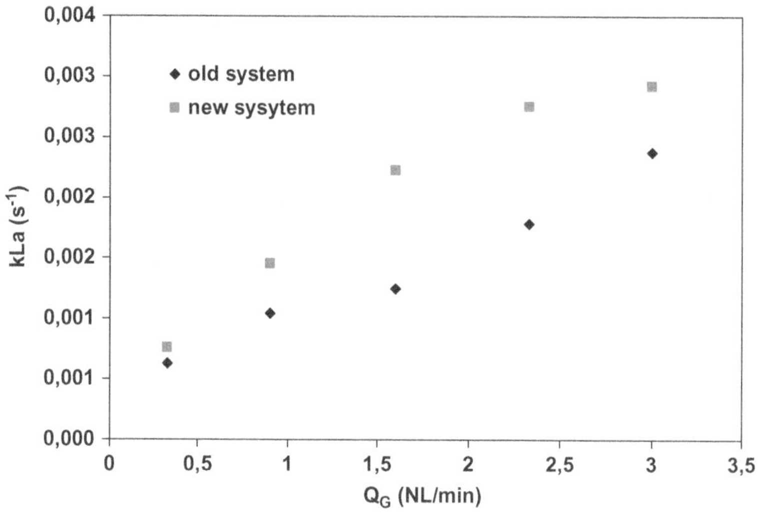

Figure 2.6 Gas-liquid mass transfer coefficient (mean value) versus air flow rate: comparison between the old and the new system without agitation

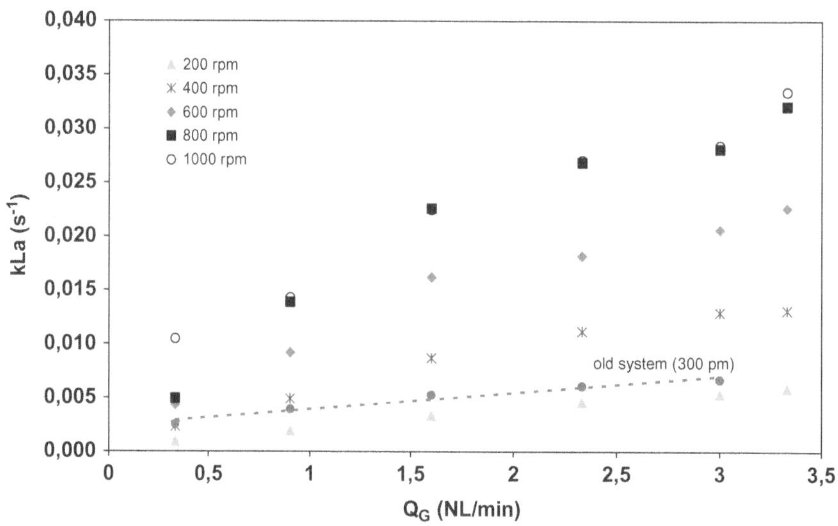

Figure 2.7 Gas-liquid mass transfer coefficient (mean value) versus air flow rate: influence of the rotation speed for the new system

These results are modelled according to the relationships usually available in literature:

$$k_La = C \cdot N^a \cdot Q_G^b \qquad \text{Eq. 1}$$

$$\frac{k_La}{N} = k \cdot Fr^e \cdot Fl_G^f \qquad \text{Eq. 2}$$

$$k_La = A \cdot \left(\frac{P_g}{\rho_L V}\right)^{\alpha} \cdot U_G^{\beta} \qquad \text{Eq. 3}$$

The associated constants are collected in Table 2.1 for both agitation systems. When compared to the well-known correlation of Van't Riet (1979) for non-ionic solutions, the exponent on U_G is found identical for the old system (0.51 against 0.5) and higher for the new system (0.77 against 0.5); the exponent on the power dissipated is smaller for the old system (0.17 against 0.39) and identical for the new system (0.39 against 0.40).

Table 2.1 Model parameters associated with Eqs (2–4)

	Eq (2)			Eq (3)			Eq (4)		
	C	a	b	k	e	f	A	α	β
Old system	0.41	0.59	0.51	0.013	0.05	0.51	0.35	0.17	0.51
New system	2.53	1.22	0.77	0.099	0.49	0.77	4.71	0.39	0.77

In order to test the axial homogeneity of oxygen mass transfer in the tank, two dissolved oxygen probes are used and the criterion (h) is applied as stated in Eq. 4 (Cabaret *et al.*, 2008).

$$h = \frac{1}{n}\sum_{n}\left(k_La_{top} - k_La_{bottom}\right)/k_La_{mean} \qquad \text{Eq. 4}$$

In Eq. 4, n represents the number of experiments (n = 15 and 30 for the old and new systems respectively, as averaged over all the Q_G). It is clear that h represents the mean deviation between k_La at the top and the bottom of the tank. For both agitation systems, a rise in rotation speeds enables the gas-liquid mass transfer to be better homogenised in the tank. No noticeable difference is observed whatever the conditions (h varying between 5 to 8 % for the lowest and highest rotation speeds respectively). Therefore, it is concluded that uniform distribution of oxygen is obtained with this system, avoiding possible local deficit of oxygen.

Assessment of better aeration and impact of pre-treatments

These new impeller, baffle system and sparger proposed by Laboratoire GEPEA of Université de Nantes, although certainly with some delay with respect to the schedule, were implemented into the bench-scale plant of Universitat Rovira i Virgili, which was then ready to assess the benefits on the ATAD performance under this better aeration design using raw sludge. Later, it was planned to check the use of pre-treated sludge (as described in Chapter 1 'Reduction and stabilisation of the excess and toxicity of sludge in anaerobic conditions') in order to overcome the usual anaerobic step applied in ATAD systems for enhancing the biodegradability of the waste sludge.

Unfortunately, the expected improvement in the operation was not possible to be substantiated because the work had to be stopped in month 32. It is not too optimistic to presume that VS removal could achieve, even surpass, the highest level provided by patented technologies, but just using current equipment and usual engineering design tools. This information was also needed for proper comparison with anaerobic digestion options in order to establish some guidelines about selection of more suited process in each situation.

Numerical study on mixing and aeration

About CFD, it must be noted that the ATAD reactor is so complex that its entire numerical simulation is nowadays very unrealistic. Therefore, only partial simulations (without mechanical agitation) were performed and served as a complementary tool for investigating the mechanisms involved and formulating a process model (scaling-up). This numerical study was begun with considering a benchmark reactor (internal loop airlift reactor), enabling thus a strategy for simulations to be defined (Talvy *et al.*, 2007).

At present, the different steps towards its application in the ATAD system are occurring: mesh definition (cell size and shape, refinement), choice of the most realistic boundary/symmetry conditions (surface deformation, sparger area), choice of the closure laws and/or models (drag coefficient, turbulence model. . .), and tests relating to the influence of various numerical parameters on results. Despite the discontinuation of the work package, this study will constitute a major chapter in the ongoing thesis carried out on this subject as the research will be mainatained.

Concluding remarks

A literature revision evidences that, even being considered an established process, not too much information is yet accessible for potential users on performance and design of ATAD systems. Most designs are proprietary technologies with limited open knowledge and, in many cases, still reporting operation problems such as insufficient aeration. Although lab-scale studies demonstrate the feasibility of this alternative for sludge sanitation and stabilisation, to date very few full-scale plants are in operation, mainly when compared to the huge amount of operational municipal WWTP. The lack of a wider implementation must probably be assigned to the fact that patented technology is needed, which increases the investment cost, and insufficient open literature makes difficult a supported decision for engineers.

In this research study, the possibility of building up bench-scale plants with open technology was demonstrated. Thus, ATAD tests over a given sludge can be conducted without too much effort in order to provide performance information for decision making purposes.

Even in the non-optimised ATAD bench-scale reactor built up as first approach, COD reduction and VS removal was found at levels comparable to full-scale plants in operation, which supports the potential of this sludge reduction option.

In addition, mixing and aeration optimisation have been demonstrated to be possible through open literature using classical mixers and established mass transfer concepts, even if experimental testing is needed. In addition, new developed protocols now allow characterising the rheology of the sludge and its impact on aeration with support for oxygen mass transfer rate, gas distribution and agitation power needed for home-made designs.

Beyond of this project, this research work will give elements at disposal for establishing a relationship between the aeration performances and the biological degradation (efficiency, thermophilic temperature, autothermal conditions) in such systems, applicable to a wide diversity of situations.

Unfortunately, the benefits of the optimised aeration and mixing system designed and later installed in the reactor could not be experimentally substantiated due to the discontinuation of the work package after the European Commission decision. Neither the expected positive effect of using pre-treated sludge, with enhanced biodegradability, has been possible to be supported through experimentation.

Nevertheless, ATAD is seen as a very attractive option for small size municipal WWTP, for which high investment in anaerobic systems for biogas production is not justified. In addition, the ATAD systems can be home-made designed in order to give proper performance, with the support of freely available knowledge and engineering tools.

IMPACT OF THE TECHNOLOGY ON ITS INDUSTRY OR RESEARCH SECTOR

Overall view

In recent years, an increased attention has been carried out to minimize the amount and toxicity of waste sludge, because it represents a rising challenge for wastewater treatment plants (WWPT) due to economic and environmental factors. For instance, the cost of sludge treatment and disposal has accounted up to 60% of the total operating cost in municipal WWTPs. Due to the expanding construction of sewage sludge plants for the biological treatment of urban and industrial effluents, the digestion of waste sludge is an essential part of wastewater treatment system.

At EU level the recycling of sludge accounts for about 40% of the overall sludge production, landfilling as well as incineration in some Member States are still the most widely used disposal outlets despite their environmental drawbacks. This situation must be seen as no longer acceptable so a total treatment/recycling of the sludge should be encouraged in the future. For this goal, agricultural landspreading can be one major use of stabilised sludge, although applications have to be developed.

Digestion/stabilisation of sludge can be carried out both aerobically and anaerobically. The anaerobic digestion is the most widespread and probably represents an environmentally friendly use of this waste because of its potential to yield biogas, which can be applied to cut carbon dioxide emissions and contribute to reach the Kyoto's objectives (Rulkens, 2008), although direct combustion has been also suggested (Werther and Ogada, 1999). However, biogas use requires adapted installations and cogeneration system to generate electricity and the investment needed for this option could not be justified in small-medium scaled WWTP. In such cases, Autothermal Thermophilic Aerobic Digestion ATAD appears to be an attractive alternative to reduce sludge amount and produce Class A biosolids that can be applied to agricultural land without restrictions.

For instance, in 1993, the USEPA (United States Environmental Protection Agency) regulation 40 CFR Part 503 (Standards for the use and disposal of sewage sludge) for sludge management came into existence. The new regulations have promulgated a higher pathogen destruction and a vector attraction reduction: such conditions are not possible by the conventional methods of sludge digestion. These pathogen requirements are based on the performance standards which define two classes of pathogen reduction: class A (PFRP, a Process to Further Reduce Pathogens) and class B (PSRP, a Process to Significantly Reduce Pathogens). Many municipalities have not access to sites for disposal of class B bio-solids. Because of the fewer restrictions to the land application of class A bio-solids, municipalities are increasingly choosing to produce class A bio-solids. Due to these new compliance parameters, it has become less beneficial from a cost and operation point of view for the municipalities to continue with their current bio-solids management techniques. That is why there is a considerable interest for the ATAD operation. In Europe, Sewage Sludge Directive 86/278/EEC, in connection to the Urban Waste Water Treatment Directive 91/271/EEC, pursues similar aims. ATAD can fulfil these requirements, but clearly the lack of information is preventing this alternative from a wider implementation.

Although the biochemical and physical steps involved in ATAD are well stated (Staton *et al.*, 2001), there is a deficit of ATAD models implementing all the steps in a unique routine using powerful tools as CFD and only some simplified model is recently available (Gomez *et al.*, 2007). Instead, even recent advances in ATAD design, designated as the second ATAD generation (Staton *et al.*, 2001), seems to be done using a 'trial-and-error' approach based in empirical experience of a number of ATAD plants (Kelly, 2003; Layden, 2007, Bartkowska and Dzienis 2007 and Layden *et al.*, 2007a,b). Thus, Kelly (2003) states 'there are no typical sludges, ...ATAD remains very much an art and universal solutions are not yet available given the variety of different sludge types and the variability of sludge character'. In addition, most of the available technology is in hands of a few owners (Fuchs, NILSA), which limits the decision capacity of the potential users and forces a subsequent dependency if implemented. As stated by Layden *et al.* (2007a), 'from the literature, it appears that the knowledge and understanding of ATAD rest with a limited number of people; better dissemination of information (i.e., research findings) is needed if ATAD is to be more widely used'.

As a result of this research work, a full set of knowledge has been generated that will allow end-users (also designers) reaching freely available elements for equipment design and operation diagnostic, which must support a wider implementation of ATAD. This set of tools will help interested end-users to evaluate the feasibility of ATAD for specific cases at relatively low-cost as no pilot-plant evaluation will be needed, and classical equipment could be used. Therefore, the approach will be conducted using the equipments, systems and protocols developed here:

- Bench-scale evaluation of ATAD performance (COD and VS reduction, OUR, etc.),
- Rheological characterisation of the sludge,
- Determination of oxygen mass transfer rates,
- Optimisation of classical mixing and aeration systems,
- Modelling with CFD of the full operation.

Applicability of ATAD

In order to assess the economic impact of application of ATAD, a digester is designed with a total residence time of six days (as it is believed that the eight days tested in the bench-scale reactor could be cut with the improvement of mixing and aeration, however not assessed because the discontinuation of the work package). In order to dehydrate the digested sludge, a press filter is used with previous addition of dilute polyelectrolyte for improving the sludge dryness.

The estimation of the operation costs was conducted for the municipal WWTP from Reus, which is an average size town (Catalonia, Spain), owned by Gestió Ambiental i Abastament, S.A. with a depuration capacity for a population of 60000 equivalent inhabitants. Its size is precisely in the frontier between the economic viability of ATAD and TAD options.

Calculation hypothesis

The calculation basis for the feed sludge volume to the digester is taken as $100\,m^3$/day. In Table 2.2, the input volatile solid and total solid data are shown. The output values are taken from experimental data or present operation costs. According to EPA (1990) the oxygen uptake could be taken as 1.42 kg O_2/kg VM destroyed.

Table 2.2 Data used in the calculations

Parameter	Input	Output
Sludge volume (m^3/day)	100	100
% DM	3.70	2.23
% VM	77.0	66.7
Dehydrated sludge dryness, %	24.5	–
Polyelectrolyte dose, %	18	–

Mass balance

Taking into account the data shown in Table 2.2 and the hypothesis listed in the previous section, the total solids and volatile solids amounts are calculated at the output of the digester. The results are shown in Table 2.3.

From Table 2.2 and Table 2.3, the necessary amount of polyelectrolyte for dehydrating the sludge and the amount of dehydrated sludge is calculated. These data are shown in Table 2.4.

Table 2.3 Mass balance summary

Parameter	Input	Output
Sludge volume (m^3/day)	100	100
kg/day DM	3700	2230
kg/day VM	2850	1487

Table 2.4 Dehydrated sludge data

Parameter	Value
Dilute polymer (kg/day)	18000
Dehydrated sludge production (kg/day)	9102

Regarding to the oxygen needs and the energy requirements for aeration and according to (EPA, 1990), it is 2.1 kgO_2/kWh. The data are shown in Table 2.5.

Table 2.5 Aeration and mixing requirements

Parameter	Value
VM destroyed (kg/day)	1363
Oxygen need (kg/day)	1935
Aeration requirement (kW)	57.6

Operation costs

The main operation costs are the dehydrated sludge management and transport and the polyelectrolyte consumption.

Experimental data obtained note that still 50% of the sludge produced does not accomplish the 86/278/CEE directive. Although, it is expected that improvement in mixing must increase COD destruction and consequently also pathogen destruction, the economic estimation will be made considering this conservative value. Then landfill management and transport costs are taken into account. The main data are shown in Table 2.6.

Table 2.6 Main operation costs

Parameter	Value
Polymer cost (€/kg)	2.94
Dilute polymer cost, 0.12% (€/t)	3.53
Agriculture sludge management (€/t)	9.40
Agriculture sludge transport (€/t)	11.2
Landfill sludge management (€/t)	50.3
Landfill sludge transport (€/t)	13.3

From Table 2.6, the overall operation costs are calculated and shown in Table 2.7 in form of ratio to the dry matter amount entering to the digester.

Table 2.7 Overall operation costs

Parameter	Cost (€/day)	Ratio (€/t DM)
Sludge management	383.4	103.6
Polymer	63.5	17.2
Total	446.9	120.8

It must be noted that this cost is competitive if compared with thermophilic processes as it is not significantly different within the estimation error. The application of pre-treatments could give about 10% saving because of the enhancement of VS.

REFERENCES

Bartkowska I. and Dzienis L. (2007). Technical and Economic Aspects of Autothermal Thermophilic Aerobic Digestion Exemplified by Sewage Treatment Plant in Gizyco. *Env. Protec. Eng.*, **33**, 17–24.

Cabaret F., Fradette L. and Tanguy P. A. (2008). Gas-liquid mass transfer in unbaffled dual-impeller mixers. *Chem. Eng. Sci.*, **63**, 1636–1647.

De la Rubia M. A., Perez M., Romero L. I. and Sales D. (2006). Effect of solids retention time (SRT) on pilot scale anaerobic thermophilic sludge digestion. *Proc. Biochem.*, **41**, 79–86.

EPA. (1990). (Environmental Protection Agency) Autothermal Thermophilic Aerobic Digestion of Municipal Wastewater Sludge; Environmental Regulations and Technology, Washington, USA.

Espinosa-Solares T., Brito-De La Fuente E., Tecante A. and Tanguy P. A. (2002). Gas dispersion in rheologically-evolving model fluids by hydrid dual mixing systems. *Chem. Eng. Technol.*, **25**, 723–727.

Fuchs Gas und Wassertechnik GmbH (www.fuchs-germany.com).

Gomez J., de Gracia M., Ayesa E. and Garcia-Heras J. L. (2007). Mathematical modelling of autothermal thermophilic aerobic digesters. *Wat. Res.*, **41**, 959–968.

Kambhu, K. and Andrews, J. F. (1969). Aerobic thermophilic process for the biological treatment of wastes – simulation studies. *J. Water Pollut. Control Fed.*, **41**, R127.

Kelly H. G. and Mavinic D. S. (2003). Auto-thermal thermophilic aerobic digestion research, application and operational experience, WEFTEC workshop W104, Thermophilic Digestion: Hot Update! Los Angeles Convention Centre, Los Angeles (Ca, USA), October 11.

Layden, N. M. (2007). An evaluation of autothermal thermophilic aerobic digestion (ATAD) of municipal sludge in Ireland. *J. Env. Eng. Sci.*, **6**, 19–29.

Layden N. M., Mavinic D. S., Kelly H. G., Moles R. and Bartlett J. (2007a) Autothermal thermophilic aerobic digestion (ATAD) – Part I: Review of origins, design, and process operation. *J. Env. Eng. Sci.*, **6**, 665–678.

Layden N. M., Kelly H. G., Mavinic D. S., Moles R. and Bartlett J. (2007b). Autothermal thermophilic aerobic digestion (ATAD) – Part II: Review of research and full-scale operating experiences. *J. Env. Eng. Sci.*, **6**, 679–690.

Nagata S. (1975). Mixing principles and applications. Kodansha scientific books, John Wiley & Sons, New York.

NILSA, Navarra de Infraestructuras Locales, S.A. (www.nilsa.com).

Paul E. L., V. A. Atiemo-Obeng, S. M. and Kresta S. M. (eds). (2003). Handbook of industrial mixing: science and practise. WileyBlackwell, New York.

Rulkens W. (2008). Sewage sludge as a biomass resource for the production of energy: overview and assessment of the various options. *Energy & Fuels*, **22**, 9–15.

Seyssiecq I., Ferrasse J.-H. and Roche N. (2003). State-of-the-art: rheological characterisation of wastewater treatment sludge. *Biochem. Eng. J.*, **16**, 41–56.

Staton K. L., Alleman J. E., Presley R. L. and Eloff J. (2001). 2nd Generation autothermal thermophilic aerobic digestion: conceptual issues and process advancements WEF/AWWA/CWEA Joint Residuals and Biosolids Management Conference Biosolids 2001: Building Public Support, Proceedings.

Talvy S., Cockx A. and Liné A. (2007). Modelling hydrodynamics of gas-liquid airlift reactor. *AIChE J.*, **53**, 335–353.

Van't Riet K. (1979). Review of measuring methods and results in non-viscous gas-liquid mass transfer in stirred vessels. *Ind. Eng. Chem. Process Des. Dev.*, **18**, 357–364.

Vlaev S. D., Valeva M. D. and Mann R. (2006). Some effects of rheology on the spatial distribution of gas hold up in a mechanically agitated vessel. *Chem. Eng. J.*, **87**, 21–30.

Werther J. and Ogada T. (1999). Sewage sludge combustion. *Prog. Ener. Comb. Sci.*, **25**, 55–116.

Xuereb C., Poux M. and Bertrand J. (2006). Agitation et mélange. Aspects fondamentaux et applications industrielles. L'Usine Nouvelle Dunod, Paris.

Chapter 3

Production of hydrogen from sludge

Alan Guwy, Richard Dinsdale, Jaime Massanet-Nicolau, Jon Maddy, Sandra Esteves and Giuliano Premier

SUMMARY

Project objectives

To investigate operating conditions for fermentation of sewage sludge giving a maximal yield of H_2 and fermentation products from anaerobic digestion:

- Design, construction and testing of a fully instrumented and controlled bioreactor producing hydrogen from pre-treated sludge and the provision of data for assessing the use of a liquid by-product.
- Production and evaluation of the data from an optimised and enhanced treatment process generating a hydrogen enriched biogas from sewage sludge.
- To evaluate the technical and economic feasibility of fermentation of sewage sludge to H_2. To make recommendations on the applicability of this technology to sewage treatment works
- Assessment of the economic and technological impact of this sludge treatment process on large sludge processing installations.

Work performed

A multistage fermentation system was designed, constructed and commissioned to carry out these objectives. The apparatus, was purposely built, and is capable of performing two-phase fermentation of sewage biosolids, producing hydrogen and methane sequentially. It is also capable of automatic continuous feeding at hourly intervals at a pre-defined hydraulic retention time, continuous real-time monitoring of environmental variables in the bioreactors and the control of these variables during fermentation.

This apparatus was used to demonstrate and optimise two-phase fermentation of sewage biosolids, producing hydrogen and methane sequentially. This involved, evaluating pre-treatment strategies for the sewage biosolids, developing methods to inhibit competing microbial activity and integrating hydrogen and methane producing bioreactors. An economic and technical feasibility of implementing the technology developed at full scale was also prepared.

METHODOLOGIES AND APPROACHES EMPLOYED

Fermentation apparatus

The apparatus consists of five main modules, a heat treatment stage, a storage and delivery stage, a hydrogen producing fermentation stage, a methanogenic fermentation stage and a waste monitoring and data collection stage. The hydrogen bioreactor has a working volume of 10 L and the methanogenic bioreactor has a working volume of 100 L. Each bioreactor is instrumented to allow the following variables to be tracked in real-time:

- pH
- Redox potential
- Temperature
- Gas production rate
- Hydrogen production (hydrogen bioreactor only)
- Methane production
- Carbon dioxide production

A schematic showing the set-up for both the hydrogen and methanogenic reactors including ancillary monitoring and control equipment is given in Figure 3.1.

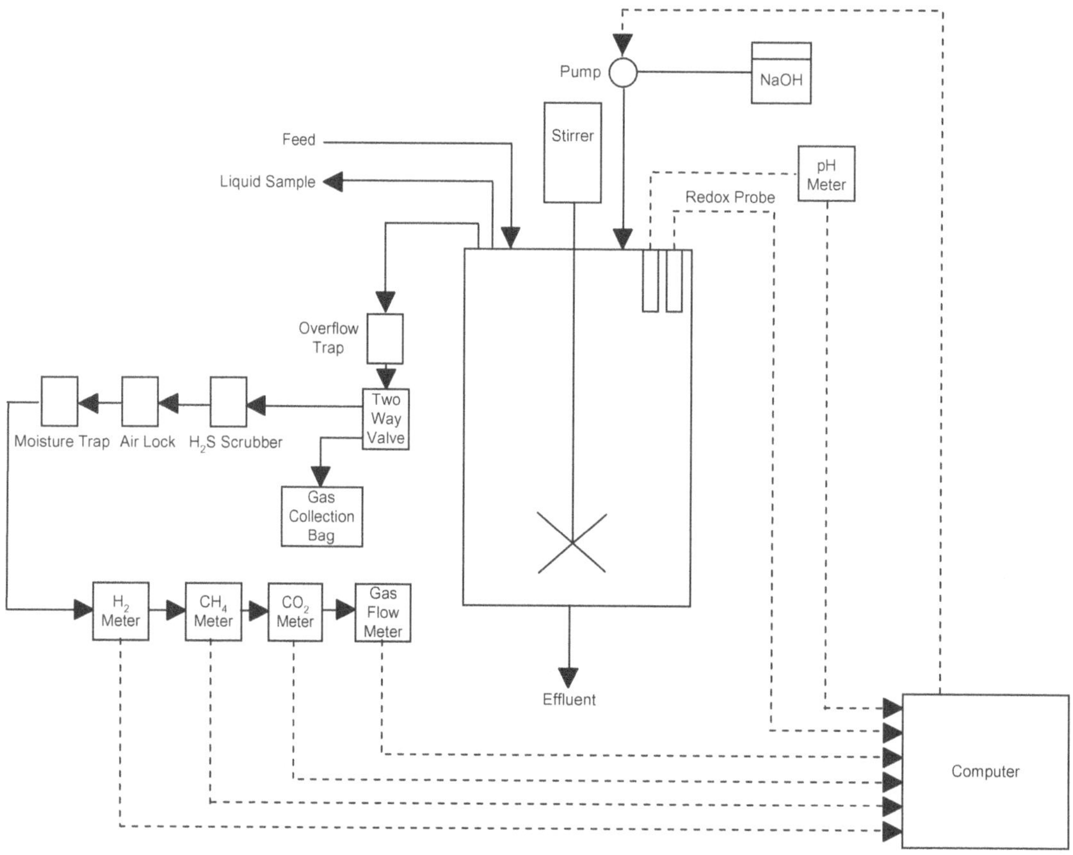

Figure 3.1 Schematic of reactor showing equipment used for monitoring and environmental control

The fermentation system also incorporates control equipment and programs which allow the following tasks to be performed automatically:

- Feeding of the bioreactors at a predefined hydraulic retention time (HRT).
- Delivery of saccharifying agents such as cellulase enzyme, at a specified rate.
- Maintaining bioreactor pH at or above a fixed point via addition of 2 M NaOH.
- Controlling the temperature of bioreactor contents using a thermostatically controlled electric heating jacket.
- Sparging of the bioreactor contents at a specified rate.

Evaluation of pre-treatment strategies

Three different pre-treatment strategies were evaluated; with a view to increasing soluble carbohydrate content in the feedstock biosolids and consequently hydrogen yield. These pre-treatments were pasteurisation combined with enzymatic saccharification, pasteurisation combined with acid hydrolysis, and thermal hydrolysis. Pasteurisation of the feedstock biosolids involved heating the biosolids to 70°C for a period of 1 hour. After pasteurisation, feedstock biosolids were kept at 2–8°C for no more than 7 days prior to being fed into the hydrogen bioreactor.

Enzymatic saccharification was carried out using a commercially available product called Cellulase 13 L which is used in the manufacture of foodstuffs, to break down structural carbohydrates such as cellulose. Cellulase 13 L is known to contain the enzyme cellulase at a concentration of 1,500 units/L (One unit of enzyme is defined as the amount of enzyme which causes the release of one micromole of glucose equivalents per minute at pH 4.6 and 40°C). In addition the product is known to contain a mixture of other enzymes capable of degrading complex carbohydrates but which are not assayed by the manufacturer. The product also contained some soluble carbohydrate as a preservative but this was removed via ultra filtration by the manufacturer prior to delivery. The Cellulase 13 L was added to the hydrogen bio-reactor at a concentration of 5% by volume at the start of the fermentation experiments and it was also added to the bioreactor during continuous feeding at a rate sufficient to maintain a 5% concentration by volume.

Acid hydrolysis was performed by reducing the pH of the feedstock biosolids to 2.0 via the addition of concentrated HCl. The acidified feedstock biosolids were then stored at room temperature for a period of 24 hours before being neutralised to pH 7.0 using 10 M NaOH.

Thermal hydrolysis of sewage biosolids was performed using a full-scale commercial thermal hydrolysis process (THP), which involves heating the sewage biosolids to 165°C at 6 bar for 20 minutes. The feedstock biosolids used when evaluating thermal hydrolysis was a mixture of primary sewage biosolids and waste activated sludge in a ratio of 70:30 by volume. This was necessary because the THP was used to treat this mixture of sewage biosolids only.

Stabilisation of hydrogen production

Pre-treating sewage biosolids using pasteurisation combined with enzymatic saccharification resulted in successful hydrogen production; however after two days hydrogen production began to decline. Two methods of preventing this decline in hydrogen production were evaluated; flushing of the hydrogen bioreactor head-space, and sparging of the bioreactor contents. In both cases the gas used was nitrogen at a flow rate of 0.2 L/min. A third fermentation experiment was conducted, again using nitrogen sparging, but at a higher rate of 0.6 L/min.

Two-phase fermentation of sewage biosolids

Having established that fermentative hydrogen production from sewage biosolids was possible in a continuously fed bioreactor; the next phase of the research involved adding a second, methane producing, fermentation stage to improve energy recovery from the sewage biosolids and to further stabilise them by reducing volatile solids and volatile fatty acids. In this two-phase fermentation system, the hydrogen bioreactor was operated under the same conditions as in the previous phase of this research; briefly, pasteurisation and enzymatic saccharification were applied to the feedstock biosolids and the bioreactor was sparged with nitrogen at 0.6 L/min. Effluent from the hydrogen bioreactor was used as the feedstock for a second methane bioreactor with a volume of 100 L which was operated at a HRT of 10 days.

Optimisation of sparging process

A methodology had been developed, that addressed the main issues in advancing fermentative hydrogen production from sewage biosolids, from laboratory scale research, towards full scale implementation. The next phase of the research involved evaluating methods that would improve the cost and practicality of the proposed fermentation methodology. To this end, the sparging process was developed further, in an effort to make it more cost and energy efficient. Two experiments were performed in this phase of the study. Firstly, the hydrogen bioreactor was operated alone with nitrogen being sparged into the bioreactor at a different rates; the aim of this experiment was to establish the optimal sparging rate in terms of hydrogen yield and production rate. The hydrogen bioreactor was started up and operated as in the previous sparging experiments and once continuous feeding had commenced, six different sparging rates of between 0.2 and 0.7 L/min were evaluated for a period 3 HRTs each.

In the second experiment, a two-phase fermentation was carried out using carbon dioxide instead of nitrogen to sparge the hydrogen bioreactor at a rate of 0.6 L/min. Start up and operation of this experiment was the same as for the two-phase fermentation experiment described in the previous phase of the research. The fermentation experiment was continued for 3 HRTs of the methane bioreactor (30 days in total).

ACHIEVEMENTS RELATED TO THE STATE OF THE ART

Current knowledge regarding fermentative hydrogen production

Background

Existing Research has shown that hydrogen can be produced from a wide array of biomass sources by fermentation (de Vrije and Claassen, 2003; Hawkes *et al.*, 2007 and Nandi and Sengupta, 1998). This process is carried out by obligately anaerobic microorganisms such as those found in the genus Clostridium. These microorganisms convert carbohydrates in the biomass to hydrogen and carbon dioxide according to the following formulae:

$$C_6H_{12}O_6 + 2 \cdot H_2O \rightarrow 2 \cdot CH_3COOH + 4 \cdot H_2 \qquad \text{Eq. 1}$$

$$C_6H_{12}O_6 \rightarrow CH_3CH_2CH_2COOH + 2 \cdot CO_2 + H_2O \qquad \text{Eq. 2}$$

As the above formulae show, the end products of this conversion process include acetic acid or butyric acid. As with conventional anaerobic digestion the process can occur at mesophilic temperatures and in the absence of oxygen, meaning that no aeration and minimal heating is required to facilitate the process.

The stoichiometric yield of hydrogen via fermentation is 4 moles H_2 per mole of hexose when acetate is the sole product and 2 moles H_2 per mole of hexose when butyrate is produced. In practice a mixture of acetate and butyrate is often produced by fermentative hydrogen producers, resulting in a typical yield of 3 moles H_2 per mole of hexose (Hawkes *et al.*, 2002; Levin *et al.*, 2004 and Nath and Das, 2004).

Fermentative hydrogen production frequently occurs in nature, in anaerobic environments including lake and river sediments and in the digestive systems of ruminants; however, hydrogen is rarely observed in these environments since it is energy rich, and rapidly consumed by other microorganisms such as methanogens, homoacetogens and sulphur reducing bacteria (Corder *et al.*, 1983 and Scholten *et al.*, 2000). The rapid consumption of hydrogen in these environments benefits the hydrogen producers since a low partial pressure makes hydrogen producing metabolism more efficient. Thus hydrogen producers and hydrogen consumers participate in a syntrophic relationship based upon interspecies hydrogen transfer (Traore *et al.*, 1983).

In addition to utilising hydrogen, some methane producers are able to utilise the by-products of fermentative hydrogen production (a mixture of acetate and butyrate) to produce methane. Since the environmental optima and growth rate of methane producing microbes are different to those of the hydrogen producing microorganisms (Fenchel and Finlay, 1995); and because methanogenic microbes may consume the hydrogen before it can be recovered, methanogenic fermentation should be carried out in a separate fermentation vessel. Thus, a two-phase fermentation system could be used to decouple interspecies hydrogen transfer, and recover both hydrogen and methane from biomass via fermentation. Two phase anaerobic digestion has been successfully used to treat sewage biosolids at both pilot and full scale with the result that the overall hydraulic retention time required to treat the biosolids was reduced (Asaadi and Marsh, 2004 and Ghosh *et al.*, 1995). However, neither of these studies reported producing hydrogen in the first stage of digestion.

Hydrogen from sewage biosolids

Sewage biosolids can be used as a source of biomass for fermentative hydrogen production. Using sewage biosolids to produce hydrogen would offer many advantages over other sources of biomass:

- It is abundant, with approximately 8 million tonnes (dry solids) being produced in Europe per annum (Gavala *et al.*, 2004).
- It is a waste product and so, unlike energy crops, no additional energy input or land use is required to produce it. Such costs can be comparatively high if a crop is grown specifically for use as a biomass source.
- The infrastructure for the collection and processing of sewage biosolods is already in place, in the form of sewers and sewage treatment works. This reduces the energy costs of transporting biomass from the point of production to the point of utilisation.

So far, very little work on fermentative hydrogen production from sewage biosolids has been conducted. Huang *et al.* (2000) have reported yields of 0.16 g H_2/kg dry solids (DS) from waste activated sludge, whereas Wang *et al.* (2003b; 2003a) reported yields equivalent to 10.08–70.56 L H_2/kg DS. However it should be noted that so far all these studies used pure cultures of bacteria as inoculum. This would be expensive and impractical to attempt at full scale since large volumes of inoculum would need to be prepared. Wang *et al.* (2003b) also demonstrated that applying different pre-treatments to sewage biosolids prior to fermentation can have a significant effect on hydrogen yield. In particular, sterilisation of the sewage sludge at 121°C consistently improved hydrogen production and reduced the production of methane which can have a negative impact on hydrogen yield. Researchers working with other substrates have successfully used pre-treatments such as enzymatic saccharification, thermal hydrolysis and acid hydrolysis to break down complex carbohydrates such as cellulose into more readily assimilated, soluble sugars (Yanez *et al.*, 2006; Zhang *et al.*, 2007 and Li and Chen, 2007). Massanet-Nicolau *et al.* (2008) demonstrated successful fermentative hydrogen production from primary sewage biosolids (PSB) pre-treated with a combination of pasteurisation and enzymatic saccharification obtaining yields of 18.1 L H_2/kg DS.

Advancement from state of the art

Objectives

To advance the production of hydrogen from sewage biosolids from the current state of the art, towards full scale application, a number of scientific and technical barriers needed to be overcome:

- Sewage sludge contains a substantial microbial flora of its own; including microorganisms which may compete with hydrogen producers, or consume the hydrogen before it can be recovered.
- The carbohydrate in sewage biosolids is predominantly present in a recalcitrant form. Results in the literature indicate that less than one percent of carbohydrate in primary sewage biosolids is present in a soluble form (Massanet-Nicolau *et al.*, 2008). The rest is present in insoluble, structurally complex forms such as cellulose.
- The rheological properties of sewage sludge are such that fermentation apparatus will be prone to blockages, settling out of solids from the liquid phase, and foaming. These issues present technical obstacles to successful research into the continuous fermentation of sewage biosolids.

The aims and objectives of the research were tailored towards overcoming the obstacles to utilising sewage biosolids as source of biomass for fermentative hydrogen production and to advancing the technology towards full scale application within the wastewater treatment industry. The goals of the research include:

- Establishing the feasibility of fermentative hydrogen production in a continuously fed mode. A continuous fermentation (as opposed to batch fermentation) is crucial for the fermentation of sewage biosolids due to the large volume of material involved. The time and cost involved in using batch fermentation to treat sewage biosolids would be prohibitive given the current production levels.
- Evaluating potential pre-treatments for sewage biosolids and assessing their impact of fermentative hydrogen production. Existing research has shown that certain pre-treatments have a positive impact on hydrogen yields from sewage biosolids; and that these pre-treatments are associated with increased solubility of chemical oxygen demand (COD) and carbohydrate.
- Assessing the stability of hydrogen production during continuous fermentation. Due to the diverse microbial population within sewage biosolids, there is potential for populations of hydrogen consuming organisms to develop during continuous fermentation. Methods of inhibiting this phenomenon will be evaluated.
- Operating a two-phase fermentation system producing hydrogen and methane sequentially from sewage biosolids. Using the by-products of hydrogen producing fermentation to feed a mesophilic bioreactor producing methane will increase the overall energy yield from the fermentation process, and stabilise the sewage biosolids further, so that they can be recycled to land.
- Optimisation of fermentation parameters. The optimal sparging rates and the use of carbon dioxide as a sparging gas will be investigated so as to reduce the economic and financial costs of producing hydrogen from sewage biosolids at full scale.

Evaluation of pre-treatment strategies

Many researchers have succeeded in producing hydrogen, fermentatively, from simple substrates such as glucose and sucrose, and even from well defined polysaccharides such as starch and cellulose (Chen and Lin, 2003; Fang and Liu, 2002 and Hussy *et al.*, 2005). However, applying these findings to the fermentation of actual waste streams such as sewage biosolids involved several difficulties. The substrate itself was more complex, and heterogeneous making it more recalcitrant than simple, defined substrates. The sewage biosolids also contain a substantial microbial population of their own, including species that will either inhibit hydrogen production or directly consume it before it can be recovered from the bioreactor. Consequently, there are relatively few accounts in the literature of fermentative hydrogen production using sewage biosolids as a substrate. Most of the research that has been published however, indicates that pre-treatment of sewage biosolids results in improved hydrogen production (Huang *et al.*, 2000; Wang *et al.*, 2003a, 2003b).

Of the pre-treatment strategies evaluated in this study, pasteurisation combined with enzymatic saccharification was clearly superior to either acid hydrolysis or thermal hydrolysis since it was the only pre-treatment which resulted in the production of hydrogen during fermentation. Recently Massanet-Nicolau *et al.* (2008) reported using a combination of pasteurisation and enzymatic saccharification to successfully produce hydrogen from PSB using batch fermentation. In the work carried out in this work package, the same pre-treatment was used to successfully produce hydrogen from PSB in a continuously fed bioreactor. Continuous, fermentative hydrogen production from sewage biosolids has not previously been reported in the literature.

Acid hydrolysis has been used to pre-treat cellulosic biomass prior to fermentation (Lloyd and Wyman, 2005) but in this study acid hydrolysis was unsuccessful at promoting fermentative hydrogen production. The failure of acid hydrolysis to result in successful hydrogen production may be due to the small amount of carbohydrate solubilisation achieved when it was used. Thermal hydrolysis has also been successfully used to enhance conventional methanogenic fermentation (Neyens and Baeyens, 2003). However despite the fact that the thermally hydrolysed sewage biosolids contained a large proportion of soluble carbohydrate, no hydrogen production was observed when it was used as a feedstock in a hydrogen bio-reactor. A direct comparison between these methods of pre-treatment with regards to their suitability for producing hydrogen from sewage biosolids has not previously been presented in the literature.

Stabilisation of hydrogen production

Although the combination of pasteurisation and enzymatic saccharification resulted in successful hydrogen production during continuous feeding, production eventually declined and at the same time a build up of acetate was observed in the bioreactor. It is hypothesized that this is due to homoacetogenesis, a biochemical pathway by which hydrogen is converted to acetate according to the formula below (Hawkes *et al.*, 2007 and Kraemer and Bagley, 2008a).

$$2 \cdot CO_2 + 4 \cdot H_2 \rightarrow CH_3COOH + 2 \cdot H_2O \qquad \text{Eq. 3}$$

In order for stable hydrogen production to be achieved, homoacetogenesis must be inhibited. All methods of inhibiting homoacetogenesis evaluated in this study successfully extended the period of hydrogen production during continuous feeding. Flushing of the headspace with nitrogen only resulted in hydrogen being produced for 4 days from the point at which continuous feeding commenced while sparging of the bioreactor contents with nitrogen at a rate of 0.2 L/min resulted in a period of hydrogen production lasting for 6 days after feeding was commenced. These findings are consistent with those of researchers working with other substrates, who also report that sparging delays the onset of homoacetogenesis (Hawkes *et al.*, 2007 and Hussy *et al.*, 2003). However, the pattern of VFA production observed in the experiments where sparging and headspace flushing were evaluated, indicate that these methods serve only to delay the onset of homoacetogenesis but, that to prevent homoacetogenesis entirely, the sparging system used in these experiments was not sufficiently effective.

It was hypothesised that headspace flushing/sparging improves the transfer of the hydrogen from the liquid to the gas phase within the bioreactor, thus making it unavailable to hydrogen consuming organisms such as homoacetogens. This process would depend, in part, on the extent of the liquid gas interface in the bioreactor. If it is assumed that, since sparging produces a greater gas/liquid interface, it removes more of the hydrogen from the bioreactor than flushing the headspace alone, then the fact that sparging was more effective than flushing at preventing homoacetogenesis may indicate that a higher sparging rate, or a more effective sparging process would be able to prevent the eventual decline in hydrogen production observed during continuous fermentation when sparging was used. Consequently, a higher sparging rate (0.6 L/min instead of 0.2 L/min) was also evaluated.

Sparging the bioreactor contents with nitrogen at a rate of 0.6 L/min successfully prevented a decline in hydrogen production for more than 12 HRTs after feeding commenced; at which point the experiment was ended. Additionally, the build up of acetic acid that accompanied the decline in hydrogen production during previous experiments was also not observed. As a result stable fermentative hydrogen production from primary sewage biosolids in a continuously fed bioreactor is reported here for the first time.

Two-phase fermentation

Having successfully developed a methodology for continuous fermentative hydrogen production from sewage biosolids, the next phase of this study was to integrate a hydrogen producing bioreactor with second a bioreactor producing methane. The methane bioreactor is fed with the effluent from the hydrogen bioreactor in a two-phase fermentation system. Such two-phase fermentation systems have already been used to treat sewage biosolids at full scale and offer many advantages including a reduced overall HRT (Asaadi *et al.*, 2004 and Ghosh, 1991).

So far however no one had shown that a continuously fed, two-phase fermentation system can be used to produce both hydrogen and methane sequentially from sewage biosolids. As well as recovering energy from the treatment of sewage biosolids, it is necessary to stabilise the fermentation residue so that it can be recycled to land. Since fermentative hydrogen production alone, does not fully stabilise sewage biosolids, a secondary treatment stage would be an essential component of any proposed methodology for producing hydrogen, biologically, from sewage biosolids. Integrating hydrogen and methane production in this manner is, therefore, a crucial step towards using sewage biosolids for fermentative hydrogen production on an industrial scale.

In this study, a two-phase fermentation was performed for a period of 30 days, during which the hydrogen and the methane bioreactors remained stable, indicating that a two-phase fermentation system is feasible for the treatment of sewage biosolids. Two-phase, continuous fermentation producing hydrogen and methane sequentially from sewage biosolids has not been reported in the literature.

In the hydrogen bioreactor, a yield of 27.76 L H_2/kg VS was obtained which is comparable to the yields obtained in previous phases of this research, and by other researchers who have conducted small scale batch fermentations of sewage biosolids (Wang *et al.*, 2003b and Huang *et al.*, 2000). In the second bioreactor, methane yields of 260.7 L CH_4/kg VS were obtained, which is less than those normally obtained from the conventional, mesophilic anaerobic digestion where typical yields are 350–400 L CH_4/kg VS (Arundel, 2000 and Tchobanoglous *et al.*, 2002) but comparable with other research into two-phase fermentation of sewage biosolids (Ghosh, 1991 and Ghosh *et al.*, 1995). It should be noted however that the HRT of the methane bioreactor in this study is only 10 days, which is less than that typically used during methanogenic fermentation of sewage biosolids (these typically operate at 15–20 days HRT). In the methane bioreactor the pH remained within the range 6.7 to 7.3 and the bicarbonate alkalinity was between 4000–5000 mg/L these values are indicative of stable methanogenic fermentation. VFA levels in the methane bioreactor remained between 1 and 2 orders of magnitude lower than in the hydrogen bioreactor; no other signs of instability were observed in either bioreactor.

The overall HRT of the two-phase process was 11 days compared with 12–20 days for conventional anaerobic digestion (Tchobanoglous *et al.*, 2002). The lower overall HRT employed in this process has significant cost implications for the anaerobic digestion of sewage biosolids. Operating at a lower HRT means that for a given volume of sewage biosolids, the cost of treatment in terms of both capital expenditure and ongoing operating costs is reduced.

Optimisation of sparging process

Earlier phases of this study established that continuous fermentative hydrogen production from sewage biosolids can be maintained for long periods if stabilised by sparging the bioreactor. In this phase of the study different rates of sparging were evaluated with a view to determining the minimum effective sparging rate. Since sparging requires energy, and adds to the cost of the fermentation process any reduction in sparging rates that can be made would improve the technical and economic feasibility of deploying this technology at full scale.

This study found that there is a marked effect on both the yield and production rate of hydrogen as sparging rate reaches 0.5 L/min. The effect is greater still at a sparging rate of 0.6 L/min but a marked improvement in hydrogen yield or production is not observed as the sparging rate increases beyond this, to 0.7 L/min. The results indicate, therefore, that 0.6 L/min is the optimal sparging rate for hydrogen production in this bioreactor. The results indicate that an optimal sparging rate does exist with regards to fermentative hydrogen production and demonstrate the potential gains in efficiency if the sparging process is correctly optimised. This is the first time that the effects of sparging rates on hydrogen yield from sewage biosolids, in continuous fermentation, have been reported (Massanet-Nicolau *et al.*, 2010).

The energetic and financial cost of sparging could also be reduced if carbon dioxide were shown to be a suitable sparging gas. Since carbon dioxide is produced by both the hydrogen and methane bioreactors during normal operation, it could be recycled and utilised as a sparging gas. When a two-phase fermentation was carried out using carbon dioxide sparging the indications were that stable hydrogen production was taking place in the hydrogen bioreactor, and that the stability of the methane bioreactor was not affected. In the hydrogen bioreactor, as well as hydrogen production being sustained throughout the experiment, n-butyric acid was also still present after 30 days of operation. The presence of n-butyric acid has been positively associated with fermentative hydrogen production by other researchers (Hawkes *et al.*, 2007 and Kyazze *et al.*, 2007).

When carbon dioxide sparging was used methane production in the second bioreactor persisted throughout the 30 experiment as did the reduction of VS, COD and carbohydrate, indicating that methanogenic fermentation was taking place. Throughout the experiment, the pH in the methane bioreactor remained between 6.7 and 7.3, and the bicarbonate alkalinity remained between 4000 and 5000 mg/L1; both of which are indicative of stable methanogenic fermentation (Arundel, 2000 and Tchobanoglous *et al.*, 2002). The use of carbon dioxide sparging to stabilise continuous fermentative hydrogen production from sewage sludge had not been previously reported in the literature.

Despite the stable operation of both bioreactors when carbon dioxide sparging was used, the production rate and yield, of both hydrogen and methane was less than when nitrogen was used as a sparging gas. Since carbon dioxide, along with hydrogen, is a substrate for homoacetogenic bacteria, it is possible that sparging with nitrogen, and thus removing carbon dioxide and hydrogen from the bioreactor, may have inhibited their growth. Sparging with carbon dioxide would only have removed hydrogen from bioreactor; consequently lessening this inhibitive effect. The effect on hydrogen production of removing carbon dioxide during hydrogen fermentation has been demonstrated by other researchers (Park *et al.*, 2005). When considering full scale implementation of fermentative hydrogen production from sewage biosolids, the case for carbon dioxide sparging, as opposed to nitrogen sparging, depends on a comparison of the energy lost through reduced yields of hydrogen and methane, and the energy cost of sparging with nitrogen instead of carbon dioxide.

IMPACT OF THE TECHNOLOGY ON ITS INDUSTRY OR RESEARCH SECTOR

The purpose of the scientific work undertaken in this study is to advance fermentative hydrogen production from sewage biosolids from the current state of the art, towards a process that could be feasibly implemented at full scale. The principle challenges are:

- Switch from the use of pure microbial cultures that are expensive and impractical to use at full scale and instead use mixed cultures, as is the current practice at sewage treatment works.
- Implement fermentative hydrogen production as a continuous fermentation process rather than a batch process. This is of critical importance if fermentative hydrogen production is to be applied to sewage biosolids, because the volumes involved in wastewater treatment make batch fermentation impractical.
- Integrate continuous fermentation into a two-phase system where the effluent from a hydrogen bioreactor is treated via methanogenic fermentation. Again this is vital if adequate stabilisation (VS destruction and VFA removal) is to be achieved.

At the commencement of this project little or no advancement towards these goals had been reported in the literature. The state of the art involved small scale (less than 500 ml) batch digesters using pure cultures of inoculum. In contrast, the research carried out in this project has resulted in development of a method for continuous, two-phase fermentation, producing both hydrogen and methane sequentially, using a mixed culture inoculum. The specific findings of this research are detailed below.

- A means of pre-treating the sewage biosolids is necessary, in order to increase the carbohydrate available to hydrogen producing microorganisms.
- Of the pre-treatments evaluated, pasteurisation, in combination with enzymatic saccharification was the most effective with regards to hydrogen yield and production rate.
- Continuous fermentative hydrogen production is possible but some means of preventing homoacetogenesis, and thus stabilising hydrogen production is required.
- Of the methods evaluated, sparging the bioreactor contents at a rate of 0.6 L/min with nitrogen was most effective at stabilising hydrogen production.
- Fermentative hydrogen production can be combined with methanogenic fermentation in a two-phase system producing hydrogen and methane sequentially.
- Carbon dioxide can be used a sparging gas during fermentative hydrogen production and results in stable hydrogen production; albeit with a slight reduction in hydrogen and methane yield. This adds to the technical and economic feasibility of the process.

These findings represent a significant step forward towards an methodology for producing hydrogen from sewage biosolids in a way that is both economically and technically feasible at full scale.

Further implications for industry relating to full scale hydrogen production from sewage biosolids have been identified in the economic and technical feasibility study undertaken in this project these include. This study discusses the various ramifications of deploying fermentative hydrogen production from sewage biosolids at full scale, including:

- The amount of sewage biosolids available for fermentative hydrogen production in Europe on a country by country basis.
- The expected hydrogen and methane yields which would result from deploying this technology at full scale.
- An assessment of the electrical or fossil fuel (diesel equivalent) energy that could be displaced.
- The potential savings arising from treating sewage biosolids at the reduced HRT made possible by this process.
- The potential markets for hydrogen and methane that are recovered.

A survey of sewage biosolids production in the 27 EU countries using data compiled by the Organisation for Economic Co-operation and Development (OECD) showed that, in the EU, over 14 million tonnes of sewage (dry solids) is produced annually.

Calculations using the yield data from the research conducted in this study, in combination with data on sewage biosolids production in the EU show that, at its current state of development, fermentative hydrogen production could be used to produce over 15,000 tonnes of hydrogen and over 1.5 million tonnes of methane annually, if used to treat all the sewage biosolids in the EU.

This data was extrapolated further to determine the energy content of the hydrogen and methane that could be produced. In the EU as a whole, fermentative hydrogen production from sewage biosolids could yield over 22,500 GWh/yr of energy from hydrogen and methane combined. In a plant treating 100 m^3 of sewage biosolids per day, energy equivalent to 791 litres of diesel (on a thermal basis) could be produced.

Based on a plant treating 100 m^3 of sewage per day, the lower overall hydraulic retention time of fermentative hydrogen production, as opposed to conventional mesophilic digestion would result in capital costs savings of 58%.

Potential markets for the hydrogen and methane produced have been evaluated and it was concluded that the most lucrative scenarios involved using these gases as fuel in a tied fleet of vehicles, or for use in stationary energy generation such as a combined heat and power unit, or a fuel cell. With incentives coming forward for the injection of biogas in the gas national grids, this option would also be of extreme value. All these scenarios are a good fit for sewage treatment plants which frequently make use of tied fleets and stationary power generation. As such there would be an immediate internal market for the hydrogen and methane produced using fermentative hydrogen production.

REFERENCES

Arundel J. (2000). Sewage and industrial effluent treatment, Wiley-Blackwell.

Asaadi M. and Marsh P. (2004). Acid phase digestion – Experience of two stage anaerobic digestion at Swindon STW in Wakefield.

Chen C. C. and Lin C. Y. (2003). Using sucrose as a substrate in an anaerobic hydrogen-producing reactor. *Advances in Environmental Research*, **7**(3), 695–699.

Corder R. E., Hook L. A., Larkin J. M. and Frea J. I. (1983). Isolation and characterization of two new methane-producing cocci: Methanogenium olentangi, sp. nov., and Methanococcus deltae, sp. nov. *Archives of Microbiology*, **134**(1), 28–32.

de Vrije T. and Claassen P. A. M. (2003). Dark hydrogen fermentations, in J. H. Reith, R. H. Wijffels and H. Barten (eds.), Bio-methane & Bio-hydrogen. Petten, Dutch Biological Hydrogen Foundation, 103–123.

Fang H. H. P. and Liu H. (2002). Effect of pH on hydrogen production from glucose by a mixed culture. *Bioresource Technology*, **82**(1), 87–93.

Fenchel T. and Finlay B. J. (1995). Ecology and evolution in anoxic worlds. Oxford, New York, Oxford University Press.

Gavala H. N., Yenal U. and Ahring B. K. (2004). Thermal and enzymatic pretreatment of sludge containing phthalate esters prior to mesophilic anaerobic digestion. *Biotechnology and Bioengineering*, **85**(5), 561–567.

Ghosh S. (1991). Pilot-scale demonstration of 2-phase anaerobic-digestion of activated-sludge. *Water Science and Technology*, **23**(7–9), 1179–1188.

Ghosh S., Buoy K., Dressel L., Miller T., Wilcox G. and Loos D. (1995). Pilot- and full-scale two-phase anaerobic digestion of municipal sludge. *Water Environment Research*, **67**, 206–214.

Hawkes F. R., Dinsdale R., Hawkes D. L. and Hussy I. (2002). Sustainable fermentative hydrogen production: challenges for process optimisation. *International Journal of Hydrogen Energy*, **27**(11–12), 1339–1347.

Hawkes F. R., Hussy I., Kyazze G., Dinsdale R. and Hawkes D. L. (2007). Continuous dark fermentative hydrogen production by mesophilic microflora: Principles and progress. *International Journal of Hydrogen Energy*, **32**(2), 172–184.

Huang C. H., Lin H. Y., Tsai Y. Y. and Hsie Y. K. (2000). 'The preliminary studies of hydrogen production from anaerobic digestion with different substrates and cultivations.' in The Twenty-fifth Wastewater Technology Conference. Yunlin, Taiwan.

Hussy I., Hawkes F. R., Dinsdale R. and Hawkes D. L. (2003). Continuous fermentative hydrogen production from a wheat starch co-product by mixed microflora. *Biotechnology and Bioengineering*, **84**(6), 619–626.

Hussy I., Hawkes F. R., Dinsdale R. and Hawkes D. L. (2005). Continuous fermentative hydrogen production from sucrose and sugarbeet. *International Journal of Hydrogen Energy*, **30**(5), 471–483.

Kraemer J. and Bagley D. (2008a). Measurement of H_2 consumption and its role in continuous fermentative hydrogen production. *Water Science & Technology*, **57**(5), 681–685.

Kraemer J. T. and Bagley D. M. (2008b). Optimisation and design of nitrogen-sparged fermentative hydrogen production bioreactors. *International Journal of Hydrogen Energy*, **33**(22), 6558–6565.

Kyazze G., Dinsdale R., Guwy A. J., Hawkes F. R., Premier G. C. and Hawkes D. L. (2007). Performance characteristics of a two-stage dark fermentative system producing hydrogen and methane continuously. *Biotechnology and Bioengineering*, **97**(4), 759–770.

Levin D. B., Pitt L. and Love M. (2004). Biohydrogen production: prospects and limitations to practical application. *International Journal of Hydrogen Energy*, **29**, 173–185.

Li D. and Chen H. (2007). Biological hydrogen production from steam-exploded straw by simultaneous saccharification and fermentation. *International Journal of Hydrogen Energy*, **32**(12), 1742–1748.

Lloyd T. A. and Wyman C. E. (2005). Combined sugar yields for dilute sulfuric acid pretreatment of corn stover followed by enzymatic hydrolysis of the remaining solids. *Bioresource Technology*, **96**(18), 1967–1977.

Massanet-Nicolau J., Dinsdale R. and Guwy A. (2008). Hydrogen production from sewage sludge using mixed microflora inoculum: Effect of pH and enzymatic pretreatment. *Bioresource Technology*, **99**(14), 6325–6331.

Massanet-Nicolau J., Guwy A., Dinsdale R., Premier G. and Esteves S. (2010). Production of hydrogen from sewage biosolids in a continuously fed bioreactor: effect of hydraulic retention time and sparging. *International Journal of Hydrogen Energy*, **35**(2), 469–478.

Nandi R. and Sengupta S. (1998). Microbial production of hydrogen: An overview. *Critical Reviews in Microbiology*, **24**(1), 61–84.

Nath K. and Das D. (2004). Improvement of fermentative hydrogen production: various approaches. *Applied Microbiology and Biotechnology*, **65**(5), 520–529.

Neyens E. and Baeyens J. (2003). A review of thermal sludge pre-treatment processes to improve dewaterability. *Journal of Hazardous Materials*, **98**(1–3), 51–67.

Park W., Hyun S. H., Oh S., Logan B. E. and Kim I. S. (2005). Removal of headspace CO_2 increases biological hydrogen production. *Environmental Science & Technology*, **39**(12), 4416–4420.

Scholten J. C., Conrad R. and Stams A. J. (2000). Effect of 2-bromo-ethane sulfonate, molybdate and chloroform on acetate consumption by methanogenic and sulfate-reducing populations in freshwater sediment. *Fems Microbiology Ecology*, **32**(1), 35–42.

Tchobanoglous G., Burton F. L. and Stensel H. D. (2002). Wastewater Engineering: Treatment and Reuse, McGraw-Hill Publishing Co.

Traore A. S., Fardeau M., Hatchikian C. E., Le Gall J. and Belaich J. (1983). Energetics of Growth of a Defined Mixed Culture of Desulfovibrio vulgaris and Methanosarcina barkeri: Interspecies Hydrogen Transfer in Batch and Continuous Cultures. *Appl. Environ. Microbiol.*, **46**(5), 1152–1156.

Wang C. C., Chang C. W., Chu C. P., Lee D. J. and Chang B. V. (2003a). Sequential production of hydrogen and methane from wastewawter sludge using anaerobic fermentation. *Journal of the Chinese Institute of Chemical Engineers*, **34**(6), 683–687.

Wang C. C., Chang C. W., Chu C. P., Lee D. J., Chang B. V. and Liao C. S. (2003b). Producing hydrogen from wastewater sludge by Clostridium bifermentans. *Journal of Biotechnology*, **102**(1), 83–92.

Yanez R., Alonso J. L. and Parajo J. C. (2006). Enzymatic saccharification of hydrogen peroxide-treated solids from hydrothermal processing of rice husks. *Process Biochemistry*, **41**(6), 1244–1252.

Zhang M. L., Fan Y. T., Xing Y., Pan C. M., Zhang G. S. and Lay J. J. (2007). Enhanced biohydrogen production from cornstalk wastes with acidification pretreatment by mixed anaerobic cultures. *Biomass and Bioenergy*, **31**(4), 250–254.

Chapter 4

Utilisation of enzymes from sludge to optimise activated sludge digestion

Debora Nabarlatz, Laura Pramparo, Agustí Fortuny Sanromà, Azael Fabregat Llangostera, Josep Font Capafons, Frank Stüber, Christophe Bengoa (Universitat Rovira i Virgili), Jaume Cabrè Fonseré, Iñaki Oriol, Raúl García (Gestió Ambiental i Abastament, S.A.), Jaume Bresó, Eva Rovira (Tratamientos y Recuperaciones Industriales S.A.), Pavel Jenicek and Jana Vondrysova (Institute of Chemical Technology of Praha)

SUMMARY

Project objectives

- Test several disintegration techniques like mechanical disintegration, ultrasound disintegration and stirring disintegration for the release of enzymes (i.e. protease) from activated sludge.
- Evaluate the influence of the addition of several compounds to the sludge according to that found in literature (i.e. cation exchange resin, non ionic detergent, etc).
- Evaluate different types of sludge for the recovery of the enzymes, time of withdrawal, preservation conditions, etc.
- Evaluate the possible pathways for the purification of the enzymatic fraction.

Waste sludge was provided by Gestió Ambiental i Abastament, S.A. The company also collaborated in the experimental task and in the data interpretation. Universitat Rovira i Virgili was responsible of the enzyme extraction procedure. Additionally, Vysok Skola Chemicko-Technologicka v Praze participated in the enzyme extraction procedure, allowing their knowledge, facilities and sludges from WWTP of Prague (Praze). Finally, Tratamientos y Recuperaciones Industriales, S.A. gave all its support in analytical procedures and economical aspects.

Work performed

The first part of the work was devoted to the extraction of enzymes (protease and lipase) from activated sludge. The extraction was performed by using stirring, mechanical and ultrasound disintegration, adding Triton X-100, cation exchange resin (CER) and Tris buffer, alone or in combination between them. The influence of the different parameters, like time of disintegration, concentration of detergent, temperature or the effect of the combination of the different

 Reduction, Modification and Valorisation of Sludge (REMOVALS). Edited by Christophe Bengoa, Azael Fabregat, Josep Font and Frank Stüber. ISBN: 9781843393450. Published by IWA Publishing, London, UK.

additives was studied. The extraction of enzymes was evaluated using sludge collected from two different wastewater treatment plants: the preliminary experiments were carried out in Prague (Czech Republic), and the study was continued in Reus (Spain). From this last sludge, the enzymes identified and quantified were: lipase, protease, alkaline phosphatase, acid phosphatase, L-leu-aminopeptidase, α-glucosidase and α-amylase. The purification of the enzymes was realised by dialysis in a tubing membrane with deionised water or by precipitation with ammonium sulphate, and after, the samples were lyophilised with a vacuum pressure of 0.5 mbar and a temperature of −40°C and stored.

End results

The results obtained showed that the VSS (volatile suspended solids) content in sludge from Prague WWTP was five to ten times higher than the content in the sludge from Reus WWTP; however, this parameter has not significant influence on the specific recovery of the enzymatic activity. The influence of the time during magnetic stirring disintegration showed that after 1 h of treatment it was possible to recover the maximal protease and lipase activity. The concentration of Triton X100 had a great influence for the extraction of protease, but it was not so significant for the extraction of lipase. The ultrasound disintegration showed that for the extraction of lipase, the ultrasonic power applied during 20 min is enough to extract more than 20 lipase units/g VSS from activated sludge (without additives), compared to 1 h that is necessary when using magnetic stirring disintegration. On the other side, the use of 2.0% v/v TX100 allows to extract near 53 protease units/g VSS at the same ultrasonic disintegration conditions. Finally, it can be concluded that both disintegration methods (ultrasonication and stirring, combined with the action of Triton X100) are suitable to extract protease and lipase from activated sludge maintaining them active.

The extraction protocol was optimized for the extraction of seven hydrolytic enzymes that are present in the activated sludge. The enzymes quantified were lipase, protease, alkaline and acid phosphatase, L-leu-aminopeptidase, α-glucosidase and α-amylase. The amount of protein released during the treatments was also measured. The lipase present in the enzymatic fraction was purified by using dialysis, precipitation with ammonium sulphate and finally lyophilisation. The method employed for the extraction of enzymes was ultrasonic disintegration of activated sludge at different disintegration times. The results showed that 30 min of disintegration applying 50 W of ultrasonic power are enough to extract the maximal amount of enzymatic activity. These results were compared with those obtained when 2% v/v Triton X100 (a detergent currently used to extract membrane proteins) was used as additive. It was observed that the use of Triton X100 increases the amount of enzymatic activity extracted of protease and acid phosphatase, but it has not a significant influence for the extraction of the other enzymes.

The recovery of enzymes from activated sludge is a promising option for the valorisation of this resource, considering the sludge as a raw material capable of generating multiple new biomolecules and products. So this study allowed us to determine the presence of a variety of hydrolytic enzymes in a standard way that could be measured, compared and scaled up in terms of enzyme units. It represents a progress beyond the state of the art because until now, only few references were present in the literature that could not be compared between them due to the broad difference in the enzymatic activity recovered and also due to the definition of enzymatic activity itself. Besides this, ultrasonication is a suitable technology to implement the extraction of enzymes at industrial scale. The final conditions for the extraction method should be optimised depending on the type of hydrolytic enzyme that will be recovered, the required yield, the economy of the process, and the purification procedure that will be necessary for the final application desired. The preliminary economical analysis indicates that the use of extracted enzymes for the pretreatment of activated sludge seems more convenient than the use of commercial ones.

METHODOLOGIES AND APPROACHES EMPLOYED

Extraction of enzymes from activated sludge

Sludge collection and handling

The activated sludge was collected from the biological reactor in the Wastewater Treatment Plant in Reus, Spain (WWTP Reus) and in Prague, Czech Republic (WWTP Prague). The samples of activated sludge were taken and transported to the laboratory in 30 min. The sludge was used as received, and it was analyzed in order to determine the content in total solids (TS) and volatile suspended solids (VSS) by gravimetric method according to standard methods.

Disintegration of sludge for the extraction of enzymes

The disintegration of activated sludge was carried out using three methods: stirring, mechanical and ultrasonic disintegration:

- Stirring disintegration: the magnetic stirring disintegration was evaluated at 500 rpm using 50 mL sludge volume, varying the disintegration time from 1 to 300 min (in Prague WWTP and Reus WWTP). In order to detect possible changes in the amount of enzymatic activity recovered when the process is scaled up, a 500 mL reactor was used in which the agitation was done by a three blades marine impeller (in Reus WWTP). The impeller rotation speed was set to 500, 1,000 and 1,500 rpm. The volume of sludge used for each test was 200 mL.

- Mechanical disintegration: it was carried out in Prague WWTP using a disintegrator ULTRA – TURRAX T25 Basic (IKA) at speed 24,000 rpm and 11,000 rpm. The experiments were developed at room temperature (due to equipment configuration) using 200 mL sludge volume and varying the disintegration time from 1 to 10 min.
- Ultrasonic disintegration: The ultrasonic disintegration in Prague WWTP was evaluated using a Sonotrode Disintegrator (ECOSON, Czech Republic), with working frequency 20 kHz and output power 180 W. All the experiments were carried out at 40% amplitude (near 72 W) at 3 different disintegration times (1, 5, 10 min). The utrasonic disintegration in Reus WWTP was evaluated using an Ultrasonic Disintegrator UP200S equipped with S14 tip (diameter 14 mm) (Hielscher Ultrasonics GmbH, Germany), with working frequency 24 kHz and output power 200 W. The experiments were carried out using 45% amplitude (near 50 W according to the fabricant) at different disintegration times between 1 and 60 min. In both cases the experiments were carried out using 200 mL sludge volume.

The temperature during the treatment was controlled by using a water-ice bath (although in mechanical disintegration was not possible). The additives used were Tris buffer for pH adjustment (10 mM at pH 7, 7.5 and 8); cation exchange resin (CER) Dowex Marathon sodium form 20–50 mesh (0.24, 0.48 and 0.8 g CER/mL sludge); and non-ionic detergent Triton X-100 (TX100) at concentrations from 0.1% to 10% v/v. Samples of sludge were taken before and after the disintegration process and were centrifuged for 10 min at 17,400 g prior to the analysis. The supernatant was used as the source of enzyme for the determination of enzymatic activity and amount of protein. All the disintegration experiments were carried out by triplicate.

Purification of the enzymatic fraction

In order to purify the fraction of enzymes present in the supernatant extracted after the centrifugation step, several purification procedures were tested:

- Dialysis: For the dialysis process, a 12–14 kDa dialysis tubing membrane (Medicell International) was used. The membrane was placed in a container having deionised water as solvent (100 mL). The equipment was placed in a water bath which was maintained at 8 ± 2ºC. To control the progress of the dialysis, the solvent was changed at least 3 times during 24 h, and the conductivity of the diffusate was measured. The process was repeated until the conductivity remained constant and near to that of original water. The samples obtained after dialysis were stored at 4ºC in the fridge for analysis or frozen (−20ºC) for lyophilisation.
- Precipitation with ammonium sulphate: The precipitation process was carried out by adding slowly 25 mL of a 100% saturated (761.4 g/L) ammonium sulphate solution to 50 mL of enzyme solution (40% saturated solution). The mixture was placed in a water bath at 8 ± 2ºC under stirring (500 rpm). After 24 h the mixture was centrifuged and the supernatant discarded. The solids (pellet) were dissolved again in 4 mL of 50 mM potassium phosphate buffer (pH 7.5 at 37ºC). The samples were stored at 4ºC in the fridge for analysis or frozen (−20ºC) for lyophilisation.
- Lyophilisation: The samples after the disintegration, precipitation and dialysis were lyophilized using an Armfield Lyophilisation Equipment (Prodel S.A., Spain) equipped with a high vacuum pump JAVAC (Brook Crompton Betts Pty.). The conditions used were a vacuum pressure of 0.5 mbar and a temperature of −40°C.

Analytical methods for the determination of the amount of protein and enzymatic activity

After all the steps of extraction and purification, the enzymatic activity and the amount of protein have been measured. The enzymes identified and quantified were lipase, protease, alkaline phosphatase, acid phosphatase, L-leu-aminopeptidase, α-glucosidase and α-amylase. The enzymatic activity was measured following standard analysis protocols by Sigma Aldrich in each case. The amount of protein was also quantified by the Bradford method. In Table 4.1 are detailed all procedures utilised for the identification and quantification of the enzymes.

ACHIEVEMENTS RELATED TO THE STATE OF THE ART

Fundamentals about the extraction of enzymes from activated sludge

As it was mentioned before, the extraction of enzymes from activated sludge collected from municipal wastewater treatment plants was just briefly explored until now, as demonstrated by the fact that less than 15 articles were found to deal with this subject until last year (2008).

Table 4.1 Analytical techniques for the determination of the enzymatic activity (Sigma-Aldrich) and amount of protein (Bradford)

Enzyme	Reagents	Experimental conditions
Lipase	Add 0.20 mL enzyme solution + 1.8 mL buffer (100 mM NaH_2PO_4 + 150 mM NaCl + 0.5% TX100, pH 7.2, 37°C) to test & blank. Add 0.02 mL of deionised water (blank) and 0.02 mL of 50 mM PNPB (test).	Substrate: p-nitrophenyl butyrate (PNPB). Absorbance measured at 400 nm during 6 min (ΔAbs400 nm/min) at room temperature. One unit of enzyme releases 1 μmol of p-nitrophenol per min.
Protease	Add 5 mL of 0.65% w/v casein in buffer (50 mM potassium phosphate, pH 7.5, 37°C) to test & blank. Add 1 mL of enzyme solution to test and incubate (10 min, 37°C). Add 5 mL of 110 mM TCA to stop reaction to both test & blank, and 1 mL of enzyme solution to blank. Incubate (30 min, 37°C). From the previous solution, take 2 mL (test & blank) and add 5 mL of 500 mM Na_2CO_3 and 1 mL of Folin & Ciocalteu's to each tube. Incubate (30 min, 37°C).	Substrate: casein. Absorbance measured at 660 nm. One unit of enzyme liberates 1 μmol of L-tyrosine per min.
Acid phosphatase	Add 0.5 mL of buffer (90 mM citrate, pH 4.8, 37°C) + 0.5 mL of 15.2 mM PNPP to test & blank. Add 0.10 mL of enzyme solution to test and incubate (10 min, 37°C). Add 4 mL of 100 mM NaOH (test & blank) and 0.10 mL of enzyme solution to blank.	Substrate: p-nitrophenyl phosphate (PNPP). Absorbance measured at 410 nm. One unit of enzyme hydrolyzes 1 μmol of PNPP per min.
Alkaline phosphatase	Add 1.35 mL (test) and 1.4 mL (blank) of buffer (1 M diethanolamine + 0.5 mM $MgCl_2$, pH 9.8, 37°C). Add 0.15 mL of 150 mM PNPP to both tubes. Add 0.05 mL of enzyme solution to test and incubate (10 min, room temp).	Substrate: p-nitrophenyl phosphate (PNPP). Absorbance measured at 405 nm during 10 min (ΔAbs400 nm/min). One unit of enzyme hydrolyzes 1 μmol of PNPP per min.
Leucine aminopeptidase	Add 2.8 mL buffer (50 mM Na_2HPO_4, pH 7.2, 37°C) and 0.10 mL 24 mM LPN in CH_3OH to both test & blank. Add 0.10 mL deionised water (blank) and 0.10 mL enzyme solution (test).	Substrate: L-leucine-p-nitroanilide (LPN). Absorbance measured at 405 nm during 7 min at room temperature (ΔAbs400 nm/min). One unit of enzyme hydrolyzes 1 μmol of L-leucine-p-nitroanilide per min.
α-amylase	1 mL 1% w/v starch solution in buffer (20 mM Na_2HPO_4 + 6.7 mM NaCl, pH 6.9, 20°C) to test & blank. Add 1 mL enzyme solution to test and incubate (3 min, 20°C). Add 1 mL colour reagent (8 mL of 1.5 g/mL sodium potassium tartrate in 2 M NaOH, + 20 mL 96 mM 3,5-dinitrosalicylic acid solution, diluted to 40 mL with deionised water) to both test & blank. Add 1 mL of enzyme solution to blank and incubate (15 min, 100°C). Add 9 mL deionised water to test & blank.	Substrate: starch. Absorbance measured at 540 nm. One unit of enzyme liberates 1 mg of maltose from starch in 3 min.
α-glucosidase	Add 5 mL buffer (67 mM K_2HPO_4, pH 6.8, 37°C) and 0.20 mL 3 mM glutathione reduced sol., to both test & blank. Add 0.20 mL deionised water (blank) and 0.20 mL enzyme solution (test). Add 0.5 mL 10 mM PNG to both tubes and incubate (20 min, 37°C). Take 2 mL from the previous solution and add 8 mL of 100 mM Na_2CO_3.	Substrate: p-nitrophenyl α-D-glucoside (PNG). Absorbance measured at 400 nm. One unit of enzyme liberates 1 μmol of D-glucose per min.
Protein	5 mL Coomasie blue brilliant reagent (Bradford colour reagent, diluted 1:5). Add 0.10 mL deionised water (blank) and 0.10 mL enzyme solution (test). Incubate (1 h, room temp).	Absorbance measured at 595 nm. Samples with 2% v/v TX100 were diluted 1:20. Calibration with BSA, range 0.01–1 mg/mL.

The extraction of enzymes from activated sludge was started in 1995 (Frolund 1995). In this study, it was detected the extracellular enzyme activity of esterase, lipase, leucine aminopeptidase, α and β glucosidase, chitinase and β-glucuronidase. Several authors reported these and other enzymes, as overviewed in Table 4.2. As it can be observed, 13 enzymes have been reported as they are present in activated sludge collected from urban wastewater treatment plants.

However, there is a wide range of enzymatic activities recovered that can vary up to three orders of magnitude depending on the author. The definition of enzymatic activity is arbitrary and depends strongly on the substrate used and the conditions of the reaction used to determine it. For this reason, it is necessary to establish formal protocols with well defined methods suitable to determine the enzymatic activities because this is a key parameter in the extraction process.

Table 4.2 List of enzymes extracted from activated sludge in municipal wastewater treatment plants

Enzyme	Amount detected	Substrate	Unit definition	Author
Esterase	0.035	4-methylumbelliferyl stearate	μmol MUF/h/mg $_{VS}$	(Frolund, 1995)
	140	p-nitrophenyl esters (chain length 4–6)	μmol p nitrophenol/min/L	(Boczar, 2001)
Lipase	0.005	4-methylumbelliferyl stearate	μmol MUF/h/mg $_{VS}$	(Frolund, 1995)
	0.002	p-nitrophenyl-palmitate	μmol p nitrophenol/min/g $_{MLSS}$	(Jung, 2002)
	350	p-nitrophenyl-palmitate	μmol p nitrophenol/min/g $_{VSS}$	(Gessesse, 2003)
	1000	4-methylumbelliferyl oleate	μmol MUF/h/g $_{TS}$	(Schade, 2005)
	90	p-nitrophenyl-palmitate	μmol p nitrophenol/L/h	(Li, 2006)
L-leu-aminopeptidase	0.015	L-leucine-4-methyl-7-coumarinylamidehydrochloride	μmol MCA/h/mg $_{VS}$	(Frolund, 1995)
	9	L-leu-p-nitroanilide	μmol p nitroaniline/min/g $_{VS}$	(Cadoret, 2002)
	38	L-leu-p-nitroanilide	μmol p nitroaniline/L/h	(Li, 2006)
Protease	0.1	Azocasein, 440 nm	1 mg azocasein/h/mg $_{MLVSS}$	(Goel, 1998)
	80	Casein	μmol L tyrosine/min/mg $_{protein}$	(Jung, 2001)
	2210	Casein	μmol L tyrosine/min/g $_{MLSS}$	(Jung, 2002)
	8	Azocasein, 340 nm	ΔAbs/min/g $_{VS}$	(Cadoret, 2002)
	4000	Azocasein, 440 nm	0.01 ΔAbs/g $_{VSS}$	(Gessesse, 2003)
	28	Casein	μmol L tyrosine/min/g $_{VSS}$	(Yu, 2007)
	4	Casein	μmol L tyrosine/min/g $_{VSS}$	(Yu, 2008a, 2008b)
α-glucosidase	0.005	4-methylumbelliferyl-α-D-glucoside	μmol MUF/h/mg $_{VS}$	(Frolund, 1995)
	2.5	p-nitrophenyl-α-D-glucopyranoside	μmol p nitrophenol/min/g $_{VS}$	(Cadoret, 2002)
	300	–	μmol glucose/min/g $_{VSS}$	(Yu, 2007)
	4	–	μmol glucose/min/g $_{VSS}$	(Yu, 2008a, 2008b)
β-glucosidase	0.009	4-methylumbelliferyl-β-D-glucoside	μmol MUF/h/mg $_{VS}$	(Frolund, 1995)
	0.002	p-nitrophenyl-β-D-glucopyranoside	units/g $_{MLSS}$	(Jung, 2002)
	40	p-nitrophenyl-β-D-glucopyranoside	μmol p-nitrophenol/L/h	(Li, 2006)
β-glucuronidase	0.005	4-methylumbelliferyl-β-D-glucuronide	μmol MUF/h/mg $_{VS}$	(Frolund, 1995)
Endoglucanase	0.002	Carboxymethyl cellulose	units/g $_{MLSS}$	(Jung, 2002)
α-amylase	2.7	Amylose azure	μmol Remazol brilliant blue/min/g $_{VS}$	(Cadoret, 2002)
	14.5	–	μmol glucose/min/g $_{VSS}$	(Yu, 2007)
	150	–	μmol glucose/min/g $_{VSS}$	(Yu, 2008a, 2008b)
Alkaline phosphatase	0.2	p-nitrophenyl phosphate	μmol p nitrophenol/h/mg $_{MLVSS}$	(Goel, 1998)
	38	p-nitrophenyl phosphate	μmol p nitrophenol/L/h	(Li, 2006)
	60	p-nitrophenyl phosphate	μmol p nitrophenol/h/mg $_{MLVSS}$	(Yu, 2008a, 2008b)
Acid phosphatase	0.5	p-nitrophenyl phosphate	μmol p nitrophenol/h/mg $_{MLVSS}$	(Goel, 1998)
	38	p-nitrophenyl phosphate	μmol p nitrophenol/h/mg $_{MLVSS}$	(Yu, 2008a, 2008b)
Dehydrogenase	0.35	Iodonitrotetrazolium (INT)	μmol INT-formazan/h/mg $_{MLVSS}$	(Goel, 1998)
Chitinase	0.005	4-methylumbelliferyl-N-acetyl-&-D-glucosaminide	μmol MUF/h/mg $_{VS}$	(Frolund, 1995)

The economy of an industrial process for the extraction of enzymes will be determined by the efficiency of the extraction and the enzymatic activity extracted. Also the yield during purification will be evaluated according to the enzymatic activity preserved after the whole process, and finally in order to compare and make some economical comparison with existent commercial enzymes, a precise definition of enzymatic activity should be stated.

Definition of enzymatic activity used for each type of enzyme

The definition used for expressing the enzymatic activity recovered after the disintegration process was chosen according to the standard protocols used for the determination (by Sigma-Aldrich). However this, the definition of the enzyme unit is expressed in terms of the substrate used and they can be easily reproduced.

- Lipase activity. One unit of enzyme will release 1.0 μmol of p-nitrophenol per minute.
- Protease activity. One unit of protease will hydrolyze casein to produce colour equivalent to 1.0 μmol of L-tyrosine per minute.
- Acid phosphatase activity. One unit of enzyme hydrolyze 1 μmol of p-nitrophenyl phosphate per minute at pH 4.8 at 37°C.
- Alkaline phosphatase activity. One unit of enzyme hydrolyze 1 μmol of p-nitrophenyl phosphate per minute at pH 9.8 at 37°C.
- Leucine aminopeptidase activity. One unit of enzyme hydrolyze 1 μmol of L-leucine-p-nitroanilide per minute at pH 7.2 at 37°C.
- α-amylase. One unit of enzyme will liberate 1 mg of maltose from starch in 3 min at pH 6.9 at 20°C.
- α-glucosidase activity. One unit of enzyme will liberate 1 μmol of D-glucose per minute at pH 6.8 at 37°C.

Extraction of enzymes by stirring disintegration: evaluation of the disintegration time effect and the use of additives on the protease and lipase recovered

Figures 4.1a and 4.1b show the results for the enzymatic activity recovered after magnetic stirring disintegration using activated sludge from WWTP Reus, expressed as units of enzymatic activity per gram VSS recovered as function of the extraction time, when 0.5% v/v TX100 is used as additive.

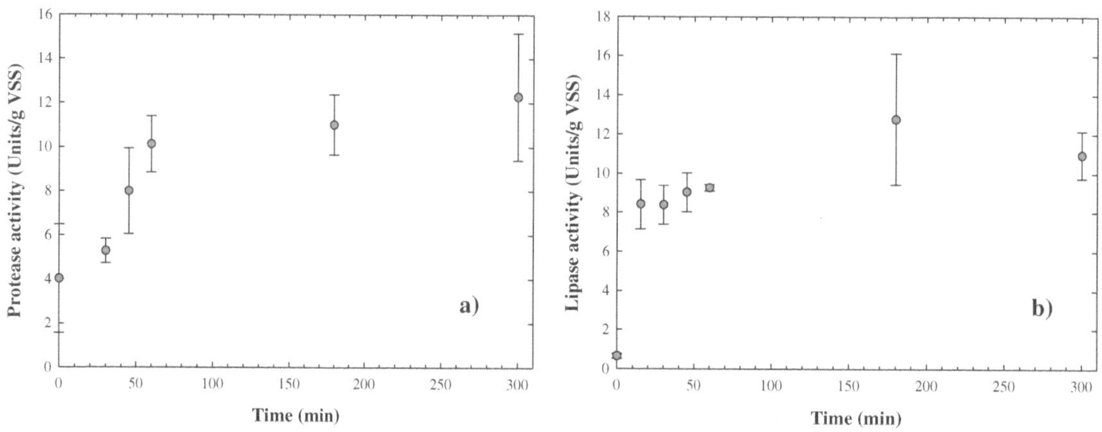

Figure 4.1 Enzymatic activity recovered by magnetic stirring disintegration expressed as function of the disintegration time for protease (a) and lipase (b). Conditions: sludge collected from Reus WWTP, T = 7°C, 500 rpm, 0.5% v/v TX100

As it can be observed in the figures, after 1 hour of disintegration it was possible to extract almost the maximal protease activity (near 10 protease units/g VSS). Approximately the same behaviour was observed for the extraction of lipase, because it was possible to recover up to 9.5 lipase units/g VSS. However, the tendency shows that at shorter times lipase extraction is faster than protease extraction, because after 15 min of extraction it is possible to recover more than 8 lipase units/g VSS (more than 85% of the maximum) being the recovery of protease almost negligible at the same time.

Figure 4.2a and 4.2b show the recovery of protease and lipase as function of the amount of TX100 added during magnetic stirring disintegration carried out at 7°C during 1 h, using sludge from WWTP Reus. As it can be observed in Figure 5.2a for the original sample, the yield in protease activity increases rapidly for TX100 concentration higher than 0.5% v/v, reaching a maximum near 65 protease units/g VSS. However, when the concentration of TX100 in the solution is higher than 1.0%, the determination of the protease activity presented several phenomena, like the appearance of a yellow precipitate during the colour reaction with Folin reagent. In the experiments for the extraction of enzymes from sludge found in literature the maximal concentration of TX100 used was 2.0% v/v (Gessesse, 2003).

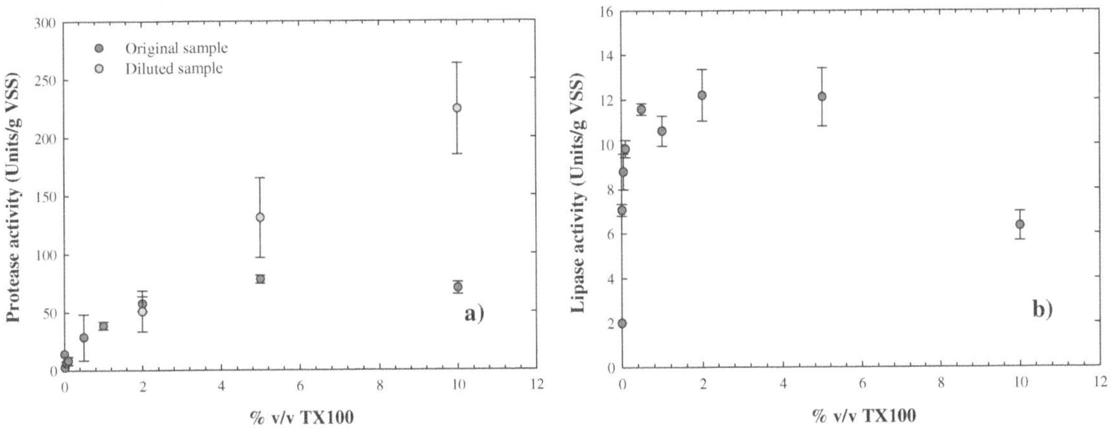

Figure 4.2 Enzymatic activity recovered by magnetic stirring disintegration expressed as function of the % v/v TX100 for protease (a) and lipase (b). Conditions: sludge collected from Reus WWTP, T = 7°C, 500 rpm, t = 1 h

The phenomena observed, together with the lack of literature data using higher concentrations of TX100, suggests that probably high concentrations of detergent causes interference in the reaction with Folin reagent, yielding a false result. In order to validate the standard analysis used to determine the protease activity when high concentrations of TX100 are used, the colour reaction with Folin reagent was done using a standard concentration of L-tyrosine and adding the detergent in the same concentrations as results from the extraction of enzymes.

The results presented in Figure 4.2a (diluted sample) show that the real activity for the samples having more than 2.0% v/v TX100 is higher than that obtained when the sample is analyzed without dilution. These results allowed us to trust in the results obtained using concentrations of TX100 lower than 2.0% v/v.

On the other side, the extraction of lipase using different concentrations of TX100 showed a different behaviour. As it can be observed from Figure 4.2b, the maximal activity recovered was near 11 lipase units/g VSS, reached when using between 0.5%–1.0% v/v TX100. After 2.0% v/v TX100 the amount of lipase activity recovered decrease, showing that probably higher concentrations of detergent partially deactivate the enzyme. For this reason for the next experiments the maximal concentration of TX100 used was 2.0% v/v.

Figure 4.3 shows the absorbance of a solution having 0.375 μmol L-tyrosine combined with different detergent concentrations. In this figure it can be observed that using concentrations lower than 2.0% v/v TX100, the colour reaction is affected in less than 3%. However, when the concentration of detergent is increased until 5 and 10% v/v, the colour intensity decreases in 19% and 52%, respectively. For this reason, when the concentration of TX100 was 2.0% or higher, the sample was diluted to corroborate which is the real activity of the protease when using higher concentrations of detergent.

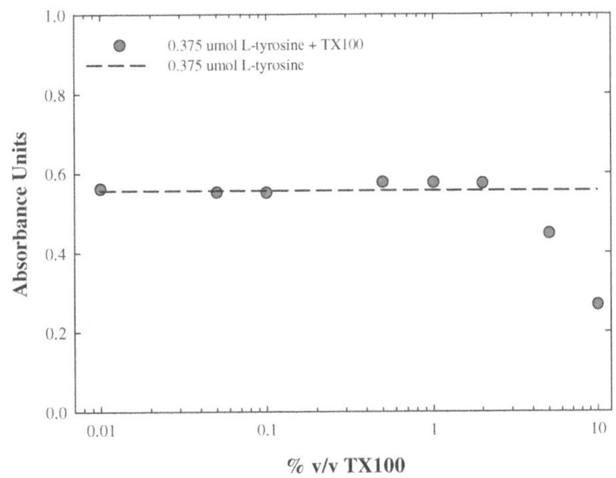

Figure 4.3 Determination of the influence of % v/v TX100 in the colour reaction of L tyrosine with Folin & Ciocalteu's Phenol reagent. Absorbance detected at 660 nm

In order to detect changes in the enzymatic activity recovered during stirring disintegration at a higher volume scale the experiments were carried out using 200 mL of sludge collected from Reus WWTP and processed in a stirred reactor at three different velocities (500 rpm, 1,000 rpm and 1,500 rpm) using 2.0% v/v TX100 as additive. The results showed that the enzymatic activities recovered using 500 rpm as stirring velocity are comparable to those obtained in Figures 4.2a and 4.2b for both enzymes.

When the stirring velocity increased, the protease activity recovered decreased and the lipase activity recovered is maintained nearly constant, which demonstrated that the use of 500 rpm is sufficient as stirring velocity for enzymatic recovery. Figures 4.4a and 4.4b show the influence of different types of additives during the experiments using magnetic stirring disintegration of sludge from WWTP Reus at 7°C during 1 h.

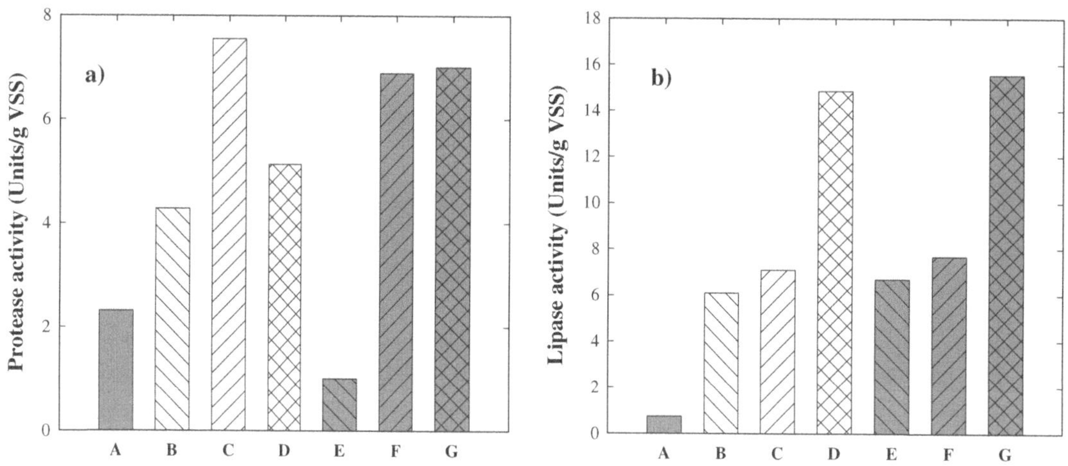

Figure 4.4 Enzymatic activity recovered by magnetic stirring disintegration expressed as function of the different additives used for (a) protease and (b) lipase. Conditions: sludge collected from Reus WWTP, T = 7°C, 500 rpm, t = 1 h. Legend: A: 10 mM Tris pH 7.5; B: 0.48 g CER/mL; C: 0.5% v/v TX100; D: 0.5% v/v TX100 + 0.48 g CER/mL; E: 10 mM Tris pH 7.5 + 0.48 g CER/mL; F: 10 mM Tris pH 7.5 + 0.5% v/v TX100; G: 10 mM Tris pH 7.5 + 0.5% v/v TX100 + 0.48 g CER/mL

As it can be observed in Figure 4.4a for the extraction of protease, the addition of Tris buffer pH 7.5 alone does not affect significantly, while the addition of 0.48 g CER/mL sludge improves slightly its extraction. But definitely the addition of TX100 alone or in the combination TX100 + Tris, or TX100 + CER yields the maximal protease activity recovered, making possible to extract up to 7.5 protease units/g VSS. This suggests that the additive that affects significantly the extraction of protease is the use of a non ionic detergent like Triton X-100.

On the other side, the results for the recovery of lipase in Figure 4.4b are slightly different: again the addition of Tris buffer pH 7.5 alone does not affect significantly the extraction of lipase, while the addition of 0.48 g CER/mL sludge improves it slightly. However, in this case the combination of TX100 + CER means a significant improvement in the extraction process, because it was possible to recover up to 6.5 and 7.5 lipase units/g VSS (when CER and TX100 were used alone), while it was possible to recover up to 15.5 lipase units/g VSS when TX100 and CER were used together. That means that probably there is a synergic effect between the detergent and the resin which helps to extract lipases, suggesting that these enzymes are retained by a combination of ionic and hydrophobic interactions to the cells in the sludge.

Extraction of enzymes by ultrasonication: evaluation of the disintegration time and the ultrasonic power on the protease and lipase recovered

According to the results optimized by stirring disintegration, the following experiments were carried out using ultrasonication. Figure 4.5a shows the results for the extraction of protease using ultrasound disintegration. It can be observed that when the sludge from Reus WWTP was disintegrated by ultrasound using 0.5% v/v TX100 during 5 min of treatment, it was possible to recover near 6.7 protease units/g VSS.

If the extraction time is increased, the maximal yield recovered was near 12.6 protease units/g VSS using 60 min of treatment. As it was observed using magnetic stirring disintegration, the increment in the concentration of TX100 from 0.1% to 2.0% increases markedly the recovery of protease for both types of sludge used. The maximal amount of enzymatic activity recovered using 2.0% v/v TX100 was near 52 protease units/g VSS. If we compare the results using sludge from WWTP Reus for both disintegration methods we can observe that it was possible to recover the same amount of protease activity (near 55 protease units/g VSS) using 2.0% v/v TX100, but it was necessary 1 h of disintegration when using magnetic stirring, while in just 10 min of operation it was possible to obtain the same result using ultrasound disintegration.

The extraction of lipase by ultrasound disintegration was carried out in the same way, and the results are presented in Figure 4.5b. The results showed that after 20 min of operation it is possible to extract almost the maximal lipase activity (near 23 lipase units/g VSS). As it can be observed from the graph, the influence of the addition of detergent is not so significant and it seems that the main factors influencing the extraction are the effect of ultrasound treatment and the disintegration time. Again, if we compare the results obtained with those using stirring, it can be observed that while the maximal lipase extracted is 12 lipase units/g VSS after 1 h of disintegration (2.0% v/v TX100), it is possible to obtain near 25 lipase units/g VSS after 30 min of disintegration using the same amount of detergent as additive.

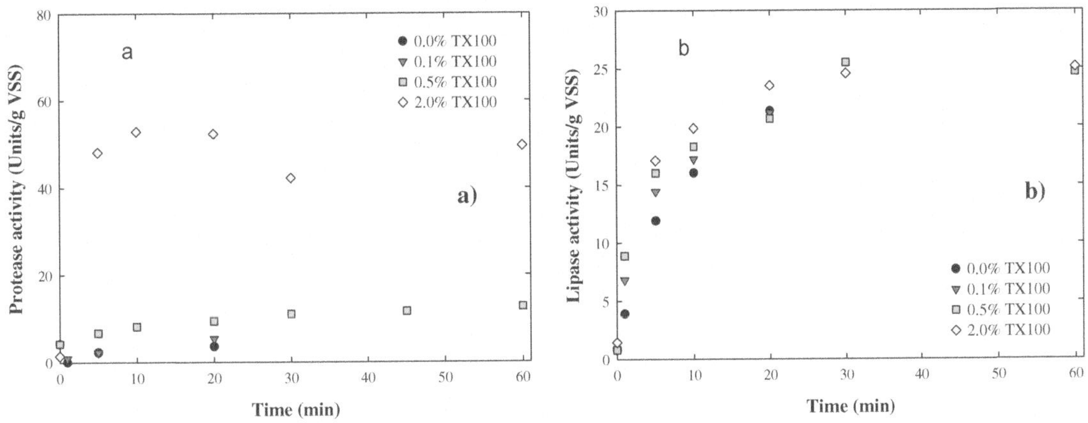

Figure 4.5 Enzymatic activity recovered by ultrasound disintegration of activated sludge collected from (a) Protease and (b) Lipase from Reus WWTP, using different types of additives. T = 5°C, 50 W Power

Finally, the extraction of enzymes by ultrasound disintegration using 2.0% v/v TX100 as additive was evaluated as function of the ultrasonic power employed (results not shown). It was demonstrated that 45% amplitude (near 50 W of ultrasonic power according to the fabricant) are enough to extract almost the maximal amount of lipase and protease, and the increase in the ultrasonic power does not increase the amount of enzymatic activity recovered.

From the results obtained, it can be concluded that the optimal procedure to extract the maximal amount of enzymatic activity will depend on the enzyme that wants to be recovered. For the extraction of lipase, 50 W of ultrasonic power applied during 20 min is enough to extract more than 20 lipase units/g VSS from activated sludge (without additives).

On the other side, the use of 2.0% v/v TX100 allows to extract near 52 protease units/g VSS at the same disintegration conditions. The use of ultrasonication allows reducing the disintegration time from 1 h necessary for stirring to 20 min for ultrasonication. Beside this, the choice of the extraction procedure will depend on the type of hydrolytic enzyme that will be recovered, the required yield, the economy of the process, and on the purification procedure that will be necessary for the final application desired.

Extraction of enzymes by ultrasonication: extension of the study to the detection of seven hydrolytic enzymes

As it was already mentioned, the objective of the present study was to determine the presence of several hydrolytic enzymes in activated sludge collected from an urban wastewater treatment plant. According to the better results found for the recovery of protease and lipase, the study was extended to detect another five enzymes, which are also well known of participating in the degradation process. There are some previous studies dealing with this subject, but it is difficult to compare them and extract useful results due to several reasons, between them, the lack of detail about the definition of enzymatic activity itself and the differences in the protocol and the substrate used to measure the enzymatic activity. For this reason, with the purpose of extract the enzymes and to use them for some specific applications in mind, it is necessary to make a detailed study about the availability of each type of enzyme, its stability and the optimal protocol for its extraction maximizing the amount of enzyme activity recovered. The enzymes studied are acid phosphatase, alkaline phosphatase, L-leu-aminopeptidase, α-amylase, protease, lipase, and α-glucosidase. Its presence in the activated sludge give us an idea of the variations in the substrate composition, microbial population, etc., depending on the presence of the different macromolecules in the activated sludge that the microorganisms are able to degrade (Voet, 2006). Lipase hydrolyzes the lipids and esters present in activated sludge, while acid and alkaline phosphatases hydrolyze different types of phosphate esters. The generic enzyme called protease hydrolyzes different types of proteins, while leucine aminopeptidase is more specific and cuts the bond between leucine and different peptides. Finally, α-amylase and α-glucosidase hydrolyze starch and maltose from cellulose, respectively.

Figure 4.6 shows the results for the enzymatic activity obtained and the amount of protein recovered when using ultrasound treatment (without TX100) at 20°C at different disintegration times. The enzymes studied are lipase, protease, alkaline phosphatase, acid phosphatase, L-leu-aminopeptidase, α-glucosidase and α-amylase.

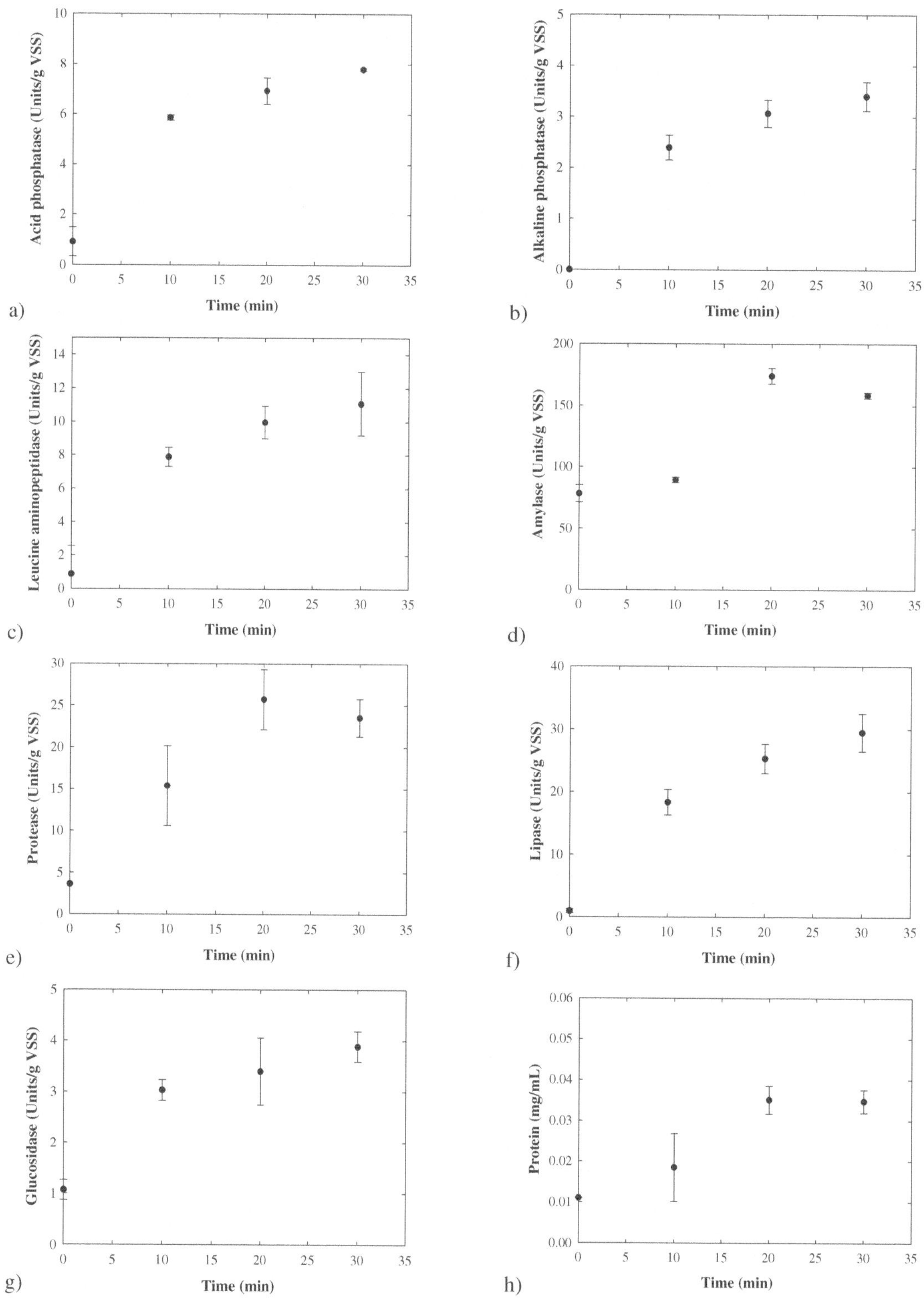

Figure 4.6 Variation in the enzymatic activity of acid phosphatase (a), alkaline phosphatase (b), leucine aminopeptidase (c), α-amylase (d), protease (e), lipase (f), α-glucosidase (g) and protein (h) as function of the disintegration time at the following conditions: 50W ultrasonic power, 20°C, 0% v/v TX100

As it can be observed, 30 min of disintegration time at the ultrasonic power used (50 W) seems to be adequate to extract the maximal amount of enzyme present in the activated sludge for all the enzymes. However, there are some differences in the amount of the enzymatic activity extracted during the time of disintegration. The amount of enzymatic activity recovered per gram of VSS for acid and alkaline phosphatase, leucine aminopeptidase, α-glucosidase and lipase has an increasing tendency that seems to continue beyond 30 min of disintegration.

On the other side, α-amylase and protease showed a maximal amount of enzymatic activity at 20 min of disintegration time, and beyond this point, the activity started slightly to decrease. This could be an effect of degradation of the enzyme during the ultrasonic process, probably also related to the own stability of the enzyme during this process.

Acid and alkaline phosphatase yielded a maximum of 7.8 and 3.4 units enzyme/g VSS (0.01 and 0.003 units/mL) respectively when using 30 min of disintegration time. These values are near to that found in the literature (0.0125 units/mL and 0.005 units/mL for acid and alkaline phosphatase, respectively) which were expressed in similar units (Goel, 1998). Yu reported between 10–17 units/g VSS for alkaline phosphatase and between 3–12 units/g VSS for acid phosphatase, respectively (Yu, 2008b, 2009).

L-leucine aminopeptidase showed a maximal activity of 11 units/g VSS, which is slightly higher than the value found in literature (around 9 units/g VSS) (Cadoret, 2002). The results for α-amylase showed a higher enzymatic activity, because it was found that it is possible to extract up to 174 units/g VSS, which is in the range reported (between 15–150 units/g VSS) (Yu, 2008a, 2008b, 2009).

The activity of α-glucosidase extracted after 30 min of disintegration (3.9 units/g VSS) was higher than some values found in the literature (2 units/g VSS) (Cadoret, 2002). However, Yu *et al.* had extracted between 2–305 units/g VSS (Yu, 2007, 2008a, 2008b). There are several results in the literature for the extraction of lipase, but only few of them are comparable. In the present study it was possible to extract up to 29.5 units/g VSS, which is a lower value than that found in literature (335 units/g VSS) (Gessesse, 2003). On the other side, protease yielded up to 25.7 units/g VSS, which is in the range of the results found in the literature (between 2.5–28 units/g VSS) (Yu, 2007, 2008a, 2008b, 2009). Finally, the maximal amount of protein present in the supernatant after centrifugation is around 0.035 mg/mL, which represents near 4% of the volatile suspended solids present in the activated sludge. If we observe the amount of enzyme recovered per mg of protein, it is detected a maximum in the amount extracted at 10 min of disintegration, and then a steep decrease. However, observing the amount of protein extracted from activated sludge it increases until reaching a maximum at 20 min of extraction. That indicates that even if the amount of protein extracted increases with disintegration time, the effect of the ultrasonication affects the enzyme probably by deactivating it.

According to the results showed previously, the use of Triton X-100 as additive in a concentration 2% v/v improves the efficiency of the extraction of protease. For this reason, the disintegration experiments were carried out without any additive (0% TX100) and they were compared with the results obtained when 2% TX100 was used. The disintegration experiments were carried out during 30 min at 20°C. The Figure 4.7 presents the results of these experiments.

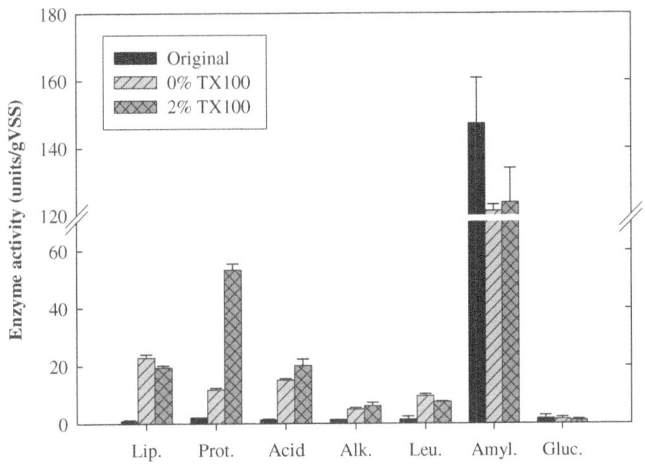

Figure 4.7 Enzymatic activity recovered after the disintegration experiments using 0% v/v and 2% v/v TX100 at the following conditions: 50 W ultrasonic power, 20°C, and 30 min disintegration time. Abbreviations: Lip. (lipase); Prot. (protease); Acid (acid phosphatase); Alk. (alkaline phosphatase); Leu. (leucine aminopeptidase); Amyl. (α-amylase); Gluc. (α-glucosidase)

As it can be observed in the figure, only protease and acid phosphatase showed an increment of its activity. The activity of protease increased from 11.7 to 53.3 enzyme units/g VSS (356% increase) when 0% and 2% of detergent was used. The activity of acid phosphatase increased from 15 to 20 enzyme units/g VSS in the same conditions (33% increase).

The other enzymes were recovered with near the same activity when the experiments were carried out with or without detergent, indicating that the use of TX100 does not improve significantly its recovery. On the contrary, the activities of lipase or leucine aminopeptidase even descend with the use of TX100 (it decreases between 15–20%). As it was already mentioned, the use of Triton X100 as additive will be effective depending on the type of enzyme that will be extracted. Its use in this case is helpful for the extraction of protease and acid phosphatase, but it may generate problems related to the purification procedure that should be avoided in the subsequent steps of the process. For this reason, the extraction process should be optimized for all the enzymes present in the activated sludge that want to be recovered, taking into account the degree of purity and the specific enzymatic activity required at the end of the process.

As it was already mentioned, 50 W of ultrasonication are enough to extract the maximal amount of enzymatic activity for all the enzymes studied. Beside this, the use of Triton X100 as additive will be effective depending on the type of enzyme that will be extracted. Its use in this case is helpful for the extraction of protease and acid phosphatase, but it has no beneficial effects on the extraction of the other enzymes. However, its use also may generate problems related to the purification procedure that should be avoided in the subsequent steps of the process. For this reason, the extraction process should be optimized for all the enzymes present in the activated sludge that want to be recovered, taking into account the degree of purity and the specific enzymatic activity required at the end of the process.

Purification of the lipase fraction

In order to evaluate the efficiency of the purification process, the lipase activity and the amount of protein were measured in all the steps of the process purification. Table 4.3 shows the results that were obtained for the analysis of the samples after disintegration and purified by two different methods: precipitated with ammonium sulphate and dialyzed.

Table 4.3 Lipase activity and amount of protein recovered after the different purification procedures. The original samples were obtained after ultrasound disintegration at the following conditions (50 W power, 20°C, 30 min disintegration time)

	Lipase (Units/L)		Protein (mg/L)	
	Treatment (% v/v TX100)			
	0%	2%	0%	2%
After ultrasound disintegration (original)	21.0 ± 1.0	17.9 ± 0.6	83.8 ± 16.03	268.8 ± 24.6
After precipitation with ammonium sulphate	53.2 ± 4.7	29.1 ± 1.1	403.4 ± 16.2	386.0 ± 56.2
% recovered of total	25.3%	26.1%	48.1%	23.6%
After dialysis	23.5 ± 2.6	18.4 ± 1.7	81.5 ± 0.0	297.8 ± 0.0
% recovered of total	109.0%	100.2%	88.6%	116.2%

The percentages recovered after the purification procedure related to that of the original activity are also shown. As it can be observed, the precipitation with ammonium sulphate allowed concentrating the sample doubling the specific lipase activity in both cases (0% and 2% TX100). But the total enzymatic activity recovered was around 25–26% of the original, indicating almost a 75% loss of the activity.

If we relate the enzymatic activity with the amount of protein recovered, we can observe that the loss of activity after precipitation is much more marked when using 0% TX100 (0.13 lipase units/mg protein recovered compared with the original 0.25 lipase units/mg protein) than using 2% TX100 (0.075 lipase units/mg protein recovered compared with the original 0.066 units/mg protein). On the other side, the dialysis process allowed to recover 100% of the lipase activity (the values over 100% could be due to the variations in the volume recovered). That indicates that this process allowed removing low molecular weight material maintaining constant the enzymatic activity.

Table 4.4 shows the enzymatic activity of the different samples obtained after disintegration and purification process by lyophilisation. As it can be observed, lipase is very sensitive to the overall cycle of frozen and lyophilisation because there is a significant loss on the enzymatic activity for almost all the samples.

For the original sample after disintegration, it was possible to recover between 17–20% of the original (for 2% and 0% TX100 respectively). The samples obtained after ultrasound disintegration using 0% TX100 and precipitated with ammonium sulphate maintained around 74% of its original enzymatic activity, while when using 2% TX100 allowed recovering just around 5% of the original enzymatic activity. On the other side, the dialysis process allowed to recover around 45% of the enzymatic activity for the samples obtained after disintegration using 0% TX100, compared with 12.7% recovered when the samples contained 2% TX100. If we observe at the specific enzymatic activity per gram of solid recovered, we can conclude that the ultrasound disintegration with 0% TX100 allowed recovering the higher specific enzymatic activity, around 23 lipase units/g solid.

Table 4.4 Lipase activity and amount of protein in the different purified samples, recovered after lyophilisation. The original samples were obtained after ultrasound disintegration at the following conditions (50 W power, 20°C, 30 min disintegration time)

	Lipase (Units/g solid)	
	Treatment (% v/v TX100)	
	0%	2%
After ultrasound disintegration (original)	3.5 ± 0.8	0.5 ± 0.0
% recovered of total	20.0%	17.1%
After precipitation with ammonium sulphate	2.3 ± 1.6	0.2 ± 0.0
% recovered of total	73.9%	4.8%
After dialysis	22.9 ± 0.8	0.6 ± 0.0
% recovered of total	44.9%	12.7%

Similar enzyme concentrations (between 10–23 lipase units/g solid) have been used in the literature to pretreat wastewater having high concentrations of oil and fats (Cammarota, 2006), indicating that lipases obtained after dialysis could be used for this application. However, we consider that for specific applications (like production of esters) a purification and concentration procedure by classical methods must be stated in order to can compete with enzymes produced at industrial scale.

As a main conclusion, the use of dialysis followed by lyophilisation allowed to recover the maximal amount of lipase, when there is no detergent in the sample. So this confirms that the use of detergent should be carefully considered when a certain enzyme wants to be recovered. The recovered lipase could be used as pretreatment for sludge degradation; however, we consider that for specific applications (like production of esters) a purification and concentration procedure by classical methods must be stated in order to can compete with enzymes produced at industrial scale.

IMPACT OF THE TECHNOLOGY ON ITS INDUSTRY OR RESEARCH SECTOR

Research impacts

One of the main impacts of this part of this work was the optimization of the extraction of seven enzymes from activated sludge using a quite simple process. Just applying 30 min ultrasonication is enough to extract the maximal amount of enzymatic activity for all the enzymes studied. The use of detergent improves the extraction of certain enzymes. The preliminary purification procedure employed for the recovery of lipase seems to be very effective to recover almost the half of the enzymatic activity in powder form. This is an interesting result in case that the enzymes will be used as pretreatment of activated sludge, or for any other application: they could be concentrated, dried and recovered with maintenance of the enzymatic activity with reasonable yield. Finally, the application of an enzymatic pretreatment will increase the biogas production and reduce the amount of sludge to be disposed off, and it can be installed in any already existing wastewater treatment plant.

Social and environmental impacts

The annual production of sewage sludge exceeds 10 millions tons a year in the European Union, and it is becoming a serious issue due to several factors like the growth of population, its accumulation in large cities, and the increment in the amount and complexity of the industrial activity. The amount of activated sludge increases annually, and the final disposal ways (landfilling and composting) together with the lack of disposal lands, should be reconsidered due to a more strict environmental legislation. For this reason, the development of new technologies is necessary to reduce the amount of sludge and make possible its reuse. The recovery of valuable products from sludge that could be used in the sludge degradation is promising. In this sense, the main social and environmental impact of the extraction of enzymes from activated sludge will be the following:

- The use of hydrolytic enzymes in an enzymatic pretreatment prior to anaerobic digestion allowed to reduce the amount of solids that should be disposed off, improves the solids dewaterability and increase the biogas production (Cammarota, 2001; Jeganathan, 2007a, 2007b; Leal, 2006; Recktenwald, 2008; Valladao, 2007 and Wawrzynczyk, 2007).
- The disintegration by ultrasonication of sludge is effective for the digestion of aerobic activated sludge itself, due to that it destroy the flocs and releases extracellular polymeric compounds and enzymes. So the use of this technology could have additional effects for the improvement of the degradation of activated sludge.

- The hydrolytic enzymes extracted, with further purification and separation, could have very promising applications in the industry that should be explored.

Industrial and economical impacts

The industrial interest in this case is related to the economical impact caused by the applicability of an enzymatic pretreatment step prior to anaerobic digestion. In this sense, a preliminary economical analysis was done. As it can be observed from Table 4.5, after 1 year of operation, even though the investment cost and the operational costs due to electricity are high, the cost for the extraction of lipase (and all the accompanying enzymes) from activated sludge and its use for the pretreatment of sludge prior to anaerobic digestion is comparable to the use of the commercial lipase available (400 €/day and 300 €/day, respectively). However, it should be considered that the cost for the use of extracted enzyme was calculated taking into account an amortization period of four years for the equipments. After this period, the costs for equipments are paid, so the cost of the extraction leads almost to zero if we consider that the electricity employed is generated during the process. On the other side, the commercial enzyme is lost during the process, and the use of immobilized commercial enzyme increases considerably the cost (Novozymes 435, which is an immobilized lipase, costs near 1000 €/kg having 10000 units/g). So the use of commercial enzyme after 4 years will become more costly than the use of the extracted ones.

Table 4.5 Economical analysis for the enzymatic pretreatment of sludge (Calculation basis 100 m^3 sludge/day)

Costs	Per 1 day	Per 1 year
Electricity spent (MWh)	0.77	280
Electricity generation (MWh)	4.64	1700
Enzyme necessary (units)	1.6 × 107	600 × 107
Volume of sludge to disintegrate (m^3)	658	240000
Electricity cost (considering 0.1123 €/kWh)	86	32000
Ultrasonication equipment amortization (€) (2 × 160000€)	219	80000
Digester equipment amortization (€) (151236 €)	104	38000
Total cost extracted lipase(€) (Ultras. + Digester + Electricity)	409	150000
Cost of the use of commercial Lipase Calb by Novozymes (€)	198	72100
Total cost commercial lipase (€) (Digester + enzyme)	302	110100

The spent in electricity (0.8 MWh/day) is compensated by the generation of electricity due to the increase in the production of biogas (50% more, which means 4.64 MWh/day generated), so we consider that this cost could be eliminated. However, it was taken into account in order to see the real cost in case that you should buy electricity, and to can compare with the treatment when using commercial enzymes. Considering that no additional source of energy is required, the overall cycle of sludge digestion could be improved without any addition of external sources of energy.

Besides this, the extraction of enzymes from activated sludge by ultrasound disintegration has other advantages:

- The disintegration by ultrasonication of sludge has proven to be effective for the digestion of aerobic activated sludge digestion itself, due to that it destroy the flocs and releases extracellular polymeric compounds and enzymes. Ultrasonication reduced the final amount of VSS from 20% in the control to 35% after ultrasonication after 10 days activated sludge digestion (Yu, 2008b). So the application of ultrasonication could improve by itself the degradability of the sludge, enhancing also the biogas production.
- The enzymes extracted from activated sludge are not just lipase, but it is a cocktail of hydrolytic enzymes. So its application as a cocktail of enzymes for the pretreatment of sludge will improve the degradation of the whole sludge, and not just of the fraction of lipids present. Of course this should be considered also for the use of the commercial enzyme, because it will be necessary to make a cocktail of 5–7 enzymes, which was not taken into account in the present analysis.
- The pretreatment stage and the use of the enzymes could be optimised by the purification and concentration of the enzymatic fraction, and the immobilization of the enzymes over different supports, in order to can separate and reutilise the enzyme fraction.
- The hydrolytic enzymes extracted, with further purification and separation, could have very promising applications in the industry that should be explored.

REFERENCES

Boczar B. A., Forney L. J., Begley W. M., Larson R. J. and Federle T. W. (2001). Characterization and distribution of esterase activity in activated sludge. *Water Research*, **35**, 4208–4216.

Cadoret A., Conrad A. and Block J. C. (2002). Availability of low and high molecular weight substrates to extracellular enzymes in whole and dispersed activated sludges. *Enzyme and Microbial Technology*, **31**, 179–186.

Cammarota M. C. and Freire D. M. G. (2006). A review on hydrolytic enzymes in the treatment of wastewater with high oil and grease content. *Bioresource Technology*, **97**, 2195–2210.

Cammarota, M. C., Teixeira and G. A. and Freire D. M. G. (2001). Enzymatic pre-hydrolysis and anaerobic degradation of wastewaters with high fat contents. *Biotechnology Letters*, **23**(19), 1591–1595.

Frolund B., Griebe T. and Nielsen P. H. (1995). Enzymatic-activity in the activated-sludge floc matrix. *Applied Microbiology and Biotechnology*, **43**, 755–761.

Gessesse A., Dueholm T., Petersen S. B. and Nielsen P. H. (2003). Lipase and protease extraction from activated sludge. *Water Research*, **37**, 3652–3657.

Goel R., Mino T., Satoh H. and Matsuo T. (1998). Enzyme activities under anaerobic and aerobic conditions inactivated sludge sequencing batch reactor. *Water Research*, **32**, 2081–2088.

Jeganathan J., Nakha G. and Bassi A. (2007a). Oily wastewater treatment using a novel hybrid PBR-UASB system. *Chemosphere*, **67**(8), 1492–1501.

Jeganathan J., Nakhla G. and Bassi A. (2007b). Hydrolytic pretreatment of oily wastewater by immobilized lipase. *Journal of Hazardous Materials*, **145**(1–2), 127–135.

Jung J., Xing X. H. and Matsumoto K. (2001). Kinetic analysis of disruption of excess activated sludge by Dyno Mill and characteristics of protein release for recovery of useful materials. *Biochemical Engineering Journal*, **8**, 1–7.

Jung J., Xing X. H. and Matsumoto K. (2002). Recoverability of protease released from disrupted excess sludge and its potential application to enhanced hydrolysis of proteins in wastewater. *Biochemical Engineering Journal*, **10**, 67–72.

Leal M., Freire D. M. G., Cammarota M. C. and Sant'Anna G. L. (2006). Effect of enzymatic hydrolysis on anaerobic treatment of dairy wastewater. *Process Biochemistry*, **41**(5), 1173–1178.

Li Y. and Chrost R. J. (2006). Microbial enzymatic activities in aerobic activated sludge model reactors. *Enzyme and Microbial Technology*, **39**, 568–572.

Recktenwald M., Wawrzynczyk J., Dey E. S. and Norrlow O. (2008). Enhanced efficiency of industrial-scale anaerobic digestion by the addition of glycosidic enzymes. *Journal of Environmental Science and Health Part A*, **43**, 1536–1540.

Schade M. and Lemmer H. (2005). Lipase activities in activated sludge and scum – Comparison of new and conventional techniques. *Acta Hydrochimica Et Hydrobiologica*, **33**, 210–215.

Valladao A. B. G., Freire D. M. G. and Cammarota M. C. (2007). Enzymatic pre-hydrolysis applied to the anaerobic treatment of effluents from poultry slaughterhouses. *International Biodeterioration & Biodegradation*, **60**, 219–225.

Voet D., Voet J. and Pratt C. (2006). Fundamentals of Biochemistry: Life at the molecular level. Madrid, Spain, Editorial Medica Panamericana.

Wawrzynczyk J., Recktenwald M., Norrlow O. and Dey E. S. (2007). Solubilisation of sludge by combined chemical and enzymatic treatment. *African Journal of Biotechnology*, **6**, 1994–1999.

Yu G., He P., Shao L. and Lee D. (2007). Enzyme activities in activated sludge flocs. *Applied Microbiology and Biotechnology*, **77**, 605–612.

Yu, G. He P., Shao, L. and Zhu Y. (2008). Extracellular proteins, polysaccharides and enzymes impact on sludge aerobic digestion after ultrasonic pretreatment. *Water Research*, **42**, 1925–1934.

Yu G., He P., Shao L. and Lee D. (2008a). Extracellular enzymes in sludge flocs collected at 14 full-scale wastewater treatment plants. *Journal of Chemical Technology and Biotechnology*, **83**, 1717–1725.

Yu G., He P., Shao L. and Zhu Y. (2008b). Extracellular proteins, polysaccharides and enzymes impact on sludge aerobic digestion after ultrasonic pretreatment. *Water Research*, **42**, 1925–1934.

Yu G., He P., Shao L. and Zhu Y. (2009). Enzyme extraction by ultrasound from sludge flocs. *Journal of Environmental Sciences*, **21**, 204–210.

Chapter 5

Gasification of sewage sludge

Stanislaw Ledakowicz, Tomasz Bedyk, Lech Nowicki and Pawel Stolarek

SUMMARY

Project objectives

Pyrolysis and gasification are the options for revalorisation of sewage sludge by thermal treatment. Pyrolysis is the process of thermal decomposition of organic substances in an inert atmosphere, at temperatures below 800ºC. In gasification process the primary pyrolysis products, mainly char undergo secondary reactions with gasification reagent (O_2, CO_2 or H_2O) forming additional amounts of synthesis gas.

The project objectives were as follows:

- to collect data for better understanding of the thermo-chemical processing of sewage sludge (devolatilisation and gasification);
- to study the influence of the extent of pyrolysis process on subsequent gasification of the pyrolysis char (yield and kinetics);
- to establish the kinetic model of the process for different gasification reagents (O_2, CO_2, H_2O).

Work performed

In order to characterize the research objects (to search for effect of sewage sludge origin on pyrolysis behaviour) eight samples of sewage sludge were chosen for the study. The experimental set-up consisting of thermo-balance coupled with mass spectrometer and gas chromatograph (TG-MS-GC) was employed. The TG-MS-GC system provides extensive information on the conversion of solid material (TG and DTG curves) as well as the evolved substances (MS spectra, GC peaks). Three analytical methods result in a complete picture of the ongoing reactions during the thermal decomposition process. Proximate analysis of sewage sludge composition by TG showed significant differences in volatile matter and ash content.

End results

Experimental data showed that total amount of volatile products evolved from the sample during pyrolysis increased with the final temperature of the process – the higher the temperature of the pyrolysis the more complete the

 Reduction, Modification and Valorisation of Sludge (REMOVALS). Edited by Christophe Bengoa, Azael Fabregat, Josep Font and Frank Stüber. ISBN: 9781843393450. Published by IWA Publishing, London, UK.

carbonisation. The effect of heating rate on the yield and reactivity of the char was insignificant. Reactivity of char in the CO_2 gasification depends on the composition of sewage sludge in particular of ash and carbon content.

Gasification process can be divided in two main stages: devolatilisation (pyrolysis) and subsequent gasification of the pyrolysis char. These two stages were modelled separately. A visual analysis of DTG and MS profiles reveals clearly that there are some peaks that can be recognized as fractions in the sewage sludge and their thermal decomposition could be grouped into few lumps of reaction products. Proposed kinetic model of pyrolysis involving a set of parallel n-order reactions was successfully applied for different sewage sludge samples. The values of activation energy are in the expected range typical for chemically control processes.

Gasification of the char was carried out isothermally at the final pyrolysis temperature as well as in dynamic mode with constant heating rate. The kinetics of gasification strongly depends on gasifying reagent used and temperature. Gasification is usually carried out with air, steam or their mixtures, hardly ever with carbon dioxide, although the latter can be very attractive from ecological point of view. Char gasification by air is completed at about 600°C while char gasification with CO_2 does not occur significantly for temperatures below 750°C. Rates of steam gasification are faster than those with CO_2.

Kinetic modelling of gasification process was performed using experimental data collected at isothermal conditions for three oxidizing agents, by utilizing two representative gas-solid reaction models, volumetric model and shrinking core model. The shrinking core model for reaction controlled regime was found to be the best for predicting the rate of char gasification in the whole range of conversion in carbon dioxide atmosphere. The same model can be applied for oxygen gasification but only for the conversion range below 80%, while for higher conversion the shrinking core model for diffusion controlled regime should be applied. Steam gasification of sewage sludge char proceeds probably according to different mechanism and the best fitting to experimental data was observed for the first-order kinetics. The values of activation energy for different gasification agents were found to be 228 kJ/mol for CO_2, 193 kJ/mol for H_2O and 113 kJ/mol for O_2. These values agree with the literature data for biomass gasification.

The results of this work package contributed to the increasing knowledge of sewage sludge gasification. Profound understanding of the process mechanism revealed that characterization of sewage sludge is necessary for proper design and development of the gasification process. Kinetic models developed in the work enable rational designing and optimization of the developing processes of thermo-chemical treatment of sewage sludge.

ACHIEVEMENTS RELATED TO THE STATE OF THE ART

Introduction

Pyrolysis and gasification are the options for revalorisation of sewage sludge by thermal treatment. The basic technology concepts are not novel but, recently, several technologies for thermal processing of waste and biomass have been developed. There are more than 150 companies around the world that are marketing systems based on pyrolysis and gasification concepts for waste treatment (Junipier, 2001). Most of these processes are optimised for specific waste or particular scale of operation. However, many of these plants have been only operated as small scale pilot plants for limited period of time due to operational and economic problems. Although many studies, including the feasibility study made for this project, have shown that gasification and pyrolysis can be commercially feasible, the conventional alternatives or incineration are still utilized in the full scale applications.

Specific processes have been also developed for sewage sludge pyrolysis and gasification. Biomass and sewage sludge pyrolysis results in the production of condensable (tars), noncondensable volatiles (gas) and char which can be used as a source of chemicals and fuels (Hossain *et al.*, 2009). Furthermore, all heavy metals are safely enclosed in a residual mineral matrix. The pyrolysis process can be optimized to maximize production of one group of products. However, such an optimization must be based on a good understanding of the fundamentals and mechanisms of the thermal decomposition.

State of the art and research objectives

Pyrolysis is the process of thermal decomposition of organic substances in an inert (oxygen free) atmosphere, at temperatures below 800°C. During the pyrolysis sewage sludge undergoes a sequence of physical and chemical steps, starting with the drying at about 100°C. Above this temperature a series of chemical and thermal reaction occur. The major products that are formed after thermal degradation of the sludge in an inert atmosphere are the following:

- non-condensable gas (at normal conditions) containing mainly carbon monoxide, carbon dioxide, hydrogen, methane and several other hydrocarbons in smaller concentrations,
- tar and/or oil (liquid fraction),
- char, which is mainly carbon with small amounts of inert materials.

In gasification process the primary pyrolysis products, condensable and non-condensable vapours and mainly char undergo secondary reactions with gasification reagent forming additional amounts of synthesis gas. In fact, pyrolysis and gasification are two separate or partly overlapping stages (depending on the gasifying agent used) because the conversion of char occurs at higher temperatures when most of thermal decomposition processes are finished. It is well known that gasification is influenced by the reactivity of the chars (Bridgwater, 2005), and the char reactivity depends on the process conditions during the pyrolysis stage of the thermo-chemical conversion. This problem was extensively studied for coal gasification, however, similar studies for biomass gasification are rather scarce (Cetin *et al.*, 2004). No attempts have been undertaken for sewage sludge to relate the pyrolysis conditions to reactivity of char with gasifying reagent. Thermogravimetric analysis (TGA) is a common way to study the behaviour of pyrolysis of sewage sludge (Dümpelmann *et al.*, 1991; Conesa *et al.*, 1997; Chu *et al.*, 2001 and Ledakowicz and Stolarek, 2000). Recently, the coupling of TGA and Fourier transform infrared spectrometer (FTIR), mass spectrometer (MS) or chromatography (GC) has been used on the pyrolysis of sewage sludge (Conesa *et al.*, 1998; Chen and Jeyaseelan, 2001; Otero *et al.*, 2002 and Karayildirim *et al.*, 2006) to characterize the gas- or liquid-phase products.

Thermal and chemical processes of biomass conversion consist of pyrolysis, gasification and direct combustion. Pyrolysis as the first thermo-chemical stage in gasification and combustion plays an important role because its conditions highly affect the yields of char and its reactivity (Di Blasi, 2009). The conversion of char occurs at higher temperatures when most of thermal decomposition processes are finished.

In the case of solid wastes like sewage sludge, which is not precisely defined, any proposal of the intrinsic kinetic model based on chemical mechanism of the process is almost impossible. Assuming that various substances in the sludge or in other complex substances decompose according to their own characteristics, it is reasonable to treat whole process of pyrolysis as a group of parallel and independent reactions. On the basis of literature overview one may draw conclusion that a particular independent pyrolytic reactions can be uniquely represented by one set of three kinetic parameters: activation energy, pre-exponential factor, and reaction order. Such a division into the groups of reactions can be made on the basis of mass loss or released products analysis as presented in work by Wójtowicz *et al.* (2003). The results of TG-FTIR were utilized to create a tobacco biomass pyrolysis model based on independent, parallel, first-order reactions with Gaussian distribution of activation energies. Wu *et al.* (2006) demonstrated the feasibility of modeling the pyrolysed sample as a continuous mixture, in which the Arrhenius-type kinetics is applied with activation energy, pre-exponential factor, and reaction order as continuous functions of conversion. Scott *et al.* (2006a) introduced an algorithm which provided a detailed description of the kinetics of devolatilisation of quite complex fuels from conventional thermogravimetric experiments at a several constant rates of heating. The algorithm works for both a continuous distribution of reactions and for a discrete one. The algorithm was able to identify how many parallel, first-order reactions participated in devolatilisation and evaluated the activation energy and pre-exponential factor. Next, the proposed algorithm was used to investigate pyrolysis of two dried sewage sludges. Although the algorithm assigned mass loss to 28 reactions for a digested sludge, many of these are similar and could be grouped together. The digested sludge could be modelled by only 8 reactions (Scott *et al.*, 2006b) whereas description of an undigested sludge required fare more reactions.

Yang and Jiang (2009) on the basis of results from the flue gas analyzer and the chemical analysis divided DTG curves of industrial waste activated sludge pyrolysis into five decomposition peaks corresponding to volatile substances, microbial cells, proteins, inorganic substances and fixed carbon using the Lorentz fitting method. Particular steps of pyrolysis were described by the different reaction mechanisms e.g. the random nucleus formation followed by growth in the case of pyrolysis of volatile substances and the microbial cells followed a chemical nth-order reaction mechanism. Thipkhunthod *et al.* (2007) applied model based on the assumptions that the pyrolysis reactions comply with an nth-order reaction. Although, there have been a number of sewage sludge pyrolysis kinetics models reported there is still need for comprehensive biomass pyrolysis models able to predict yields of particular volatile products as a function of substrate characteristic and process conditions.

In gasification process the primary pyrolysis products, condensable and non-condensable vapours and mainly char undergo secondary reactions with gasification reagent forming additional amounts of synthesis gas.

Since pyrolysis and gasification of biomass are termo-chemical processes, the temperature and rates of heating have pronounced effect on the weight loss of biomass. When the environment is reactive complete gasification of biomass takes place yielding gaseous fuel and ash. Reactive environment (air/oxygen) results in complete gasification of biomass while inert environment (nitrogen/argon) aids devolatilization (pyrolysis) yielding more char (Kirubakaran *et al.*, 2009). Inguanzo *et al.* (2001) proved that the reactivity of pyrolyzed sewage sludge to CO_2 is considerably lower than reactivity to air and proceeds in a more progressive way. The differences in reactivity are more affected by the heating rate than by the final pyrolysis temperature. Furthermore, it is well known that gasification is influenced by the reactivity of the chars (Cetin *et al.*, 2004), and the char reactivity depends on the process conditions during the pyrolysis stage of the thermochemical conversion. It was found (Henrich *et al.*, 1999) that chars from the pyrolysis of waste materials produced at lower and medium temperatures belong to the most reactive technical carbon materials. It is because the highly disordered carbon structure in combination with porosity is achieved.

The conversion of the char particle results from the interaction of several physical processes, among which the most important are mass transfer throught the pores within the particle and mass transfer of the gaseous reagents to the exterior of the particle, and the intrinsic kinetics of the reaction. Many reports available from the literature consider the gasification of coal char in different atmospheres. Gasification with steam seems to be more complex than CO_2 gasification. Steam gasification of carbon takes places according to two basic mechanisms indicated as the oxygen exchange mechanism and the hydrogen inhibition mechanism. The reaction of carbon and oxygen were also extensively investigated (Hurt and Calo, 2001). A huge amounts of literature has been produced on the reactivity and the heterogeneous kinetics of coal and coal char gasification and combustion (Di Blasi, 2009), but the number of studies on sewage sludge char reactivity is comparatively much smaller and the kinetic modelling of the gasification and combustion reactions has not been given significant consideration.

Gasification kinetics is of importance in the design of gasification reactors. This is because gasification is a slower process than the initial pyrolysis (Liliedahl and Sjoestroem, 1997). Gasification results in a continuous change in char composition. Hence the char reactivity tends to vary with degree of conversion or time. Therefore, the first order kinetics often hold for part of the conversion interval.

Generally, there are very few publication concerning kinetics of biomass gasification (Klose and Woelki, 2005 and Henrich *et al.*, 1999). As far as we know there are no publications on gasification kinetics of char from sewage sludge. Reactivity of pyrolyzed sewage sludge in air and carbon dioxide was studied by Inguanzo *et al.* (2001).

METHODOLOGIES AND APPROACHES EMPLOYED

The Mettler-Toledo TGA/SDTA851 LF (temperature range 20 ÷ 1100°C, sample mass up to 5000 mg) thermobalance was coupled with Balzers ThermoStar Mass Spectrometer QMS 200 and gas chromatograph Varian 3800. The gas supplying system consisting of three calibrated mass flow meters (Brooks model 4850S) was set to control the feed gas flow rate and gas composition. The content of C, H, N, S and O in the raw samples of sewage sludge and for samples on different stage of the process was determined by elemental analyser CE Instruments NA 2500. Figure 5.1 shows the scheme of our experimental set-up.

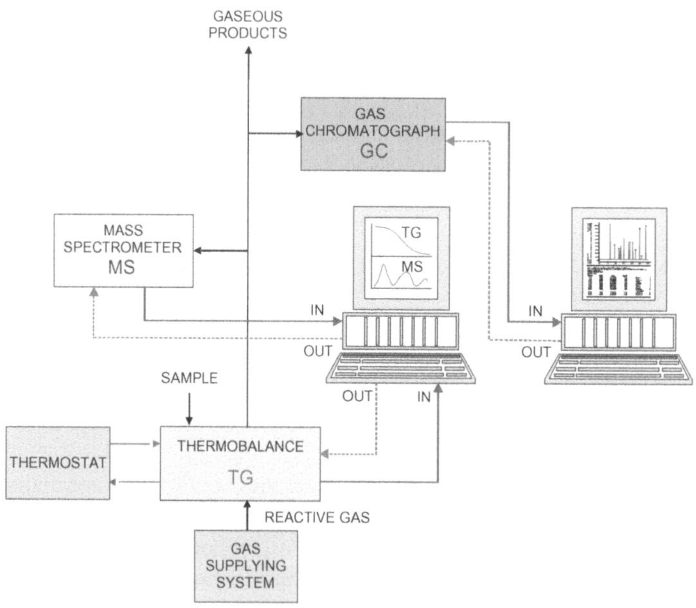

Figure 5.1 Scheme of the TGA- MS system

The TG-MS-GC system provides extensive information on the conversion of solid material as well as the evolved substances. Three analytical methods result in a complete picture of the ongoing reactions during the thermal decomposition process. In order to use results obtained from the system for quantitative analysis of reaction products the MS profiles recorded as relative intensity must be converted into composition of evolved gases. Two calibration methods of MS signals were proposed:

1. External standard – two component mixture of inert gas (Ar) and CO or CO_2 or H_2 or CH_4 was prepared outside of the system by using precise mass flow meters set on specified flow rates for each component.

2. Internal standard – the calibration gas was produced or consumed by chemical reaction occurring in the thermobalance. On the basis of sample mass loss measured in TG the molar fraction of produced or consumed gas can be calculated and compared with MS signal. For internal standard method reaction of CuO reduction in CO or H_2 containing atmosphere was performed. The amount of evolved or consumed gas can be easily changed by using different amount of oxide. The concentrations of CO_2, H_2O and change in concentration of CO or H_2 was calculated from the mass loss determined from TG measurement and stoichiometric balance for the chemical reaction occurred.

The main tasks for gas chromatograph were to confirm the results of product gas analysis obtained by mass spectrometer and more accurate analysis of light hydrocarbons in produced gas. Gas chromatograph was calibrated with standard gas mixture containing C_1-C_4 hydrocarbons, carbon monoxide, carbon dioxide, nitrogen and hydrogen.

In order to describe particular group (lump) of reactions, distinguished on the basis of gaseous products and DTG curve deconvolution, it is essential to identify so called 'kinetic triplet' activation energy, pre-exponential factor, and reaction model. A solution of an non-linear inverse problem allows for identification of the above mentioned parameters on the basis of the obtained experimental data.

A numerical interactive environment MATLAB was applied to prepare our computing program. The following optimization methods were engaged in our application and can be chosen alternatively:

- lsqnonlin: a function for solving non-linear least-squares (non-linear data-fitting) problems, which uses the Levenberg-Marquardt method with a line search or alternatively the Gauss-Newton method.
- fmincon: a function, which uses of the Kuhn-Tucker equations as the basis for sequential quadratic programming (SQP) methods.
- ga: which searches for a minimum of a function using the genetic algorithm. The genetic algorithm (GA) works on a population using a set of operators that are applied to the population. A population is a set of points in the design space. The initial population is generated randomly by default. The next generation of the population is computed using the fitness of the individuals in the current generation.
- patternsearch: the pattern search algorithms find a sequence of points, x_0, x_1, x_2,... that approaches the optimal point. The value of the objective function either decreases or remains the same from each point in the sequence to the next.

The best approach to the identification of the parameters of the differential equation system used by us is application of a hybrid scheme to optimize a function combining the genetic algorithm or pattern search method with another optimization method. Our problem with fast and slow changing variables is numerically stiff. Stiff deterministic solvers are the best choice. In our application we can use alternatively the following MATLAB functions:

- ode15s (stiff/NDF) – Variable order solver based on the numerical differentiation formulas.
- ode23s (stiff/Mod. Rosenbrock) – One-step solver based on a modified Rosenbrock formula of order 2.
- ode23t (Mode. stiff/Trapezoidal) – An implementation of the trapezoidal rule using a 'free' interpolant.
- ode23tb (stiff/TR-BDF2) – An implementation of an implicit Runge-Kutta formula.

The parameters of the model (activation energies, pre-exponential factors, reaction order and w_i – a fraction of total weight loss connected with the particular reaction) were calculated from the experimental data by minimizing of the objective function. Our objectives were DTG and TG fitting and so called 'peaks pulling' (a location of temperature of maximum peaks, at DTG curve, of the numerical functions in comparison with the experimental values for the individual reactions). The applied weighted sum strategy converts the multiobjective problem of minimizing the vector into a scalar problem by constructing the weighted sum of the all objectives. Such approach to the objective function is very innovative. The computing program created by us is universal and after small changes can be applied to identification of the kinetic parameters of different processes.

Characterization of research objects – effect of sewage sludge origin on pyrolysis behaviour

Eight samples of sewage sludge were chosen for the study. List of all samples and their origin are presented in Table 5.1.

For proximate analysis in TG samples of sewage sludge were heated with constant heating rate to the temperature 1000ºC in the inert atmosphere of argon and then samples were cooled to the temperature of gasification (900ºC) and in isothermal conditions gaseous atmosphere was changed to carbon dioxide. On the basis of this experiments amount of moisture, volatiles, char and ash was determined. DTG (differential of TG curve) profiles for all studied sewage sludge samples were collected for further kinetic analysis and modelling.

Proximate analysis of sewage sludge composition (see Table 5.2) showed significant differences in volatile matter and ash content owing to a more or less complete digestion process and the original waste water. The lowest ash content was obtained for CON and SLO samples. The CON sample was obtained from the treatment of food industry waste waters and SLO comes from a small rural waste water treatment plant. The latter was not subjected to any stabilization process. Low ash content means high heating value of the sample, although this parameter was not determined in this study.

Table 5.1 List of samples

Sample	Type of WTP	Type of sludge
GOS-Lodz (GOS) Poland	municipal	digested
Kostrzyn (KOS) Poland	industrial (paper factory)	raw
Slonsk (SLO) Poland	rural	digested
Constanti (CON) Spain	industrial (food industry)	digested
Reus (REU) Spain	municipal	digested
Praha (PRA) Czech Republic	municipal	digested
Kalisz (KAN) Poland	municipal	digested
Kalisz (KAW) Poland	municipal	digested, limed

Table 5.2 Proximate analysis of samples

Sample	Moisture (%)	Volatiles (%)	Char (%)	Ash (%)
GOS	5.3	44.6	7.4	42.7
KOS	6.8	52.4	7.2	33.5
SLO	5.4	58.9	12.4	23.3
CON	5.0	62.1	10.8	22.1
REU	5.2	53.0	8.1	33.7
PRA	4.2	42.3	4.7	48.8
KAN	4.3	52.6	9.3	33.8
KAW	2.3	48.8	4.7	44.2

The content of C, H, N, S and O in the raw samples of sewage sludge and in solid residue after pyrolysis process were determined (Table 5.3). On the basis of these data balance of basic elements for all investigated samples and approximate composition of produced tars in pyrolysis process were calculated.

Table 5.3 Elemental analysis of the raw samples (0) and after pyrolysis (P)

Sample		N	C	H	S	O*	Ash	Total	H/C	O/C
		% of raw sample mass							Mole fraction	
GOS	0	2.8	26.0	3.5	1.1	23.9	42.7	100	0.81	0.69
	P	0.2	6.5	0.1	0.7	0.0		50.1	0.06	0.00
KOS	0	1.9	25.5	4.2	0.2	34.7	33.5	100	0.98	1.02
	P	0.1	6.4	0.3	0.1	0.3		40.8	0.29	0.04
SLO	0	5.8	37.0	5.4	0.9	27.7	23.3	100	0.88	0.56
	P	0.5	10.2	0.1	0.2	1.4		35.7	0.08	0.10
CON	0	5.6	34.6	5.1	0.8	31.8	22.1	100	0.88	0.69
	P	0.0	10.6	0.1	0.1	0.0		32.9	0.03	0.00
REU	0	4.0	31.0	4.7	0.6	26.0	33.7	100	0.90	0.63
	P	0.2	7.5	0.0	0.3	0.0		41.8	0.02	0.00
PRA	0	2.7	22.9	3.4	0.7	21.6	48.8	100	0.89	0.71
	P	0.3	6.8	0.2	0.4	0.0		53.5	0.18	0.00
KAN	0	3.7	32.6	5.0	2.1	24.6	32.0	100	0.92	0.57
	P	0.2	9.5	0.1	0.5	0.0		43.1	0.03	0.00
KAW	0	2.7	24.7	3.7	0.7	25.9	42.3	100	0.89	0.79
	P	0.1	5.6	0.2	0.1	0.0		48.9	0.25	0.00

**Calculated by difference.*

Because of the absence of reactive atmosphere, total mass of evolved volatile products during the process should be equal to the mass loss observed in thermobalance. By comparing the rate of total mass loss with mass of evolved gases the rate of tar formation was estimated. To make the comparison of DTG and MS-GC data possible the latter were converted to the same units using the calibration results. A visual analysis of DTG and MS profiles (Figure 5.2) reveals clearly that there are some peaks that can be recognized as fractions in the sewage sludge and consequently their thermal decomposition could be grouped into few lumps of reaction products. The position of characteristic peaks for all samples is very similar meaning that all sewage sludges contain the same components which undergo similar groups of decomposition reaction. Differences in the intensity of individual peaks may be attributed to different fractional composition of samples and their different reactivity.

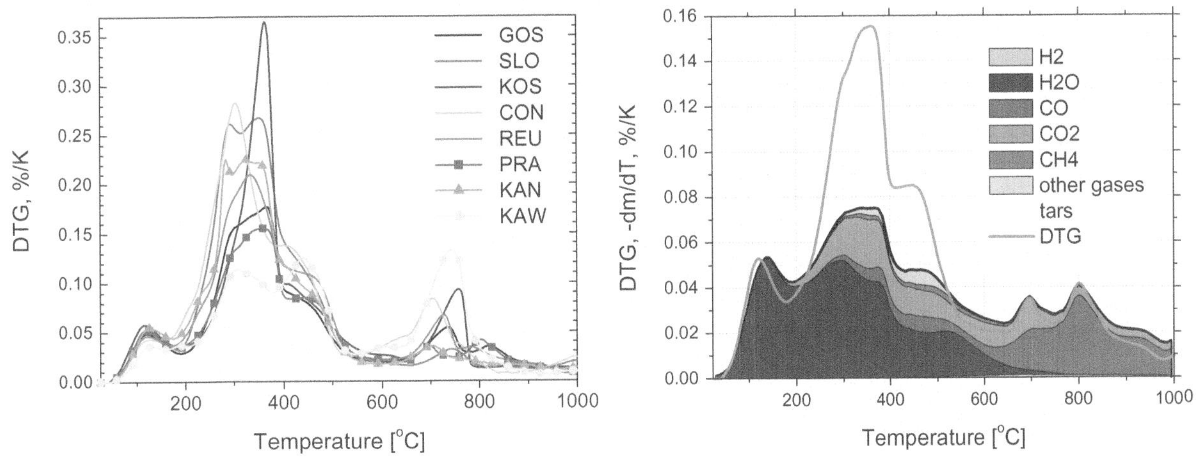

Figure 5.2 DTG curves for pyrolysis for all samples of sewage sludge and converted MS profiles for PRA

The first step that can be observed during pyrolysis appears at temperature range from 100 to 150°C corresponds to evaporation of moisture not removed completely during the drying in the oven. From the MS signals converted to the rate of formation of gas products one can see that water is the main product formed around 150°C. Dried sewage sludge sample tested in our laboratory contained 2–5% of moisture.

The main mass loss steps occurred between about 200 and 400°C. A double peak can be observed in this temperature range, the first shoulder with the maximum at about 290°C can be identified with biodegradable materials and undigested organics (Moldoveanu, 1998). The side group reactions with water elimination of organic biopolymers are perhaps represented in the second shoulder with the maximum at about 350°C as for cellulose pyrolysis – this reaction takes place at this temperature range (Otero *et al.*, 2002). Water and CO_2 with smaller amounts of CO, CH_4 and other hydrocarbons were the main components detected by mass spectrometer in the outlet gas in the discussed temperature range.

The next pronounced step occurs around 460°C and it is overlapped with the lower temperature region. Weight losses in this step may be produced by depolimerisation reactions of compounds like cellulosic material. Emission of CO, CO_2, H_2O and CH_4 is related with reactions occurring in this temperature region.

Weight loss above 600°C during pyrolysis may be produced by decomposition of calcium carbonate, devolatilisation reactions of unreacted source material and more likely of carbonaceous material (char) and tars formed at the earlier stages of the process. The characteristic peak with maximum at 720°C that can be seen on DTG and MS curves was identified as the decomposition of calcium carbonate (Bedyk *et al.*, 2007). The main product of this reaction is CO_2, which can easily oxidize char to CO above 700°C. The next peak with maximum at 800°C corresponds to CO as a main product of char gasification.

Except of gas and char some tars are always formed during the pyrolysis. Tars are defined as an organic pyrolysis products excluding hydrocarbons and benzene. Amount of mass of sample transformed to tars during the pyrolysis of sewage sludge was calculated from DTG and MS-GC data. The formation of tars occurred after 200°C and ceased at 500°C.

Influence of the pyrolysis extent on gasification process

The results of TG-MS-GC analysis of GOS sewage sludge heated in argon with the rate of 20°C/min are presented in Figure 5.3a. As can be seen fast devolatilisation of the sample occurs from 200 to 500°C. In this temperature range all liquid products (tars and oil) and char are formed. Above 600°C a slow and gradual mass loss of the sample is observed, which is due to oxidation of char by carbon dioxide and water released from the minerals contained in ash (autooxidation of char) (Bedyk *et al.*, 2007 and Caballero *et al.*, 1997). Carbon oxides which are involved in the autooxidation process are main gaseous products formed in this temperature region (see Figure 5.3b).

Table 5.4 summarizes the results of elemental analysis of dried sewage sludge and solid residue. Samples of sewage sludge were heated with constant rate of 20°C/min up to the final temperature listed in the first column of the Table. As can be seen from the data the solid product of pyrolysis obtained at temperatures above 500°C contains mainly minerals (ash), atomic carbon and some amount of oxygen, which is a component of the mineral layer (ash). Sulfur, nitrogen and hydrogen are also present. Above 500°C the atomic carbon content in the solid residue does not change significantly with the final pyrolysis temperature, although some carbon is released in the form of carbon oxides. The main mass loss of the sample at these temperatures is caused by releasing of oxygen as water and carbon dioxide (see Figure 5.3b).

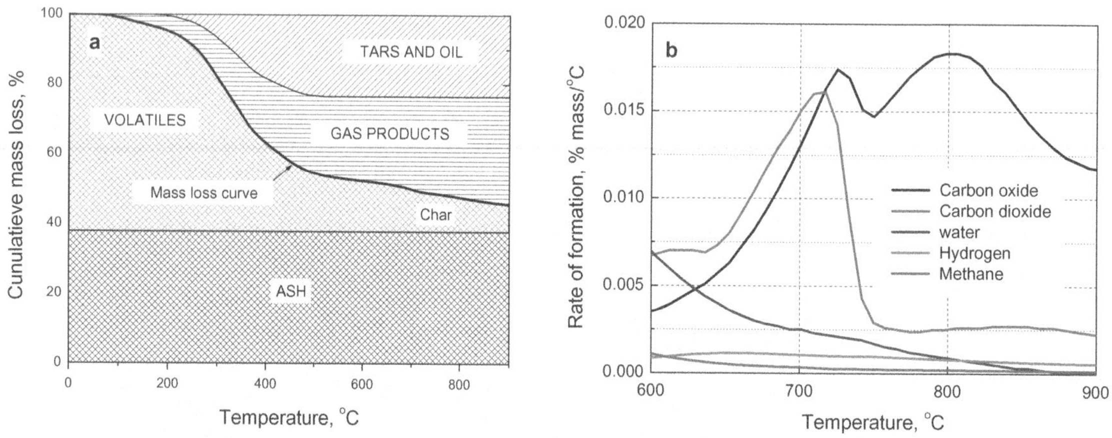

Figure 5.3 The results of TG-MS-GC analysis of GOS sample pyrolysed at different temperatures (a) – composition of solid and volatile products, (b) – formation rate of the main gaseous products

Table 5.4 Characterization of sewage sludge and solid residue after pyrolysis at different temperatures

Sample[a]	C	H	N	S	O	Other elements	Solid residue	Char[b]
	% of raw sample mass							
0	30.73	4.36	3.71	0.94	27.18	33.08	100.0	
200	30.47	2.11	3.64	0.74	24.34	32.90	94.2	
300	23.87	2.45	2.49	0.58	17.17	32.74	79.3	
400	15.50	1.25	1.28	0.45	11.64	31.88	62.0	
500	11.70	0.69	1.34	0.42	8.90	31.55	54.6	16.8
600	11.06	0.54	0.93	0.29	7.82	31.86	52.5	14.7
700	10.38	0.34	0.49	0.34	6.19	31.26	49	10.2
800	10.36	0.25	0.57	0.29	4.68	31.15	47.3	9.5
900	7.79	0.21	0.52	0.36	3.36	31.06	44.8	7.0

[a] *Final heating temperature of sample.*
[b] *Volatile mutter subjected to gasification in CO_2 at 900°C.*

Reactivity of char in CO_2 depends on the origin of sewage sludge. This is attributed to different amount of ash and carbon content in the raw sample. The more amount of ash contains sewage sludge the higher rate of gasification reaction is observed. The effect of heating rate on the yield and reactivity of the char was insignificant.

Calcium, in the form of oxide or hydroxide, is often applied in the final stage of sewage sludge processing for hygienisation (e.g. KAW in Table 5.1). The factors which influence the biocidality of a liming process are a high alkaline reaction and an increased temperature caused by the exothermal reaction of calcium oxide hydration (Tingyu *et al.*, 2000). Liming of sludge also enables stabilization and enhances dehydration (Khan, 1987). Moreover, it has been demonstrated that, similarly to the pyrolysis of coal, the addition of CaO has also a great impact on the thermal decomposition of sewage sludge.

In our study a sample of dried (about 97% dry mass) and grinded sewage sludge (GOS) was mixed in different proportions with calcium oxide and heated with a constant heating rate (20°C/min) in thermobalance in the inert atmosphere of Ar to the temperature 1000°C. With an increase in CaO content in the initial sludge sample, an increase in water vapour emission in the process of pyrolysis is observed, which cannot be attributed to the hydration and dehydration reactions of CaO. Simultaneously, a decrease in tars production and an increase of H_2 emission are observed. Calcium oxide acts not only as a stabilizing agent for sewage sludge, but it is also a great catalyst in processes of pyrolysis and gasification of this waste. It was observed that increasing content of CaO in the sample we can achieve:

- higher conversion of carbonaceous material into gas fuels,
- increase of beneficial gaseous products such as CO and H_2,
- decrease of unbeneficial liquid products (tars).

Except of CaO dolomite $CaMg(CO_3)_2$ was also tested. Dolomite which is consider as a typical tar breaking catalyst in pyrolysis process appears to be less active than calcium oxide (Bedyk *et al.*, 2009).

Gasification reagent impact on the yield and kinetics

The pyrolysis char used in gasification experiments had been previously produced outside the TG-MS-GC system, in a fixed bed reactor by heating dried sewage sludge sample (GOS) in 1000°C for 4 hours in argon flowing through the reactor.

DTG profiles obtained under dynamic conditions (heating rate of 10 K/min) for inert gas (Ar), CO_2/Ar mixture (1/1 molar ratio) and water vapour atmosphere (0.5 g/h H_2O, 50 ml/min of Ar) are compared in Figure 5.4a. Experiments on steam gasification of sewage sludge were conducted using special water vapor generator, being integral part of Netzsch STA 409 thermobalance. As can be seen from the figure, all curves are similar below temperature of 700°C, where pyrolysis occurs only. Above this temperature gasification reaction of char formed at earlier stages of the process takes place in reactive atmospheres.

The DTG profiles recorded for oxygen gasification at different gas composition (Figure 5.4b) differ significantly from those observed for CO_2 and H_2O. However the main reaction steps in the lower and middle temperature range are the same irrespective of gas type. Excluding drying period the process of devolatilisation under the atmosphere containing oxygen begins about the same temperature as for pyrolysis in an inert atmosphere. However, the rate of the process is much faster now. The maximum rate of the second step (the first shoulder on the double peak in argon) continues to rise as oxygen content increases and the temperature where this maximum is reached shifts toward lower values. For 20% O_2 this temperature falls to about 260°C, however, in this case incineration rather than gasification occurs.

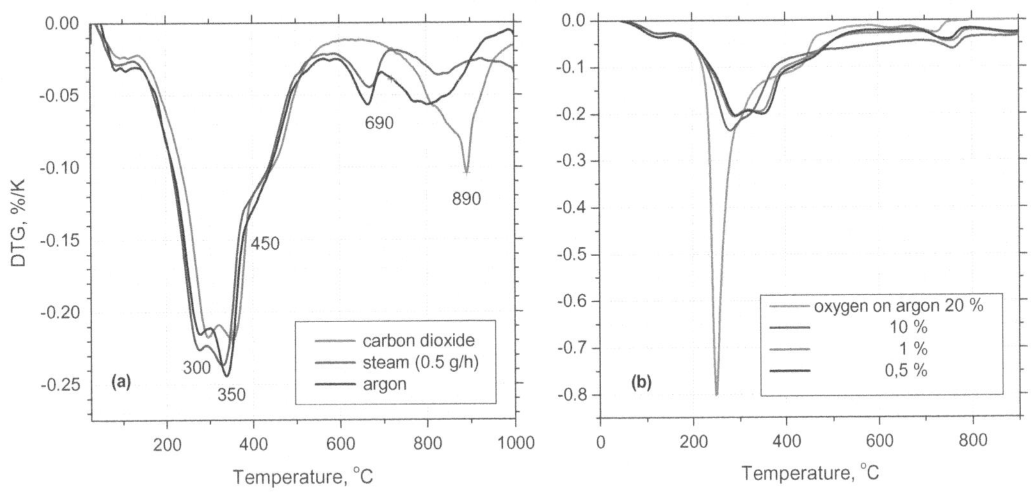

Figure 5.4 Effect of oxidizing agents on gasification of sewage sludge char

Separate sets of experiments concerning char gasification with different oxidizing atmospheres at isothermal conditions was also performed to generate data for kinetics modelling. The experiments at isothermal conditions were performed with samples weighing about 20 mg, which were placed in a alumina crucible and heated at 50°C/min from ambient to final oxidizing temperature in an inert atmosphere of argon. After stabilization of thermal conditions of the sample, flow of the oxidizing reagent was initiated and mass sample loss measured and recorded.

Kinetics of sewage sludge pyrolysis and gasification

In technical applications pyrolysis and gasification processes are often separated especially when carbon dioxide or steam are used as oxidizing agents (Quaak *et al.*, 1999). In this case the pyrolysis process should be optimised towards maximization of pyrolysis char yield. Therefore, our kinetic studies were divided into two parts. In the first part kinetic models of pyrolysis process were developed for better understanding of the complex mechanism and to get a general tool for quantitative analysis of the pyrolysis process. This tool was next used to compare kinetic behaviour of sludges of different origin in the pyrolysis process. The discrimination and identification of kinetic models for pyrolysis char gasification with different oxidising agents was the subject of the second part of modelling.

According to Santacesaria (1999) two main difficulties arise in undertaking the kinetic study of a chemical reaction:

- the development of suitable experimental techniques to collect reliable reaction rates data in different conditions, and
- the definition of an appropriate strategy of elaboration of the data collected to evaluate the unknown of the problem, that is the kinetic law, a functional relationship, able to simulate all the data collected.

The difficulty increases when few reactants originate many products as a consequence of many reactions taking place. In this case, the first step of the kinetic approach will be the determination of an acceptable scheme for the occurring reactions. Then, the study can be oriented to define the kinetics of each reaction in the pattern. The kinetics of the pyrolysis decomposition by thermogravimetric analysis was studied by different authors. This technique is also a common way to study the pyrolysis kinetics of sewage sludge (Dümpelmann *et al.*, 1991; Conesa *et al.*, 1997; Chu *et al.*, 2001; Ledakowicz and Stolarek, 2001; Scott *et al.*, 2006; Font *et al.*, 2005; Casajus *et al.*, 2009; Thiphunthod *et al.*, 2006 and Petersen and Werther, 2005).

The reaction scheme of sewage sludge pyrolysis is very complex. The decomposition of sludge should be described with the models involving many reactions. In typical approach, the simplified lumped reaction kinetic model is used, with the detailed principles arising from visual analysis of DTG curve. In our study the deconvolution of DTG curves as well as MS spectra performed according to Fraser-Suzuki asymmetric profile were the basis for identification of the number of lumps and their contribution to the overall mass loss. This is the new approach that has not been applied yet. As a first attempt, the following six lumps were taken into account:

- at the temperature range up to 180°C – dewatering and devolatilisation,
- at the temperature range from 180°C up to 600°C – three basic groups of reaction characteristic for the decomposition of biomass,
- at the temperature range from 600°C to 1,000°C – two groups of reactions connected with degradation of mineral residues, and decomposition of char or nonbiodegradable matter, which was very hard to degrade at lower temperatures.

To obtain mathematical description of the process a scheme of six co-current independent reactions was considered:

$$
\begin{aligned}
&A_1 \xrightarrow{k_1} s_1 \cdot B_1 + (1 - s_1) \cdot G_1 \\
&A_2 \xrightarrow{k_2} s_2 \cdot B_2 + (1 - s_2) \cdot G_2 \\
&\vdots \\
&A_6 \xrightarrow{k_6} s_6 \cdot B_6 + (1 - s_6) \cdot G_6
\end{aligned}
\qquad \text{Eq. 1}
$$

where A_i denotes a substrate that takes part in the i-th reaction, B_i – solid product of this reaction, and G_i – volatile product, while the stoichiometric coefficient s_i denotes the ratio of solid product mass to the mass of reacted substrate.

To describe the rate of the pyrolysis occurring in a solid sample it was assumed that the process is controlled mainly by chemical decomposition and internal mass and heat transfer limitation could be negligible at the conditions selected (low heating rates, initial mass of samples). The model corresponding to the above assumptions can be written as the following set of ordinary differential equations:

$$
\begin{aligned}
&\frac{d\alpha_1}{d\tau} = k_{01} \cdot \exp\left(-\frac{E_1}{R \cdot T(\tau)}\right) \cdot (1 - \alpha_1)^{n_1}, \alpha_1(0) = 0 \\
&\frac{d\alpha_2}{d\tau} = k_{02} \cdot \exp\left(-\frac{E_2}{R \cdot T(\tau)}\right) \cdot (1 - \alpha_2)^{n_2}, \alpha_2(0) = 0 \\
&\vdots \\
&\frac{d\alpha_6}{d\tau} = k_{06} \cdot \exp\left(-\frac{E_6}{R \cdot T(\tau)}\right) \cdot (1 - \alpha_6)^{n_6}, \alpha_6(0) = 0
\end{aligned}
\qquad \text{Eq. 2}
$$

where α_i is the conversion degree of i-th reaction defined as $\alpha_i = (m_{0i} - m_i)/m_{0i}$; m_i is mass of the substrate A_i and m_{0i} – initial mass of A_i (total mass loss in i-th reaction), which is related to the total weight loss of the sample Δm given by the equation $m_{0i} = \Delta m \cdot w_i$, in which w_i – fraction of total mass loss associated with i-th reaction.

The parameters (activation energies, pre-exponential factors, reaction order and the fractions of w_i of the reactions in the total weight loss) appearing in the model were calculated from the experimental data by minimizing a target function.

Preliminary information about the contribution of each lump into the total weight loss can be obtained by deconvolution of DTG profiles into individual peaks. More details of employing various deconvolution methods of DTG curves was given by Stolarek and Ledakowicz (2003). The deconvolution allows also to determine the temperature maximum T_{max} position at which the reaction rate achieves maximum value. These data are necessary to use ASTM E 698 method (Kissinger, 1956, 1957), which is a relatively simple way to obtain starting values of kinetic parameters for non-linear regression calculations. If the lumps are assumed to be non-interacting and evolved by independent parallel reactions the first-order kinetic parameters such as activation energy and pre-exponential factor can be determined from mathematical evaluation of DTG curves according to Kissinger's law method.

The weight loss represented by TG and DTG curves recorded during the pyrolysis of GOS sewage sludge sample at four heating rates (5, 10 15 and 20 K/min) were used to the quality of the proposed kinetic model. Figure 5.5 shows the comparison of experimental and calculated DTG curves based on the set of kinetic parameters which minimizing the target function for all investigated heating rates. As it can be seen the fitting is very good for all the experiments. That means that parameter values obtained (see Nowicki *et al.*, 2008) were independent on heating rate in the range from 5 to 20 K/min. Although, the high reaction orders (up to 5.3) obtained from calculations may be unacceptable from physical and chemical point of view, we have to remember that individual lamp represents a set of simple (may be first order) reactions taking place in the same temperature region. Describing such a process by single n-th order rate equation results in high values of n. Stolarek and Ledakowicz (2005) believe that high reaction orders are typical for lumped models and can be interpreted as a group of many first order reactions with gradually increase activation energies.

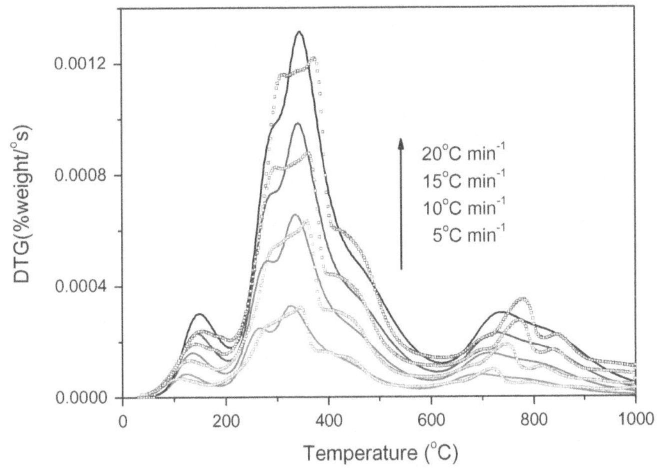

Figure 5.5 Comparison of experimental (solid lines) and calculated DTG profiles (points)

Proposed kinetic model involving six parallel n-order reactions was successfully applied for different sewage sludge samples. The fitting of the kinetic model to the experimental data is very good. The values of activation energy are in the expected range typical for chemically control processes, except for the first lump, which is mainly the result of evaporation of residual water. As far as reaction order is concerned again high values of those parameters were obtained. This is probably the result of the lumping of many elemental reactions and its complex behaviour can be modelled by n-th order kinetics.

Kinetic modelling of gasification process was performed using experimental data collected at isothermal conditions for three oxidizing agents. Raw data were recalculated to express conversion as a function of time. The kinetics of char gasification was modelled by utilizing two representative gas-solid reaction models, volumetric model and shrinking core model. For the first-order kinetics, the conversion in an isothermal batch reactor can be expressed by the following equation (in an integrated form):

$$-\ln(1-\alpha) = k \cdot t \qquad \text{Eq. 3}$$

where k is the reaction rate constant.

In the next step the shrinking-core model was examined to interpret the conversion-time data. With the model in the chemical reaction controlled regime, the following equation may be applied:

$$\Phi(\alpha) = 1-(1-\alpha)^{1/3} = \frac{t}{\tau} \qquad \text{Eq. 4}$$

where the reaction time τ for complete conversion of the reaction A(g) + bB(s) ″ products is defined as:

$$\tau = \frac{C_{B0} \cdot d_0}{2 \cdot b \cdot k \cdot C_A} \qquad \text{Eq. 5}$$

C_{B0}, d_0, k and C_A are the initial carbon concentration, initial sample particle diameter, surface reaction rate constant, and CO_2 gas concentration, respectively.

With the model in the diffusion controlled regime, the following equation may be applied:

$$\Phi(\alpha) = 1 - 3 \cdot (1-\alpha)^{2/3} + 2 \cdot (1-\alpha)^{2/3} = \frac{t}{\tau} \qquad \text{Eq. 6}$$

where the reaction time τ for complete conversion of the reaction A(g) + bB(s) products is defined as:

$$\tau = \frac{C_{B0} \cdot d_0^2}{24 \cdot b \cdot D_{ef} \cdot C_A} \qquad \text{Eq. 7}$$

D_{ef} is a diffusion coefficient.

The shrinking core model for reaction controlled regime was found to be the best for predicting the rate of char gasification in the whole range of conversion in carbon dioxide atmosphere (Figure 5.6a). The same model can be applied for oxygen gasification but only for the conversion range below 80%, while for higher conversion the shrinking core model for diffusion controlled regime should be applied (Figure 5.6b).

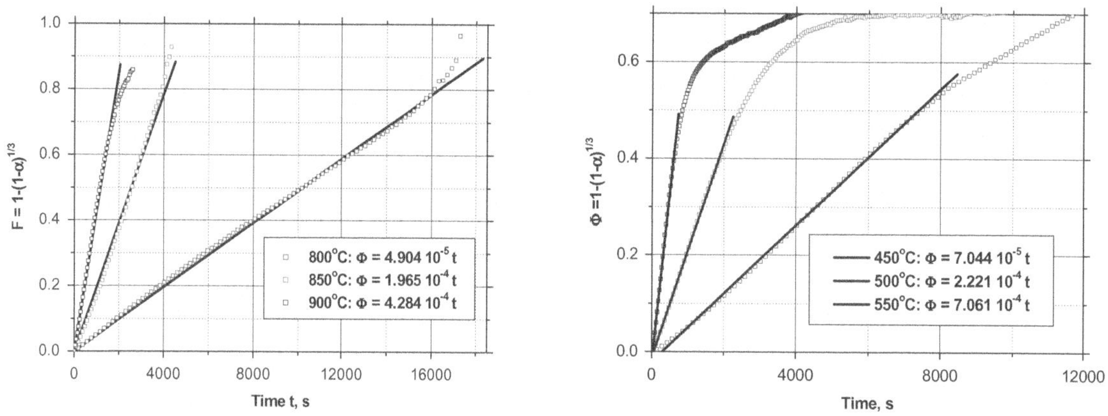

Figure 5.6 Conversion data based on the for the shrinking-core model in the chemical reaction controlled regime for gasification with carbon dioxide (a) and for oxygen (b)

Steam gasification of sewage sludge char proceeds probably according to different mechanism and the best fitting to experimental data was observed for the first-order kinetics. The values of activation energy for various gasification reactions were found to be 228 kJ/mol, 193 kJ/mol and 113 kJ/mol, respectively. These values agree with the literature data for biomass gasification.

IMPACT OF THE TECHNOLOGY ON ITS INDUSTRY OR RESEARCH SECTOR

Sewage sludge gasification has the potential to offer a major contribution to meeting the targets of the European Union for CO_2 mitigation and proper waste management. It opens the possibility to utilize sewage sludge from traditional landfilling, or low-efficient and costly incineration. Production of synthesis gas enables to change heating and electricity by steam cycles to high-efficient bio-energy production by gas engines and combined gas- and steam turbine cycles. Moreover the gasification product – synthesis gas has potential to be used as substrate for liquid fuel production via Fischer-Tropsch reaction or methanol synthesis, or fertilizers depending on social or industry demand. The pyrolytic gasification of sewage sludge is an alternative to the sewage sludge incineration system. It is possible to reduce more than half the greenhouse gas volume emitted from the sewage treatment plant. Moreover, gasification plant will be better accepted by society than incineration plant, which is still associated with dioxin pollution problems, while gasification produces synthesis gas that should be cleaned by standard methods. This holds promise of a very major emission reduction and energy saving effect.

The results of this work contributed to the increasing knowledge of sewage sludge gasification. The construction and implementation of experimental set-up consisting of thermo-balance coupled with mass spectrometer and gas chromatograph TG-MS-GC enabled characterization of sewage sludge of different origins or any biomass. Profound understanding of the process mechanism revealed that characterization of sewage sludge is necessary for proper design and development of the gasification process. Kinetics of pyrolysis and subsequent gasification step have been modeled by various kinetic approaches and the identification of kinetic parameters enables to design and optimize the developing processes of thermo-chemical treatment of sewage sludge. As the sewage sludge pyrolysis kinetics has been studied for many years the kinetic data of sewage sludge gasification is very scarce and therefore our contribution is beyond the state-of the art in this field of research.

The achievements of this work can be accounted by the publications that were written and submitted to publication or have been already published during the realization of the project:

1. L. Nowicki, P. Stolarek, T. Olewski, T. Bedyk, S. Ledakowicz, Mechanism and Kinetics of Sewage Sludge Pyrolysis by Thermogravimetry and Mass Spectrometry. Analysis. Chemical and Processing Engineering Vol. 29, 813–25, 2008.
2. L. Nowicki, P. Stolarek, T. Olewski, T. Bedyk, S. Ledakowicz, Kinetics of sewage sludge pyrolysis. European Conference on Sludge Management, ECSM 08, 1–2 Sep. 2008, Book of Abstracts, p. 18, University of Liege, Belgium.
3. T. Bedyk, L. Nowicki, P. Stolarek, S. Ledakowicz, Effect of MgO, CaO or dolomite addition on pyrolysis of sewage sludge. European Conference on Sludge Management, ECSM 08, 1–2 Sep. 2008, Book of Abstracts, p. 21, University of Liege, Belgium.
4. L. Nowicki, T. Bedyk, P. Stolarek,, S. Ledakowicz, Thermo-chemical processing of sewage sludge under different gaseous atmosphere. European Conference on Sludge Management, ECSM 08, 1–2 Sep. 2008, Book of Abstracts, p. 58, University of Liege, Belgium.
5. M. Markowski, T. Bedyk, L. Nowicki, P. Stolarek, S. Ledakowicz, Quantitative analysis of thermo-chemical processing of sewage sludge by TG-MS-GC system. European Conference on Sludge Management, ECSM 08, 1–2 Sep. 2008, Book of Abstracts, p. 59, University of Liege, Belgium.
6. T. Bedyk, L. Nowicki, P. Stolarek, S. Ledakowicz, Effect of CaO and dolomite additive on the thermal decomposition of sewage sludge in an inert atmosphere. Journal of Residuals Science & Technology, 2009, 6, No. 1, p. 3–10.
7. O. Piddubniak, S. Ledakowicz, L. Nowicki, New approach to a problem of heat transfer with chemical reaction in a cylinder of finite dimensions, submitted to International Journal of Heat & Mass Transfer, 2011, 54, 338–344.
8. A. Klepacz-Smółka, P. Stolarek, L. Nowicki, T. Bedyk, T. Olewski, S. Ledakowicz, Wyznaczanie mechanizmu i kinetyki pirolizy osadów ściekowych za pomocą analizy termograwimetrycznej i spektrometrii masowej, (Determination of mechanism and kinetics of sewage sludge pyrolysis by TG and MS analysis), Ogólnopolskie Sympozjum 'Reaktory Wielofazowe i Wielofunkcyjne dla Procesów Chemicznych i Ochrony Środowiska', (II Polish Symposium 'Polyphase and Polyfunctional Reactors for Chemical Processes and environment protection') Ustroń-Jaszowiec 19 22.10.2009. Book of Abstracts, p. 18.
9. L. Nowicki, A. Antecka, T. Bedyk, P. Stolarek, S. Ledakowicz: The kinetics of gasification of char derived from sewage sludge. Journal of Thermal Analysis and Calorimetry, 2010 accepted.

REFERENCES

Bedyk T., Nowicki L., Stolarek P. and Ledakowicz S. (2007). CaO as a catalyst for pyrolysis of sewage sludge. *Proceedings of ECOpole*, **1**, 85–90.

Bedyk T., Nowicki L., Stolarek P. and Ledakowicz S. (2009). Effect of CaO and dolomite additive on the thermal decomposition of sewage sludge in an inert atmosphere. *J. Res. Sci. Techn.*, **6**, 3–10.

Bridgwater A. V. (2005). The technical and economic feasibility of biomass gasification for power generation. *Fuel*, **74**, 631.

Caballero J. A., Font R., Marcilla A. and Conesa J. A. (1997). Characterization of sewage sludge by primary and secondary pyrolysis. *J. Anal. Appl. Pyrol.*, **40–41**, 433–450.

Casajus C., Abrego J., Marias F., Vaxelaire J., Sanchez J. L. and Gonzalo A. (2009). Product distribution and kinetic scheme for the fixed bed thermal decomposition of sewage sludge. *Chem. Eng. J.*, **145**, 412–419.

Cetin E., Moghtaderi B., Gupta R. and Wall T. F. (2004). Influence of pyrolysis conditions on the structure and gasification reactivity of biomass chars. *Fuel*, **83**, 2139.

Chen X. and Jeyaseelan S. (2001). Study of Sewage Sludge Pyrolysis Mechanism and Mathematical Modeling. *J. Environ. Eng.*, **127**(7), 585–593.

Chu C. P., Lee D. J. and Chang C. Y. (2001). Thermal pyrolysis characteristics of polymer flocculated waste activated sludge. *Water Research*, **35**, 49–56.

Conesa J. A., Marcilla A., Prats D. and Rodriguez-Pastor M. (1997). Kinetic study of the pyrolysis of sewage sludge. *Waste Management & Research*, **15**, 293.

Conesa J. A., Maecilla A., Moral R., Moreno-Caselles J. and Perez-Espinoza A. (1998). Evolution of gases in the primary pyrolysis of different sewage sludges. *Thermochim. Acta*, **313**, 63–73.

Di Blasi C. (2008). Combustion and gasification rates of lignocellulosic chars. *Progress in Energy and Combustion Science*, **35**, 121–140.

Dümpelmann R., Richarz W. and Stambach M. R. (1991). Kinetic studies of the pyrolysis of sewage sludge by TGA and comparison with fluidized beds. *Can. J. Chem. Eng.*, **69**, 953–963.

Font R., Fullana A. and Cinesa J. (2005). Kinetic models for the pyrolysis and combustion of two types of sewage sludge. *J. Anal. Appl. Pyrolysis*, **74**, 429–438.

Henrich E., Buerkle S., Meza-Renken Z. I. and Rumpel S. (1999). Combustion and gasification kinetics of pyrolysis chars from waste and biomass. *J. Anal. Appl.Pyrolysis*, **49**, 221–241.

Hossain M. K., Strezov V. and Nelson P. F. (2009). Thermal characterisation of the products of wastewater sludge pyrolysis. *Journal of Analytical and Applied Pyrolysis*, **85**, 442–446.

Hurt R. and Calo J. (2001). Semi-global kinetics for char combustion modeling. *Combust Flame*, **125**, 1138–1149.

Inguanzo M., Menendez J. A., Fuente E. and Pis J. J. (2001). Reactivity of pyrolyzed sludge in air and CO2. *J. Anal. Appl. Pyrolysis*, **58–59**, 943–954.

Juniper. (2001). Pyrolysis & Gasification of Waste. (vol. 2) Technology & Business Review. Juniper Consultancy Services Ltd.

Karayildirim T., Yanik J., Yuksel M. and Bochorin H. (2006). Characterisation of products from pyrolysis of waste sludges. *Fuel*, **85**, 1498–1508.

Khan M. R. (1987). Significance of char active surface area for appraising the reactivity of low- and high-temperature chars. *Fuel. Sci. Technol. Int.*, **5**(2), 185–231.

Kirubakaran V., Sivaramakrishnan V., Nalini R., Sekar T., Premalatha M. and Subramanian P. (2009). A review on gasification of biomass. *Renewable and Sustainable Energy Reviews*, **13**, 179–186.

Kissinger H. E. (1956). Variation of peak temperature with heating rate in diffferential thermal analysis. *J. of Research of the National Bureau of Standards*, **57**, No. 4, October.

Kissinger H. E. (1957). Reaction kinetics in differential thermal analysis. *Analytical Chemistry*, **29**, 1702–1706.

Klose W. and Woelki M. (2005). On the intrinsic reaction rate of biomass char gasification with carbon dioxide and steam. *Fuel*, **84**, 885–892.

Ledakowicz S. and Stolarek P. (2000). Wyznaczanie paremetrów kinetycznych pirolizy odpadów za pomocą analizy termograwimetrycznej. *Inż. Chem. Proc.*, **21**, 345–357.

Liliedahl T. and Sjoestroem K. (1997). Modelling of char-gas reaction kinetics. *Fuel*, **76**, 29–37.

Moldoveanu S. C. (1998). Analytical pyrolysis of natural organic polymers, Amsterdam, NY, Tokio, Elsevier.

Nowicki L., Stolarek P., Bedyk T. and Ledakowicz S. (2008). Mechanism and kinetics of sewage sludge pyrolysis by thermogravimetry and mass spectroscopy analysis. *Chem. Proc. Eng.*, **29**, 813–825.

Otero M., Diez C., Calvo L. F., Garcia A. I. and Moran A. (2002). Analysis of the co-combustion of sewage sludge and coal by TG-MS. *Biomass and Bioenergy*, **22**, 319–329.

Petersen I. and Werther J. (2005). Three-dimensional modeling of a circulating fluidized bed gasifier for sewage sludge. *Chem. Eng. Sci.*, **60**, 4469.

Quaak P., Knoef H. and Stassen H. (1999). Energy From Biomass: A Review Of Combustion And Gasification Technologies. World Bank technical paper no. 422. Energy series.

Sanrascaria E. (1999). Fundamental chemical kinetics: the first step to reaction modelling and reaction engineering. *Catalysis Today*, **52**, 113–123.

Scott S. A., Dennis J. S., Davidson J. F. and Hayhurst A. N. (2006a). An algorithm for determining the kinetics of devolatilisation of complex solid fuels from thermogravimetric experiments. *Chem. Eng. Sci.*, **61**, 2339–2348.

Scott S. A., Dennis J. S., Davidson J. F. and Hayhurst A. N. (2006b). Thermogravimetric measurements of the kinetics of pyrolysis of dried sewage sludge. *Fuel*, **85**, 1248–1253.

Stolarek P. and Ledakowicz S. (2003). Krytyczna ocena metody ASTM 698E wyznaczania parametrów kinetycznych rozkładu termicznego. *Inż. Chem. Proc.*, **24**, 701–715.

Stolarek P. and Ledakowicz S. (2005). Pyrolysis kinetics of chitin by non-isothermal thermogravimetry. *Thermochim. Acta*, **433**, 200–208.

Thiphunthod P., Meeyoo V., Ransungsunvigit P., Kityanan B., Siemanond K. and Rirksomboon T. (2006). Pyrolytic characteristics of sewage sludge. *Chemosphere*, **64**, 955–962.

Thipkhunthod P., Meeyoo V., Rangsunvigit P. and Rirksomboon T. (2007). Describing sewage sludge pyrolysis kinetics by a combination of biomass fractions decomposition. *J. Anal. Appl. Pyrolysis*, **79**, 78–85.

Tingyu Z., Shouyu Z., Jiejie H. and Yang W. (2000). Effect of calcium oxide on pyrolysis of coal in a fluidized bed. *Fuel Processing Technology*, **64**, 271–284.

Wójtowicz M. A., Bassilakis R., Smith W. W., Chen Y. and Carangelo R. M. (2003). Modelling the evolution of volatile species during tobacco pyrolysis. *J. Anal. Appl. Pyrolysis*, **66**, 235–261.

Wu R. M., Lee D. J., Chang C. Y. and Shie J. L. (2006). Fitting TGA data of oil sludge pyrolysis and oxidation by applying a model free approximation of the Arrhenius parameters. *J. Anal. Appl. Pyrol.*, **76**, 132–137.

Yang X. and Jiang Z. (2009). Kinetic studies of overlapping pyrolysis reactions in industrial waste activated sludge. *Bioresource Technology*, **100**, 3663–3668.

Chapter 6

Production of organic substrate from sludge for enhancement of nutrient removal

Pavel Jenicek, Jana Vondrysova, Dana Pokorna, Jana Koubova, Pavel Fialka, Jana Zabranska and Michal Dohanyos

SUMMARY

Project objectives

The desired method of optimisation of existing WWTPs technology is improving of COD/N (COD:P) ratio in the biological reactor. The addition of an external carbon source causes mostly unacceptable increase of operating costs. Therefore the main objective of this work-package was to find an internal carbon source and the way of its optimal utilization. The disintegration methods were until now successfully used for enhancement of anaerobic digestion of excess activated sludge. The use of such techniques for production of degradable organic substrate to improve denitrification or phosphate uptake was evaluated. Efficient nutrient (N,P) removal is today one of the most problematic tasks of the wastewater treatment. Biological methods of nutrient removal are very efficient and sustainable; however lack of organic substrate in wastewater often limits the process efficiency. Addition of external carbon source is possible but expensive solution. Another possible way is to search for internal carbon source and excess activated sludge can be one of them. Different methods of excess sludge disintegration and treatment were studied and tested. The aim of the sludge treatment was the production of organic substrate in optimal quantity and quality for nutrient removal and minimisation of surplus sludge production. The disintegration of cell walls of sludge and release of so called 'cell lysate' can even stimulate the activity of different trophic group of microorganisms.

Work performed

During the project four disintegration methods of activated sludge were tested to organic substrate production, for comparison also primary and activated sludge fermentation was performed. The quantity of produced substrate was evaluated according the specific soluble COD increase during disintegration. The quality of produced substrate was then evaluated according denitrification rate and phosphorus uptake rate. The denitrification tests were performed also with nitrite in parallel. The same testing was performed also with excess activated sludge from biological membrane reactor. Two disintegration methods: mechanical (by lysate centrifuge) and thermal by rapid thermal reactor were tested also in full scale condition. Long term operation of sequencing batch reactors (SBR) was carried out to decide on optimal use of organic substrate produced by disintegration and to find limiting parameters and risk factors of the proposed technology such as stability of nitrification and denitrification activity of microorganisms, separation properties and others.

End results

It was shown, that quantity and quality of substrate produced by sludge disintegration can be controlled by choice of sludge type and by condition of process. The degree of disintegration was evaluated by the soluble COD increase after mechanical, ultrasound, thermal and thermo-chemical disintegration. The evaluation of quality of produced substrates shows that substrates prepared by disintegration of activated sludge improve the denitrification rate significantly. The tests show also that the quality of substrate is comparable with ethanol often used as external organic substrate. It was proved by long term operation of SBR that the main advantage of the sludge disintegration is minimized sludge production which is achieved at comparable effluent concentration of nitrogen. The improvement of nitrogen removal efficiency was not confirmed sufficiently because the amount of disintegrated sludge is limited by requirement to keep the satisfactory sludge concentration. The amount of disintegrated sludge and the degree of minimization has to be related to stable condition of activated sludge process as regards TSS, VSS concentration and nitrification capacity. Sludge disintegration methods are cost effective only in the case of high sludge-disposal costs and of interest in solving of operational problems such as foaming and bulking of sludge.

METHODOLOGIES AND APPROACHES EMPLOYED

Disintegration

The Mettler-Toledo TGA/SDTA851 LF (temperature range 20–1100°C, sample mass up to 5000 mg) thermobalance was coupled with Balzers ThermoStar Mass Spectrometer QMS 200 and gas chromatograph Varian 3800. The gas supplying system consisting of three calibrated mass flow meters (Brooks model 4850S) was set to control the feed gas flow rate and gas composition. The content of C, H, N, S and O in the raw samples of sewage sludge and for samples on different stage of the process was determined by elemental analyser CE Instruments NA 2500. Figure 6.1 shows the scheme of our experimental set-up.

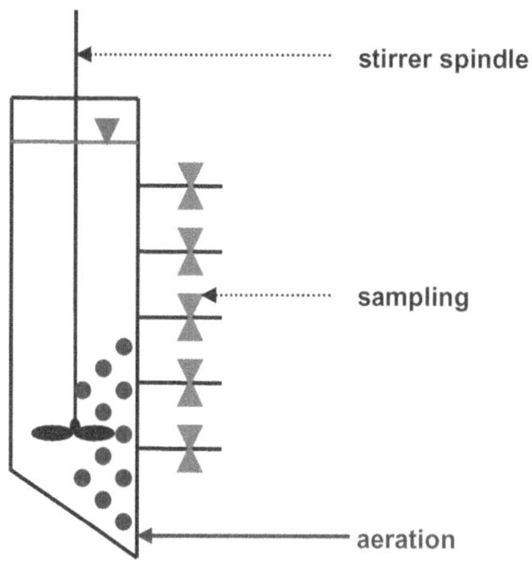

Figure 6.1 The scheme of sequencing batch reactor

Mechanical disintegration

For mechanical disintegration was used disintegrator ULTRA – TURRAX T25 Basic (purchased from IKA, Germany). The input of energy was 500 W. The disintegrator made it possible to change the rotation speed from 11,000 to 24,000 rpm. Due to construction of disintegrator it was not possible to use water ice bath to control temperature during disintegration.

Ultrasound

Sludge samples were treated by a probe ultrasonic disintegrator SONOTRODE DESINTEGRATOR (purchased from ECOSON, Slovakia) with input power 180 W, which had 20 mm tip and emitted ultrasound at 20 kHz. The method of sonication was under normal pressure. For each sonication experiment, 500 mL sludge was filled in an 800 mL glass beaker with open end. The ultrasonic probe was dipped 2.5 cm above the beaker bottom into the sludge at the middle of beaker.

Thermal disintegration

The sludge was heated to 95 ± 1°C under normal pressure. The time interval was 10 and 20 minutes. The 500 mL of sludge was filled in 1000 mL glass beaker with open end. The beaker was placed to the magnetic stirrer with temperature control. The disintegration time measurement started when the required temperature was reached.

Thermo-chemical disintegration

Sodium hydroxide and sulfuric acid were used in the thermo-chemical pre-treatment. The sludge was mixed with alkali or acid in 1:1 ratio. The concentration of chemicals was 0.25; 0.50; 1.0; 5.0 and 10.0 mol/L. The mixture was heated to 95 ± 1°C under normal pressure. The time interval was 10 minutes. The 250 mL of sludge and 250 mL of chemical was filled in 1000 mL glass beaker with open end. The beaker was placed to the magnetic stirrer with temperature control. The disintegration time measurement started when the required temperature was reached.

Fermentation

The primary, raw and excess activated sludge for the fermentation experiments originated from the central WWTP of Prague. Primary sludge (Prim) was taken in the primary sedimentation tank, excess activated sludge (ES) was sludge coming from thickened sludge tank after centrifugation. This sludge was added to primary sludge in the digester, the mixture is called raw sludge (Raw).

The sludge for fermentation was closed in 1,000 mL unmixed bottles and put in thermo-boxes at three different temperatures: psychrophilic temperature (20°C), mesophilic temperature (35°C) and thermophilic temperature (55°C). In order to evaluate the biogas production and composition, the fermentation was carried out in 120 mL serum bottles closed with rubber stoppers allowing the measurement of produced biogas.

Sequencing batch reactors

The effect of disintegration of activated sludge and addition of fermented primary sludge to improve denitrification was verified on sequencing batch reactors (SBR). The volume of reactor was 2 L. The scheme of the reactor is shown in Figure 6.1.

The SBR was formed from plastic tube with sloping bottom. The reactor had five sampling points. It was used coarse bubble aeration. Switch on and off of aeration was provided by timer. The effluent of sludge and water, dosing of wastewater and disintegration of sludge was manually controlled.

To follow the effect of sludge disintegration on nutrient removal, they were operated two sets of SBRs. In first period three SBRs were operated to compare thermo-chemical disintegration of activated sludge and fermentation of primary sludge with referential SBR. In second period four SBRs were operated to compare ultrasound, mechanical and thermal disintegration of activated sludge with referential SBR.

The operating conditions of SBR were divided in two groups. The first reactors were operated only with denitrification. It means that the reactors were fed only with tap water, organic substrate and nitrate nitrogen (marked SBR – denitrification). The second reactors were operated in condition denitrification-nitrification. All reactors were fed with wastewater from CWWTP of Prague, which contained ammonia nitrogen (marked SBR – denitrification-nitrification).

ACHIEVEMENTS RELATED TO THE STATE OF THE ART

The interest in increase of efficiency of nutrient removal from wastewater is growing in the recent years, especially, when in EU whole regions and countries (the Czech Republic for example) were declared as a sensitive area with strict effluent limits of nitrogen and phosphorus.

The main limiting factor for both nitrogen and phosphorus biological removal in many municipal and industrial wastewater treatment plants is lack of organic substrate for denitrification or phosphorus release.

The activated sludge process transforms a substantial part of the organic matter present in wastewater into excess sludge. Sludge treatment and disposal are costly processes accounting for up to 60% of operating costs. Reduction of sludge production and exploitation of valuable components of sludge is a way to tackle the disposal problems and reduce the operational costs.

The process which can release part of substrate from the sludge for the use at nutrient removal is the sludge disintegration. At last decade the practical use of the disintegration technologies was related to improvement of anaerobic digestion mainly (Dohanyos *et al.*, 1997; Thiem *et al.*, 1997 and Muller *et al.*, 1998). The use of sludge disintegration in activated sludge wastewater treatment technology is exceptional till now and the influence on biological nutrient removal and process stability has to be evaluated.

The use of fermented primary sludge as substrate for denitrification and/or enhanced biological phosphorus removal is well known and in full scale verified technology (Moser-Engeler *et al.*, 1999 and Ahn and Speece, 2006). The use of excess activated sludge for the same purpose is less studied and more complicated for following reasons:

- slower hydrolysis,
- lower biodegradability,
- lower COD/N (or COD/P) ratio of produced substrate.

To eliminate above mentioned drawbacks at least partly different disintegration processes were proposed (Muller, 2000a). On the other hand the use of excess activated sludge for substrate production brings lower sludge production and consequently the decrease of sludge disposal costs as another benefit (Strunkmann *et al.*, 2006 and Dytczak *et al.*, 2007).

In the project it was shown that quantity and quality of substrate produced by sludge disintegration can be controlled by choice of sludge type and by conditions of disintegration procedure. The results of disintegration experiments showed that the kinetic of COD and N release is not equal and therefore the use of optimal type of disintegration and procedure of disintegration can be optimized as regards COD/N ratio.

Till now most of studies were focused on evaluation of substrate quality in batch tests and on the comparison with external substrates (methanol, ethanol, acetate etc.) (Muller, 2000b; Ahn *et al.*, 2002; Kampas *et al.*, 2007 and Kampas *et al.*, 2009). The complex evaluation of the impact of long term returned activated sludge disintegration including changes in residual COD concentration, separation of activated sludge, nitrification activity and increased energy costs was not published. Except some data about sludge ozonation which is primarily oxidation process (Boehler and Siegrist 2004 and Ki *et al.*, 2004) and about systems which were not optimized for nutrient removal (Strunkmann *et al.*, 2006). Also data about energy and cost analysis are rare and related to ozonation mainly. (Boehler and Siegrist, 2007).

It is vague to transfer data about specific costs from laboratory to full scale especially as regards the specific energy consumption for substrate production and sludge disintegration. However as rough guidance we can use the data of full scale installations using sludge disintegration for improvement of anaerobic digestion – Table 6.1.

Table 6.1 Comparison of technical economic data of different methods of sludge pretreatment (according to Chudoba *et al.*, 2007)

	Unit	Mechanical disintegration Lysatec	Ultrasound disintegration VTA	Thermal hydrolysis CAMBI
Investment costs	€/1000 PE	216	1400	2000–4000
Operational costs	€/1000 PE/year	16	80	?
Biogas production increase	%	15	10–30	20–25

Therefore the aim of the project was to fill the gap in knowledge about the impact of long term returned activated sludge disintegration including changes in residual COD concentration, separation of activated sludge, nitrification activity and increased energy costs with respect to implementation of disintegration technologies into the wastewater treatment plant design.

The disintegration of returned activated sludge is viable alternative of activated sludge technology in certain cases. The benefits of the technology must be evaluated for each case individually. The main criteria for implementation of the disintegration are:

a) costs of sludge disposal,
b) influent COD/N ratio and required N removal,
c) extent of filamentous bacteria presence in sludge,
d) energy requirements of disintegration method.

Costs of sludge disposal

The costs of sludge disposal differs significantly according the method of disposal, agricultural use is cheapest alternative usually (100–300 €/t dry solids) however not accepted for each sludge and not accepted in some European countries at all. The sludge incineration is most expensive between common methods of sludge disposal 600–800 €/t dry solids (German data: Haeck, Voith, 2006).

If the costs of sludge disintegration are lower than that of sludge disposal the method can be considered cost effective. Supposing the decrease of sludge production to 25–50% of original amount, the disintegration can spare 50–600 €/t dry solids of disposed sludge.

Influent COD/N ratio and required N removal

According experimental experience it is possible roughly evaluate the eligibility or potential effect of sludge disintegration on nitrogen removal according the influent COD/N (at standard quality of sewage).

COD/N	Potential of disintegration with respect to N removal improvement.
< 4	Negligible improvement can be expected, not sufficient to remove required amount of N.
4–6	Improvement of N-removal about 5–20 mg/l can be significant from technological point of view.
> 6	No significant improvement can be expected.

Extent of filamentous bacteria presence in sludge

At the excessive growth of filamentous bacteria causing filamentous bulking and foaming the sludge disintegration can destroy the filamentous structure of sludge by cutting and shortening of filaments which improves separation of sludge and effluent quality as regards suspended solids concentration. Disintegration of well settling flocculent sludge can cause a specific deterioration of sludge volume index.

Energy requirements of disintegration method

Energy requirement for disintegration is very important factor, strongly affecting the application of the disintegration technology.

Thermal methods of sludge disintegration are the most energy consuming. At nutrient removal system the need of the temperature increase is low except of coldest winter month and in fact technologically unexploitable in contrast to anaerobic digestion where the necessary heating of the digester can be accomplished by thermal pretreatment of sludge.

Mechanical more than ultrasound disintegration is promising as regards energy consumption (Figure 6.4). Special case of disintegration is thickening lysate centrifuge in which increase of energy consumption in comparison with thickening of sludge is negligible. Till now the lysate centrifuges are operated in full scale for improvement of anaerobic digestion only.

Operational cost analysis of disintegration by thickening lysate centrifuge can be expressed by increase of power consumption due to lysating knifes addition in centrifuge (Figure 6.2).

Figure 6.2 Stator of lysate centrifuge and additional knives

Comparing energy consumption by lysate centrifuge with other methods mentioned in Figure 6.3 (increase in comparison with common centrifuge) specific consumption is about 0.06 kWh/kg SS, with degree of disintegration 10–20% (expressed as ratio $COD_{released}/COD_{tot}$) – it means approximately 2–4 times better energetic efficiency.

Such energy consumption means additional costs related to sludge disintegration about 100 €/t dry solids (supposed the price of electricity 0.18 €/kWh). This is acceptable price for minimization of sludge amount to be disposed because sludge disposal costs are several times higher mostly.

Conclusions

The use of substrate produced from sludge for enhanced biological phosphorus removal is feasible, however if there is deficit of organic substrate, phosphorus can be removed by precipitation using Fe^{3+} or another precipitant and produced substrate has to be used for N removal.

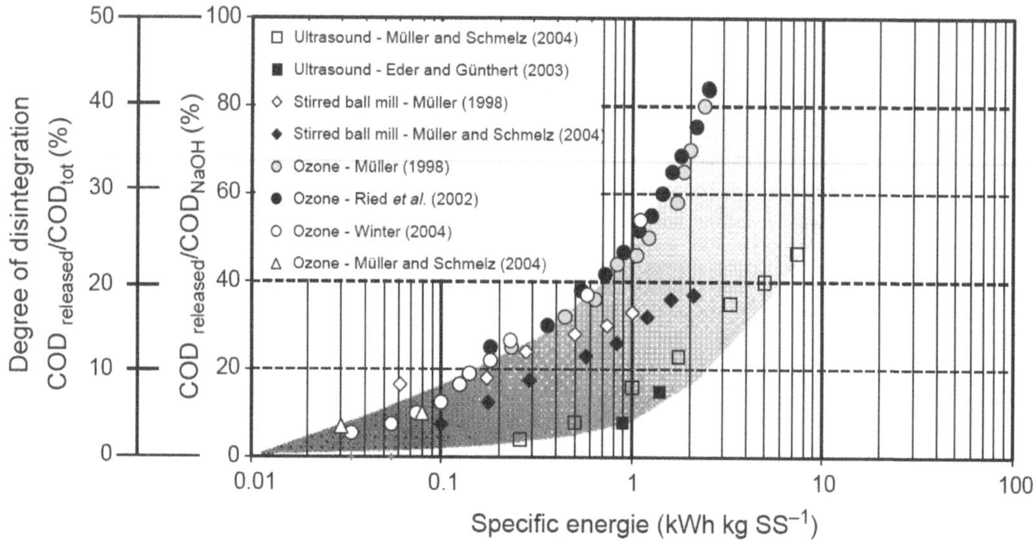

Figure 6.3 Degree of disintegration as a function of specific energy for various disintegration methods (according to Boehler and Siegrist, 2007)

The improvement of nitrogen removal obtained by sludge disintegration is limited and technologically significant seems to be in narrow range of influent COD/N ratio only.

From economical point of view more important benefit of sludge disintegration is minimization of excess sludge production at unchanged effluent quality (nutrients released at disintegration are efficiently removed).

Operational cost analysis shows that the use of lysate centrifuge is most suitable way of sludge disintegration from economical point of view, because of lowest energy consumption. Additional operational costs in such case 100 €/t dry solids can save 50–600 €/t dry solids according to specific sludge disposal costs.

Sludge disintegration methods can be cost effective only in the case of high sludge-disposal costs and in the case of interest in specific operational problems solution such as foaming and bulking.

The amount of disintegrated sludge and the degree of minimization have to respect the stable and optimal operational condition of activated sludge process as regards TSS, VSS concentration and nitrification capacity mainly. Therefore the application of sludge disintegration must be carefully evaluated for each application.

IMPACT OF THE TECHNOLOGY ON ITS INDUSTRY OR RESEARCH SECTOR

The modern wastewater treatment technologies are usually costly to be applied generally including less developed countries. Proposed technology offers achieving of high treatment efficiency and lowering the amount of produced excess sludge and costs of sludge disposal. In addition the proposed technology is more sustainable in comparison with external substrate addition, because it is not consuming chemicals (except of thermo chemical treatment) whose production is connected with energy consumption and CO_2 production. The proposed technology helps to solve the problem of increasing production of excess activated sludge – which disposal start to be in many countries environmentally questionable.

The implementation of the sludge disintegration technology in activated sludge treatment design is easy and requires only disintegration unit addition, in the case of mechanical disintegration by thickening centrifuge only modification of the apparatus.

The currently applied biological methods of nutrients removal proved to be very efficient and sustainable. However, lack of organic substrate in wastewater often limits the process efficiency. Addition of external carbon source is a possible but not really sustainable solution. Another possible way is to search for internal carbon source within the wastewater treatment process itself. Primary or excess activated sludge can be one of them. The method of volatile fatty acids production from sludge by its pre-fermentation is already well known, but the process control is not examined satisfactorily. Several new methods of excess sludge disintegration and treatment are being studied at present. The aims of such treatment processes are the production of organic substrate suitable for nutrients removal and the minimization of surplus sludge being produced. Disintegration of microbial cell walls and concurrent production of cell lysate can even stimulate the activity of different trophic group of microorganisms. In this way, a new technological process of organic substrate production from sewage sludge can be proposed, which will bring suitable internal carbon source for nutrients removal, while reducing the amount of surplus sludge produced at the same time.

The production of substrate for enhanced biological nutrient removal by primary sludge fermentation is simpler and more efficient method in comparison with excess sludge disintegration.

From economical point of view it is more important benefit of sludge disintegration the minimization of excess sludge production at unchanged effluent quality. Excess sludge disintegration methods are cost effective in the case of high sludge-disposal costs or interest in solution of specific problems with sludge quality.

REFERENCES

Ahn Y. H. and Speece R. E. (2006). Elutriated acid fermentation of municipal primary sludge. *Water Research*, **40**(11), 2210–2220.

Ahn K.-H., Yeom I.-T., Park K.-Y., Maeng S.-K., Lee Y., Song K.-G. and Hwang J.-H. (2002). Reduction of sludge by ozone treatment and production of carbon source for denitrification. *Water Science and Technology*, **46**(11–12), 121–125.

Boehler M. and Siegrist H. (2004). Partial ozonation of activated sludge to reduce excess sludge, improve denitrification and control scumming and bulking. *Water Science and Technology*, **49**(10), 41–49.

Boehler M. and Siegrist H. (2006). Potential of activated sludge disintegration. *Water Science and Technology*, **53**(12), 207–216.

Boehler M. and Siegrist H. (2007) Potential of activated sludge ozonation. *Water Science and Technology*, **55**(12), 181–187.

Chudoba P., Soukupova S. and Todt V. (2007). Possibilities of anaerobic sludge digestion intensification. Proc. of Conf. Anaerobiosis 2005, 29–30, September 2005, Trebon (in Czech).

Dohanyos M., Zabranska J. and Jenicek P. (1997). Enhancement of sludge anaerobic digestion by use of a special thickening centrifuge. *Water Science and Technology*, **36**(11), 145–153.

Dytczak M. A., Londry K. L., Siegrist H. and Oleszkewicz J. A. (2007). Ozonation reduces sludges production and improves denitrification. *Water Research*, **41**(3), 543–550.

Haeck M. and Voith U. (2006). Mechanical sludge dewatering and sludge thickening – cost reduction by applying process automation. Proceedings of the IWA Specialized Conference – Sustainable sludge management: state of the art, challenges and perspectives, 54–59, Moscow, 29–31, May 2006.

Kampas P., Parsons S. A., Pearce P., Ledoux S., Vale P., Cartmell E. and Soares A. (2009). An internal carbon source for improving biological nutrient removal. *Bioresource Technology*, **100**(1), 149–154.

Kampas P., Parsons S. A., Pearce P., Ledoux S., Vale P., Churchley J. and Cartmell E. (2007). Mechanical sludge disintegration: Providing an alternative carbon source for nutrient removal. *Environmental Technology*, **28**(4), 471–477.

Ki Y. P., Jae W. L., Ahn K.-H., Sung K. M., Jong H. H. and Song K.-G. (2004). Ozone disintegration of excess biomass and application to nitrogen removal. *Water Environment Research*, **76**(2), 162–167.

Moser-Engeler R., Kühni M., Bernhard C. and Siegrist H. (1999). Fermentation of raw sludge on an industrial scale and applications for elutriating its dissolved products and non-sedimentable solids. *Water Research*, **33**, 16, 3503–3511.

Muller J. (2000). Disintegration as a key-step in sewage sludge treatment. *Water Science and Technology*, **41**(8), 123–130.

Muller J. A. (2000). Pretreatment processes for the recycling and reuse of sewage sludge. *Water Science and Technology*, **42**(9), 167–174.

Müller J., Lehne G., Schwedes J., Battenberg S., Näveke R., Kopp J., Dichtl. N., Scheminski A., Krull R. and Hempel D. C. (1998). Disintegration of sewage sludges and influence on anaerobic digestion. *Water Science and Technology*, **38**(8–9), 425–433.

Pokorna E., Vondrysova J., Lacaze E. and Jenicek P. (2007). Improvement of organic substrate produced by sludges hydrolysis using natural zeolite. Proc. of Conf. The solution of extreme wastewater treatment requirement, 51–58, Boskovice, Czech Republic, in Czech.

Strünkmann G. W., Müller J. A., Albert F. and Schwedes J. (2006). Reduction of excess sludge production using mechanical disintegration devices. *Water Science and Technology*, **54**(5), 69–76.

Tiehm A., Nickel K. and Neis U. (1997). The use of ultrasound to accelerate the anaerobic digestion of sewage sludge. *Water Sci. Technol.*, **36**(11), 121–128.

Vondrysova J. and Jenicek P. (2007). Composition of biogas during fermentation of biomass, Poster in Conf. 'Biogas 2007', 23–24 April 2007, Ceske Budejovice, Czech Republic.

Vondrysova J. and Jenicek P. (2008). Comparison of activated sludge from membrane biological reactor and conventional system – biodegradability and pretreatment by disintegration CD – Proc. of 5th Int. Conf. on Anaerobic Digestion of Solid Wastes, Poster 2, 25–28 May 2008, Hammamet, Tunisia.

Vondrysova J., Jenicek P., Pokorna E., Vondrysova J. and Lacaze E. (2006). Production of organic substrate by sludge fermentation with possibility to minimize N_{ammon} concentration by zeolites. Proc. of Int. Conf. Wastewaters 2006, 34–40, Tatranske Zruby, 18–20 October 2006, Slovak Republic (in Czech).

Vondrysova J., Koubova J. and Jenicek J. (2008). Using activated sludge and primary sludge as a substrate for denitrification, In proceedings Kaly a odpady 2008, 12–13, March 2008, Bratislava, 50–57 (in Czech).

Vondrysova J., Koubova J. and Jenicek P. (2009). Improving denitrification by disintegration of activated sludges – Proc. of 2nd IWA Specialized Conference on Nutrient Management in Wastewater Treatment Processes 6–9, September 2009, Krakow.

Vondrysova J., Koubova J., Grymova K. and Jenicek P. (2009). Using sewage sludge for preparation of organic substrate for denitrification, In proceedings Řešení extrémních požadavků na čistění odpadních vod, 26–27, February 2009, Blansko, 38–48 (in Czech).

Vondrysova J., Pokorna E., Bouskova A. and Jenicek P. (2007). Optimization of sludge treatment methods with respect to production of organic substrate for enhancement of nitrogen removal. Proc. of Int. Conf. Wastewater Biosolids Sustainability, 1051–1056, 24–27, June 2007, Moncton, Canada.

Vondrysova J., Pokorna E., Lacaze E. and Jenicek P. (2006). Sewage sludge as a source of organic substrates for biological nutrient removal. Water Management, 56, 11, Wastewater Letters IV–VII (in Czech).

Vondrysova J., Slamova K., Koubova J and Jenicek P. (2007). Comparison of different sludge disintegration methods with respect to production of organic substrate for enhancement of nitrogen removal, Proc. of 7th Int. Conf. Wastewater 2007, 139–142, 18–20, September 2007, Brno, Czech Republic.

Wang F., Lu S. and Ji M. (2006). Components of released liquid from ultrasonic waste activated sludge disintegration. *Ultrasonics Sonochemistry*, **13**, 334–338.

Chapter 7

Advanced technology for the biological nitrogen removal of the reject water from the sludge dewatering systems

Julián Carrera, Albert Bartrolí, Eduardo Isanta, Irene Jubany, Josep A. Torà, Juan Antonio Baeza and Julio Pérez

SUMMARY

Project objectives

The aim of this work was the development of novel engineering alternatives to improve technical and economically the biological nitrogen removal process to treat rich-ammonium wastewaters, especially the reject water from the dewatering of the digested sludge of a wastewater treatment plant (WWTP). These novel alternatives were based on biological nitrogen removal via nitrite. The first stage of this process is the oxidation of ammonium to nitrite, the so-called partial nitrification. Two different technologies were tested with the aim to obtain higher treatment capacities than the obtained with existing industrial systems. These technologies were: (i) a nitrifying activated sludge system and (ii) a nitrifying biofilm airlift reactor. Finally, the last partial objective was the development, calibration and validation of a mathematical model able to simulate the performance of both technologies.

Work performed

Two pilot plants were used to treat a high strength ammonium wastewater (with an ammonium concentration ranging from 1 to 3 g $N\text{-}NH_4^+/L$): (i) a nitrifying activated sludge system with three continuous stirred tank reactors (CSTRs) in series and (ii) a nitrifying biofilm airlift reactor with activated carbon (AC) as biocarrier. We implemented two automatic control loops to achieve a full and stable partial nitrification process in both systems. These loops control the nitrogen loading rate applied to each system using as manipulated variable the inflow rate. The main difference between both control strategies applied to each system was the in-line measured variable: (i) the Oxygen Uptake Rate (OUR) in the activated sludge system with three CSTRs in series and (ii) the ammonium concentration in the reactor for the biofilm airlift reactor. A stable operation of the partial nitrification process was achieved during 15 months in both systems. This period is enough to demonstrate the robustness and reliability of the developed technologies. During this period in both systems, the effluent ammonium concentration was lower than 20 mg $N\text{-}NH_4^+/L$ whereas the nitrite accumulation was always around 98% of the inlet ammonium. The temperature of operation was 30°C during the first 12 months and it was decreased to 20°C for a period of 3 months without any change in the percentage of ammonium oxidation and nitrite accumulation, showing therefore the feasibility of the technologies to be applied at room temperature. A second goal

 ISBN: 9781843393450. Published by IWA Publishing, London, UK.

in the present study was the development, calibration and successful validation of a mathematical model for describing the performance of the partial nitrification process in both systems previously described. All kinetic and stoichiometric parameters related to nitrifying biomass were calibrated with specific experiments. It is important to outline that such a specific calibration for all kinetic and stoichiometric parameters is lacking in the literature, since the vast majority of models developed for this process include a set of parameters assumed which are not well supported by experimental evidence. The developed and calibrated mathematical model was validated for both nitrifying systems (activated sludge and biofilm) in different scenarios: (i) the controlled start-up of the nitrifying activated sludge system using biomass from a municipal WWTP as inoculum, (ii) the stable achievement of the partial nitrification process in the nitrifying activated sludge system and (iii) the stable achievement of the partial nitrification process in the nitrifying biofilm airlift reactor.

End results

Both developed systems represent an improvement of the existing technology for partial nitrification of high-strength ammonium effluents because the applied control strategies warrant the achievement of almost 100% of partial nitrification with high efficiency and high stability on long-term operation. Moreover, the developed control strategies allow achieving higher nitrifying biomass concentration in the reactors than the patented SHARON® process and consequently, they allow treating higher loading rates (or reducing the volume of the reactors). The loading rates achieved with our both alternatives (5–6 kg N/m^3 · d at 30°C) are, to the best of our knowledge, the highest values reported in the literature. Additionally, the technologies developed in present research study allow achieving 100% of stable partial nitrification at lower temperatures (20°C) than SHARON® process (which operates at 35°C). This is a very interesting added value for the developed technologies, since they are feasible therefore for partial nitrification of ammonium-reach industrial wastewaters at mild temperature. Additionally, the developed, calibrated and validated mathematical model is able to properly predict the time course of soluble compounds (ammonium, nitrite and nitrate) and biomass concentration in both partial nitrification systems. This achievement is completely new since available models do not compare the biomass concentration predicted with the model with experimental values, especially in biofilm systems.

METHODOLOGIES AND APPROACHES EMPLOYED

Partial nitrification pilot plants

Nitrifying activated sludge system

This pilot plant consisted of three aerobic reactors (named R1, R2 and R3) with a working volume of 26 L followed by a 25 L settler (Figure 7.1).

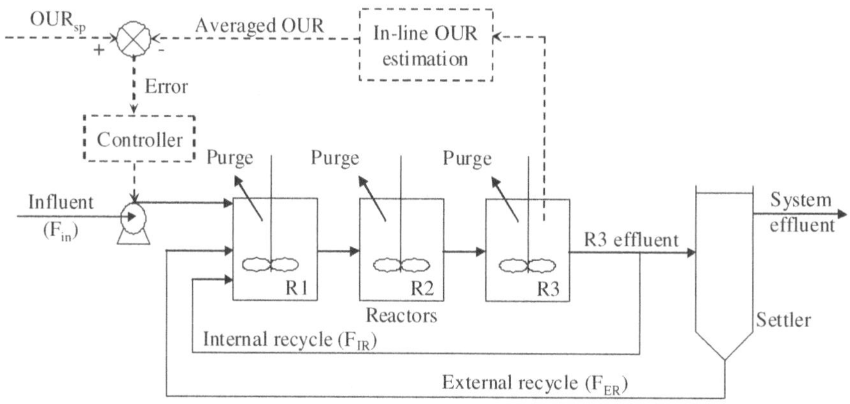

Figure 7.1 Nitrifying activated sludge system and loading rate control loop based on the in-line OUR measurement

The reactors were connected in series and they worked under completely mixed conditions. A fraction of R3 effluent was recycled to R1 to increase the dynamics of the system and to improve the mixing between reactors. Mixed liquor was withdrawn daily from the three reactors in order to keep a desired SRT. Each reactor was equipped with DO (WTW

Oxi 340i CellOx 325), pH (Crison pH 52-03) and temperature probes. They were connected to a probe controller equipment (Crison pHrocon18) which actuated as a pH on-off controller in R1, R2 and R3 by the addition of solid sodium carbonate through solid dispensers. The probe controller equipment was connected to a process computer which was in charge of controlling the key process parameters (DO, temperature and flow-rates). The DO control was based on a digital proportional-integral-derivative (PID) algorithm and operated by manipulating pneumatic control valves which modified the airflow supplied through air diffusers placed at the bottom of the reactors. The temperature control, based on an on-off control, was implemented in R1 and it operated by switching an electric heating device. Automatic in-line OUR estimation was implemented in each reactor. The OUR measurement was performed every 10 minutes and it was based on the DO decrease in the liquid phase with no air inlet. The OUR measurement followed these steps: deactivation of the DO PID controller, aeration valve shut down, DO measurement every 4 seconds until it decreased by 1.0 mg O_2/L and calculation of OUR as the slope of acquired data.

Nitrifying biofilm airlift reactor

A biofilm airlift reactor with a working volume of 112 L was used in this study. The internal diameter of the downcomer was 26.7 cm. The riser had a height of 127 cm and an internal diameter of 18.5 cm. It was positioned at a distance of 10 cm from the bottom of the downcomer. Compressed air was supplied through an air diffuser placed at the bottom of the reactor at a superficial air velocity of 0.62 cm/s. The airflow rate was 166.7 Ncm^3/s. The temperature of the reactor was controlled using an electric resistance and a temperature controller. The pH control loop used solid $NaHCO_3$ to maintain a constant pH and to supply the alkalinity required for nitrification. Ammonium concentration in the bulk was determined by means of an on-line probe (NH4D sc probe with Cartrical® cartridge, HACH LANGE, Germany). The biofilm airlift reactor was inoculated with activated sludge from a municipal WWTP (25 L from the main biological reactor of the Manresa municipal WWTP, Barcelona, Spain). The reactor was also loaded with 2.5 kg of AC, with a mean diameter of 1.2 mm and wet density of 1.25 g/mL (provided by Chemviron Carbon Limited, UK).

Methodologies

Total Ammonium Nitrogen (TAN = $N\text{-}NH_4^+ + N\text{-}NH_3$) was analyzed by means of a continuous flow analyzer based on potentiometric determination of ammonia. Total Nitrite Nitrogen (TNN = $N\text{-}NO_2^- + N\text{-}HNO_2$) and nitrate were measured with Ion Chromatography (ICS-2000 Integrated Reagent-Free IC System, DIONEX). Volatile Suspended Solids (VSS) and total suspended solids (TSS) were determined according to Standard Methods (1995).

Fluorescence in situ hybridization (FISH) technique coupled with confocal microscopy was used to determine the predominant nitrifying species. A Leica TCS SP2 AOBS confocal laser scanning microscope at a magnification of x63 (objective HCX PL APO ibd.B1 63x1.4 oil) equipped with two HeNe lasers with light emission at 561 and 633 nm was used for biomass quantification. Hybridizations were carried out using at the same time a Cy3-labeled specific probe and Cy5-labeled EUBmix probes. Specific probe used for AOB detection was Nso190 (Mobarry *et al.*, 1996) while for NOB detection was NIT3 (Wagner *et al.*, 1996). Detailed information about FISH quantification can be found elsewhere (Jubany *et al.*, 2009a). To apply the methodology developed for activated sludge (as described in Jubany *et al.*, 2009a) to granular biomass, 30–40 microscopic fields were analyzed and a single z-position was selected based on the highest intensity for each granule.

To analyze the biofilm onto AC particles small variations in the described methodology were implemented. Five AC particles were put into 2 mL microcentrifuge tubes, the hybridization buffer and the molecular probe working solution were added too (80 and 5 µL, respectively). After hybridization, the washing step and the following air-drying step were performed before microscopic examination following the method described in Jubany *et al.* (2009a). For CLSM observation, the AC particles were placed on a 35-mm, No. 1.5 glass-bottom Petri dish (P35G-1.5-14-C, MatTek, Ashland, MA). Three stacks of 40 to 50 optical z-sections images with a size of 2.0 µm were saved (512 × 512 pixel images) and later used for quantification.

ACHIEVEMENTS RELATED TO THE STATE OF THE ART

Automatic control loops to achieve partial nitrification

The treatment of wastewaters containing high concentration of ammonium was widely investigated in the scientific and technical literature in the last decades. Nitrification is known as a biochemical reaction to oxidize the ammonium to nitrate. Biological nitrification is produced in two steps: the ammonium oxidation to nitrite by ammonium oxidizing bacteria (AOB) and nitrite oxidation to nitrate by nitrite oxidizing bacteria (NOB). Usually, biological nitrification is used as a first part of the treatment within the nitrogen removal from wastewaters, followed by a denitrification step to remove the nitrogen as nitrogen gas. The running costs of the nitrogen removal treatments are closely related to the oxygen and carbon source requirements of the process (Turk and Mavinic, 1987). In order to minimize these costs a shortcut has been proposed and successfully applied with considerable savings with regards to the running costs of the installation. The shortcut consisted mainly in obtaining a nitrite-rich effluent from the nitrifying reactor, consequently suppressing the second oxidative step from nitrite to nitrate. This is usually known as nitrite based N-removal and it can be applied followed by a heterotrophic denitrification from nitrite (Mulder *et al.*, 2001) or by an autotrophic denitrification (or Anammox® process) (Patent WO 98/07664; Fux *et al.*, 2002). A wide number of different strategies have been described to obtain partial nitrification to nitrite with different reactor configurations, with or without biomass retention.

Among the systems with suspended biomass without biomass retention the named Single-reactor High Activity ammonia Removal Over Nitrite (or SHARON® process) (Patent EP-A-826639; Patent WO 98/07664; Hellinga *et al.*, 1998) process was developed to washout the NOB from the reactor, imposing a temperature higher than 25ºC, and usually in the range 30–40ºC, as well as a hydraulic retention time (HRT) in the reactor of about 1 day, impeding in this way the proliferation of NOB in the reactor. The main drawback of this system is that there is no retention of biomass and therefore either the treated ammonium load is low or a large reaction volume is required (Fux *et al.*, 2002). To overcome these limitations a number of systems with biomass retention have been investigated with the aim to achieve partial nitrification to nitrite with more compact reactor systems (Garrido *et al.*, 1997; Fux *et al.*, 2004a and Muñoz *et al.*, 2009). Among them, activated sludge in sequencing batch reactors (SBRs) were investigated in the past using NOB inhibition by free ammonia and low DO concentration to force the nitrite accumulation (Peng *et al.*, 2006; Li *et al.*, 2004; Fux *et al.*, 2006 and Lemaire *et al.*, 2008). On the other hand, biofilm reactors were investigated using low DO concentration to force the system to produce nitrite (biofilm reactors operating in continuous mode: Garrido *et al.*, 1997; Joo *et al.*, 2000; Bernet *et al.*, 2005; Tokutomi *et al.*, 2006 and Frijters *et al.*, 2007; biofilm reactor operating in batch mode: Antileo *et al.*, 2006).

The biofilm reactors were somehow disregarded because of a simultaneous presence of nitrite and nitrate in the effluent (Fux *et al.*, 2004a; Van Loosdrecht and Salem, 2006 and Picioreanu *et al.*, 1997), i.e. no full partial nitrification to nitrite was achieved in many of the studies reported. Nevertheless, there are also references in which stable continuous partial nitrification to nitrite have been obtained with artificial biofilms (Noto *et al.*, 1998), and natural biofilms (Tokutomi 2004 and Frijters *et al.*, 2007) and recently supported by theoretical analysis through mathematical modelling (Pérez *et al.*, 2009). The key kinetic parameter to obtain partial nitrification in oxygen limited nitrifying biofilms was identified as the oxygen affinity of the two groups of microorganisms involved in nitrification, AOB and NOB, being the influence of the possible inhibition effects of ammonia more reduced than those formerly considered (Pérez *et al.*, 2009).

Regarding the instrumentation, control and automation, specific control strategies have been described to be applied to nitrifying reactors with biomass retention with the aim to obtain and maintain partial nitrification. Those strategies were applied to two reactor configurations: (i) those working with suspended biomass reactor(s) and settler (Muñoz *et al.*, 2009); and (ii) those in which the biomass retention is attained with biofilms (Bernet *et al.*, 2005). In particular, a control strategy based on the ratio dissolved oxygen/ammonium concentration in the bulk liquid ($DO/N\text{-}NH_4^+$) has been described specifically to control the partial nitrification to nitrite by Bernet *et al.* (2005) in biofilm reactors. These authors concluded that the use of the $DO/N\text{-}NH_4^+$ ratio, rather than the DO concentration uniquely, was a better parameter to control partial nitrification to nitrite, when using as manipulated variable the air-flow supplied to the reactor. Nevertheless, the conversion results obtained with that control strategy was an effluent containing mainly nitrite and nitrate, with no more than 60% of accumulation of nitrite (Bernet *et al.*, 2005).

The achievement of a controlled partial nitrification system with a complete and stable oxidation of the ammonium to nitrite is a very interesting objective for a successful operation of a partial nitrification plus Anammox® system (Van der Star *et al.*, 2007). The lack of control in the connexion between the partial nitrification and the Anammox® stages could be a potential problem if the second step is fed with a nitrite/ammonium ratio higher than 1. This problem can be solved by: (a) integrating both stages in one reactor (Vázquez-Padín *et al.*, 2009) or (b) with a totally stable full partial nitrification system (close to 100% of nitrite accumulation) and by-passing the adequate flow of (untreated) rich-ammonium wastewater.

We developed in this study two (automatic) control systems able to achieve full partial nitrification to nitrite (i.e. nitrite accumulation higher than 98% in the effluent) in activated sludge and biofilm reactors operating in continuous mode. We used two pilot plants to treat a high strength ammonium wastewater (with an ammonium concentration ranging from 1 to 3 g $N\text{-}NH_4^+/L$): (i) a nitrifying activated sludge system with three CSTRs in series and (ii) a nitrifying biofilm airlift reactor with AC as biocarrier.

We implemented two automatic control loops to achieve a full and stable partial nitrification process in both systems. These loops control the nitrogen loading rate applied to each system using as manipulated variable the inflow rate. The main difference between both control strategies applied to each system was the in-line measured variable: (i) the OUR in the activated sludge system with three CSTRs in series (see Figure 7.1) and (ii) the ammonium concentration in the reactor for the biofilm airlift reactor (see Figure 7.2).

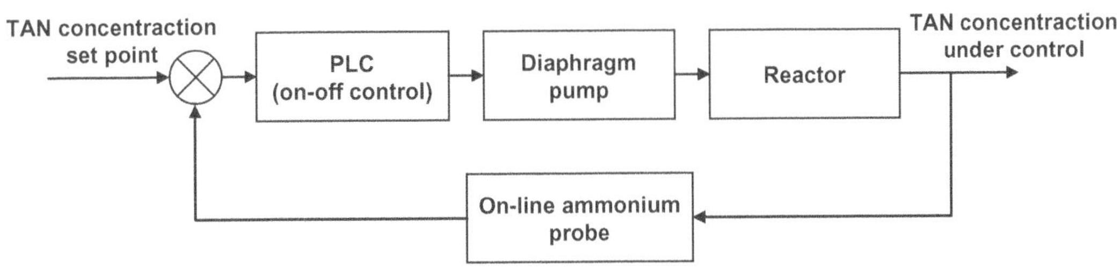

Figure 7.2 Nitrifying biofilm airlift reactor and loading rate control loop based on the in-line ammonium measurement. From Priority application (2009) No. ES200901432

Full and stable partial nitrification process in an activated sludge system

A stable operation of the partial nitrification process was achieved during 15 months in the nitrifying activated sludge system (see Jubany *et al.*, 2009b; 2009c for an extended explanation). This long period of stable operation demonstrates the robustness and reliability of the developed process. During all this period, the effluent ammonium concentration was lower than 20 mg N-NH_4^+/L while the nitrite accumulation was always around 98% of the inlet ammonium. The temperature of operation was 30ºC during the first 12 months and it was decreased to 20ºC for a period of 3 months without any change in the percentage of ammonium oxidation and nitrite accumulation. The maximum capacity of the system at 30ºC was established in 5 kg N/m^3 d. This value is clearly higher than the obtained with other industrials systems (see Table 7.1). Moreover, the maximum capacity of the system at 20ºC was established in 2.5 kg N/m^3 d. This attainment is especially significant because the most widespread partial nitrification system which is the SHARON® process can not operate at mild temperatures. Indeed, there are few references of partial nitrification process successfully operating at this mild temperature (Lemaire *et al.*, 2008) but with noticeably lower capacities. The possibility of the successful application of partial nitrification at this temperature extends the potential market of this technology to other rich-ammonium industrial wastewaters.

Table 7.1 Maximum capacity of partial nitrification for several industrial systems.

Technology	Maximum capacity (kg N/m^3 · d)	Temperature (ºC)	Reactor volume* (m^3)	Reference
SHARON®	0.6	35	416	Patent WO 98/07664
Airlift reactor	2.6	30	96	Tokutomi (2004)
SBR	1.0	36	250	Ganigué *et al.* (2009)
Swim-bed reactor	2.7	18	93	Qiao *et al.* (2010)
Activated sludge	2.5	20	100	This work
	5.0	30	50	
Biofilm airlift reactor	0.6	20	416	This work
	6.0	30	42	

**Reactor volumes are estimated considering a nitrogen mass flow of 250 kg N/d.*

The biomass concentration achieved an average value of 2.0 g VSS/L with a percentage of AOB around 85% and an undetectable percentage of NOB. The complete and stable washout of NOB ensured the stability of the partial nitrification in the long term operation. There are few references of complete and stable washout of NOB in activated sludge systems and, to the best of our knowledge, there are not examples in the literature of real-time-control for achieving partial nitrification in continuous suspended biomass systems during long-term operation at mild temperatures (less than 30ºC). On the other hand, the achieved AOB biomass concentration was significantly higher than the obtained with SHARON® reactors (Fux *et al.*, 2002; Van Hulle *et al.*, 2005 and Mosquera-Corral *et al.*, 2005). This achievement is significant because the nitrogen loading rate that can be applied in this system is higher than the applied to SHARON® reactor and this means lower reaction volume and consequently, lower investment costs. In this sense, the necessity of three aerobic reactors in series does not represent a significant increase of the construction costs. Firstly, the total reaction volume is lower than other systems (due to the higher nitrifying biomass concentration) and finally, the requirement of three reactors can be simply solved dividing the total aerobic volume with two baffles. The aeration cost of the total volume will be the same with a sole reactor or with the volume divided in three parts. The agitation of each reactor is provided by the aeration.

Full and stable partial nitrification process in a biofilm airlift reactor

Regarding the nitrifying airlift reactor, after the start-up with AC as biocarrier, the stable operation of the partial nitrification process was achieved during 12 months. This long period of stable operation demonstrates the robustness and reliability of the developed process (see Bartrolí *et al.*, 2010a for an extended explanation). During this period, the effluent ammonium concentration was steadily maintained at 20 mg $N\text{-}NH_4^+$/L through the automatic control loop while the nitrite accumulation was always around 98% of the inlet ammonium. The temperature of operation was 30°C during the first 9 months and it was decreased to 20°C for a period of 3 months without any change in the percentage of ammonium oxidation and nitrite accumulation. The maximum capacity of the system at 30°C was established in 1.3 kg N/m^3 d. This is comparable to the obtained with other industrials systems (see Table 7.1). Moreover, the maximum capacity of the system at 20°C was established in 0.6 kg N/m^3 d. As previously discussed, the possibility of the successful application of partial nitrification at 20°C extends the potential market of this technology to other rich-ammonium industrial wastewaters.

With regards to biofilm development, a set of different simultaneous processes require accurate description to fully understand the complexity of the system. The AC initially added to the airlift reactor suffered a breakage phenomenon due to its fragility. The erosion formed very small AC particles, without altering significantly the initial particle dimensions of the added AC (1.0–1.2 mm), but resulting in a bulk liquid with certain amount of very small particles (i.e. less than 0.1 mm) noticeable due to an evident blackish tinge of the bulk liquid. This process could be associated to the shear stress that suffered the particles when moving inside the reactor, particularly when passing through the raiser to the downcomer (top section of the airlift reactor) and from the downcomer to the raiser (bottom section of the airlift reactor). The small particles derived from the breakage of the AC carrier induced the formation of granular biomass, converting the reactor in a hybrid system in terms of biofilm types, given than biofilm growing on the surface of 1.0–1.2 mm particles was coexisting with large biomass flocs, the so called granules. The biomass achieved in this period was 2.8 gVS/L with the following distribution: granular biomass (72%) and biofilm developed onto the surface of the AC particles (28%).

The type of AC used as biocarrier is not expected to have a significant influence in the performance of the nitrifying biofilm airlift bioreactor. This is based on (i) the carbon particles are used as mere physical support for biofilm development and therefore activation or other previous treatments applied to obtain the carbon are not expected to be relevant; (ii) a significant fraction of the nitrifying biomass was finally developed in the bioreactor as granules and not on to the carbon particles surface; this again reduces the possible effects of a different type of carbon on the performance of the bioreactor.

After the first operation period with a hybrid biofilm system, the AC particles were removed from the airlift and the nitrifying biomass remained only in form of granules. Then, the temperature was again increased to 30°C and the retention of biomass in the airlift was improved to achieve a higher biomass concentration. In two months, the biomass concentration and the capacity of the system increased to 5 gVS/L and 6 kgN/m^3 d but maintaining 100% of partial nitrification (see Bartrolí *et al.*, 2010b for an extended explanation). This is a noteworthy achievement because it is the highest capacity of a partial nitrification system found in the literature (see Table 7.1).

Coupling the partial nitrification with a denitrification process

In general, both developed systems represent an improvement of the existing technology for partial nitrification of high-strength ammonium effluents because the applied control strategies warrant the achievement of almost 100% of partial nitrification with high efficiency and high stability on long-term operation. Moreover, the developed control strategies allow achieving higher nitrifying biomass concentration in the reactors than the patented SHARON® process and consequently, they allow treating higher loading rates (or reducing the volume of the reactors). Additionally, the technologies developed in this research study allow achieving 100% of stable partial nitrification at lower temperatures (20°C) than SHARON® process (around 35°C). This fact could be very interesting for applying partial nitrification to industrial wastewaters at mild temperature.

Both developed partial nitrification systems can be efficiently coupled to a denitrifying system. In the case of heterotrophic denitrification, an organic carbon source is required. This organic carbon source can be external (e.g. ethanol or methanol) or internal (from the excess of the sludge). In any case, the required amount of organic carbon source will be smaller than the one used in a conventional denitrification process from nitrate. In the present study, a denitrifying SBR was operated with a rich-nitrite effluent and two different organic carbon sources; ethanol (as model of a readily biodegradable organic matter) and organic substrate from the digestion of primary and secondary sludge of a municipal WWTP. The denitrification capacity achieved with ethanol at 20ºC ($0.32\,kg\,N/m^3 \cdot d$) is higher than the obtained with the organic substrates from sludge at 20ºC: $0.08\,kg\,N/m^3 \cdot d$ for the primary sludge and $0.13\,kg\,N/m^3 \cdot d$ for the secondary sludge.

If the denitrification would be autotrophic (e.g. Anammox® process), an effluent with approximately 50% of ammonium and nitrite is needed. This ratio between ammonium and nitrite concentrations, i.e. $[N\text{-}NH_4^+]/[N\text{-}NO_2^-] = 1$, in the effluent of the nitrifying step is critical for the subsequent autotrophic denitrification, and it must be controlled for a successful performance of an Anammox® reactor. Most of the existing partial nitrification systems, coupled with Anammox®, try to produce an effluent achieving $[N\text{-}NH_4^+]/[N\text{-}NO_2^-] = 1$ but sometimes is very complicated to control such a concentration ratio in the reactor influent. Both partial nitrification systems developed in this project should provide a permanent and stable effluent with a $[N\text{-}NH_4^+]/[N\text{-}NO_2^-] = 1$. Taking into account that the effluent of both systems contains a 100% of nitrite, the desired ratio required for the subsequent autotrophic denitrification (Anammox®) process could be easily obtained with a by-pass of 50% of the inflow of the partial nitrification system.

Mathematical model for describing the performance of the partial nitrification

The other main achievement of this work is the development, calibration and successful validation of a mathematical model for describing the performance of the partial nitrification process in the both systems previously described. Potential industrial applications of such a simulation tool may range from design of reactor systems for the treatment of high strength ammonium wastewaters or real-time optimization of the performance of this type of systems, to very specific development of advanced control techniques to improve performance of the system as well as plant wide optimization.

Several models have been used in the past to describe the nitrification. Most of them consider the ammonium oxidation as a one-step process and are based on the Activated Sludge Model no.1, ASM1 (Henze *et al.*, 2000). Chandran and Smets (2000) demonstrated that two-step models are needed to describe nitrification batch experiments when nitritation and nitratation rates are similar or when the nitratation process is the limiting process. There are some conditions under which this behavior is observed: high temperature (Hellinga *et al.*, 1998), low DO concentration (Chung *et al.*, 2007) and existence of inhibitors (Kim and Seo, 2006). Therefore, several studies have been published using two-step models for the description of the nitrification process (Hao *et al.*, 2002; Wett and Rauch, 2003 and Wyffels *et al.*, 2004).

The use of two-step models is specially required when dealing with high-strength ammonium wastewater because AOB and NOB can be easily inhibited by ammonium and nitrite. Anthonisen *et al.* (1976) already demonstrated the existence of substrate inhibition of AOB and NOB and the inhibition of NOB due to ammonium. In fact, they postulated and demonstrated that free ammonia and free nitrous acid concentrations, rather than ammonium or nitrite ion concentrations, inhibit nitrification. Other authors have also found inhibition of AOB due to FNA (Van Hulle *et al.*, 2007). Usually substrate inhibition is modeled with a Haldane-type of equation (Carrera *et al.*, 2004a). With respect to ammonium inhibition to NOB and nitrite inhibition to AOB, they are usually modeled as non competitive inhibitions (Pambrun *et al.*, 2006). On the other hand, a complete model to accurately describe the existing biological reactions occurring in a basically nitrifying system should evidently consider the processes related to AOB and NOB (growth and decay) but also the processes related to heterotrophic bacteria. It has been demonstrated that the fraction of heterotrophic bacteria in a system with very low chemical oxygen demand (COD) concentration in the influent is high enough to be taken into account (Vadivelu *et al.*, 2006). Heterotrophs always exist in the nitrification mixed cultures because they grow from the decay products of the nitrifiers. Therefore an extensive model should include, at least, the growth and decay of heterotrophs and the processes related to the decay products (hydrolysis, ammonification, etc.).

Previous efforts carried out in the development of models for partial nitrification were focused mainly on describing substrates removal (ammonium, nitrite and nitrate) but did not provide realistic information about biomass concentration (Dosta *et al.*, 2007; Pambrun *et al.*, 2007 and Van Hulle *et al.*, 2007). A proper description of the biomass concentration is desirable, in particular in the case of biofilm reactors, because the impact of the biomass concentration on the performance of these reactors is of special relevance. The mathematical model developed in this project considers all the processes previously described for achieving an accurately description of the dynamic and steady-state conditions of soluble compounds and biomass concentration in both partial nitrification systems: activated sludge and biofilm airlift (see Jubany *et al.*, 2008 and Isanta *et al.*, 2010 for an extended explanation of the mathematical model).

Once a mathematical model is defined, a value for each model parameter must be assigned. Usually, all the parameters are taken from literature, especially when performing theoretical studies which do not compare the results obtained by simulation with experimental data (Coelho *et al.*, 2000 and Hao *et al.*, 2002). In other cases, only some of the parameters are calibrated with specific experiments in order to fit the experimental data to the model prediction

(Moussa *et al.*, 2005 and Wyffels *et al.*, 2004). However, the widely scattered values found in the literature for each parameter makes it difficult to choose the appropriate ones to describe a specific system, especially if it is taken into account that they were usually estimated with different methods and under different conditions. In this study all the kinetic and stoichiometric parameters related to nitrifying biomass were calibrated with specific experiments (see Jubany 2007 for an extended explanation of the calibration procedure). It is important to outline that such a specific calibration for all kinetic and stoichiometric parameters is lacking in the literature, since the vast majority of models for this process include a set of parameters assumed which are not well supported by experimental evidence.

The developed and calibrated mathematical model was validated for both nitrifying systems (activated sludge and biofilm) in different scenarios: (i) the controlled start-up of the nitrifying activated sludge system using biomass from a municipal WWTP as inoculum (Jubany *et al.*, 2008). (ii) The stable achievement of the partial nitrification process in the nitrifying activated sludge system (Jubany *et al.*, 2009d). (iii) The stable achievement of the partial nitrification process in the nitrifying biofilm airlift reactor (Isanta *et al.*, 2010).

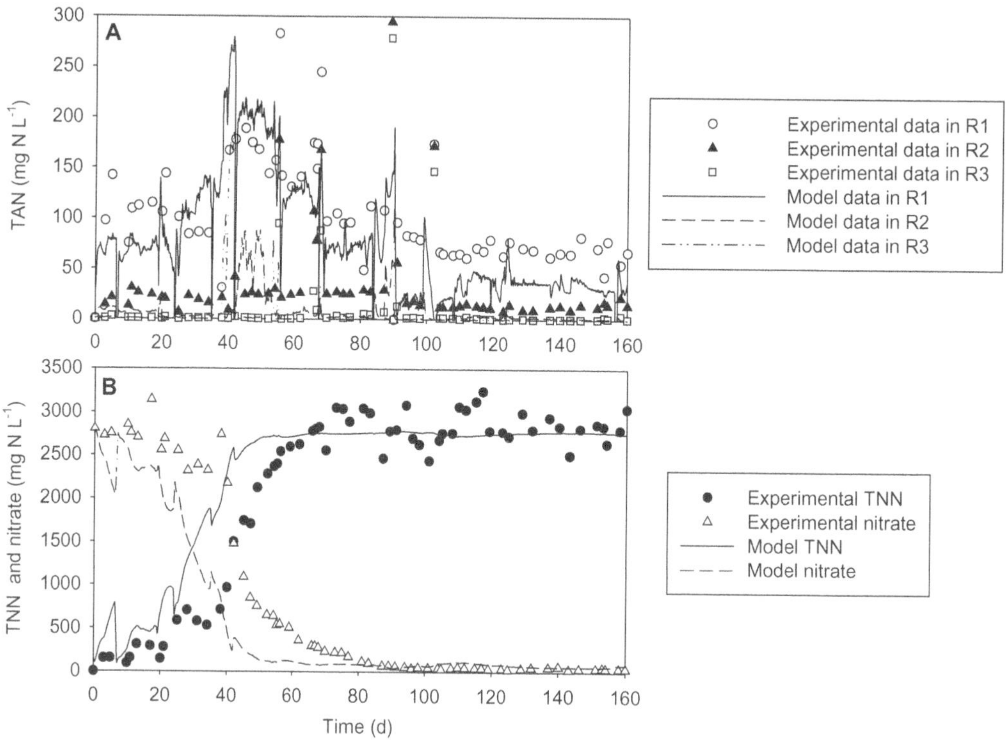

Figure 7.3 Validation of the model in the scenario of the stable achievement of the partial nitrification process in the nitrifying activated sludge system. A. Time course of TAN concentration in each reactor. B. Time course of TNN and nitrate concentrations in the third reactor (R3)

As the results presented in Figures 7.3 to 7.6 demonstrated, our model is able to properly predict the time course of soluble compounds (ammonium, nitrite and nitrate) and biomass concentration in both partial nitrification systems: activated sludge and biofilm airlift. This achievement is completely new since available models do not compare the biomass concentration predicted with the model with experimental values, especially in biofilm systems.

Finally, the mathematical model was used to compare the performance of partial nitrification process in both configurations and to verify the main reason of the achievement of nitrite accumulation. The study revealed that the NOB washout and as a result, the stable nitrite accumulation in the nitrifying activated sludge system was mainly due to the NOB inhibition by FA. This inhibition was caused by the accumulation of ammonium in the first two reactors provoked by the automatic control loop and the value of pH maintained in the system. Also, the simulation study showed that the effect of DO concentration on the nitrite accumulation was three times lower than the FA inhibition. However, the simulation study of the nitrifying biofilm airlift reactor revealed that the impact of the inhibition to the nitrite accumulation was rather weak. The main cause of the nitrite accumulation in the biofilm system was the limitation by DO. These results are significant because they could be very useful for making a decision on the technology to be selected.

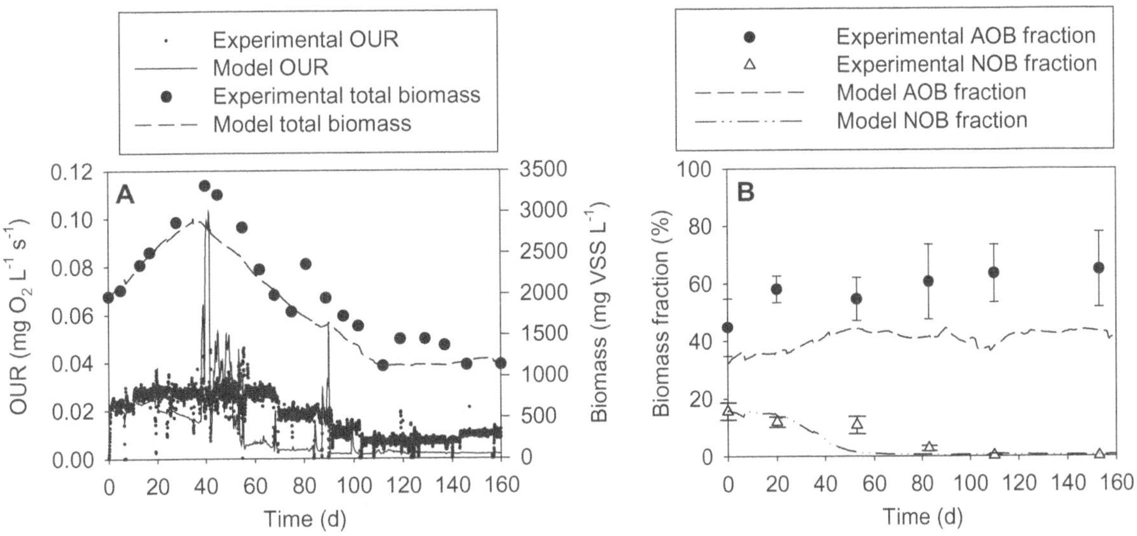

Figure 7.4 Validation of the model in the scenario of the stable achievement of the partial nitrification process in the nitrifying activated sludge system. A. Biomass concentration and OUR in the third reactor (R3). B. Biomass fractions estimated with FISH and model prediction

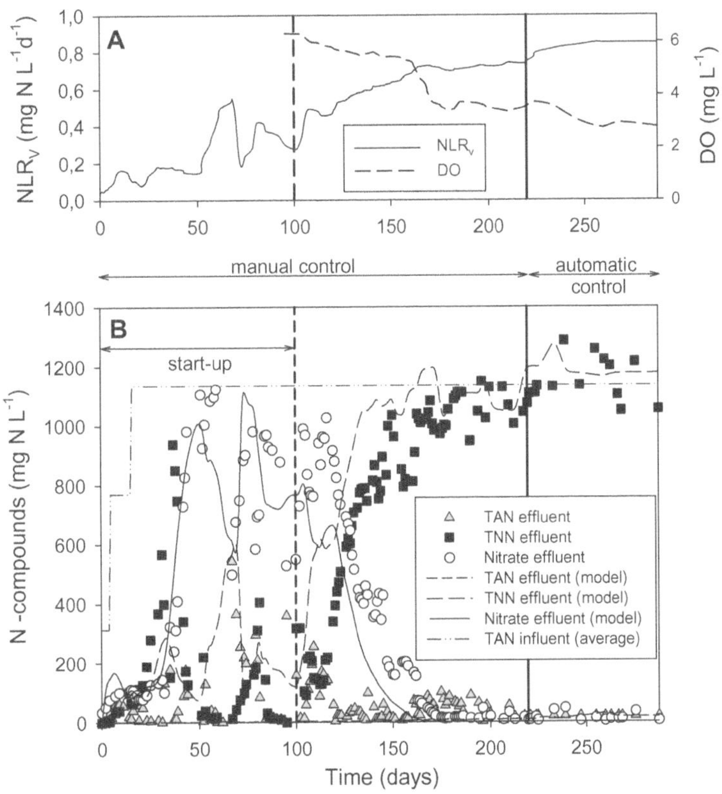

Figure 7.5 Validation of the model in the scenario of the stable achievement of the partial nitrification process in the nitrifying biofilm airlift reactor A. Variations of the applied volumetric nitrogen loading rate (NLR_V) and time course of DO concentration. B. Measured and simulated TAN, TNN and nitrate concentrations determined in the effluent of the biofilm airlift reactor during the whole period of operation (288 days). The TAN concentration in the effluent was included in the graph as an average. From Isanta *et al.*, 2010

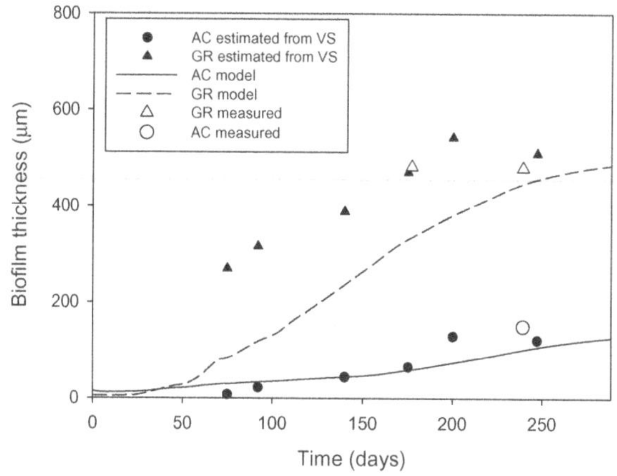

Figure 7.6 Time course of biofilm thickness for both the granular biomass and the biofilm developed onto the AC particles in the biofilm airlift reactor. Experimental values are included as scatter plots (direct measurements and estimation from VS concentrations experimentally measured), whereas model estimations are the line plots. AC = activated carbon particles; GR = granular biomass

IMPACT OF THE TECHNOLOGY ON ITS INDUSTRY OR RESEARCH SECTOR

Social, environmental and industrial benefits of the intensive treatment of the reject water

In recent years significant emphasis has been placed on reducing the amount of nutrient discharge because it results in the growth of algae, which therefore leads to eutrophication. For this reason European regulations (91/217/EEC) became more stringent for nitrogen and phosphorous. The 91/217/EEC Directive defines wastewater collection and treatment for all settlements and treatment of waste sludge. Its deadlines were phased from 1998 to 2005, depending on the size of the settlement and the characteristics of the affected area. The Urban Wastewater Treatment Directive has been operating for the last ten years, and during this time significant measures have already been undertaken to reduce pollution and many countries have improved their wastewater treatment plants thereby contributing to an enhancement in the quality of rivers. For example, biochemical oxygen demand (BOD_5) levels in European rivers have been reduced by 20 to 30% since the 1990s, phosphorous concentrations have declined by 30–40% and ammonium concentrations by about 40% (EEA, 2002).

Despite efforts to reduce nitrogen in urban wastewater, the total nitrogen concentrations in European rivers have remained high. This reflects the nitrogen impact from agriculture and at the same time the insufficient nitrogen removal by WWTPs (EEA, 2002). The regions in the South and East are especially behind with discharges of untreated or insufficiently treated wastewater still prevailing, thereby showing the great need to invest in the extension of WWTPs.

The 91/217/EEC Directive represents the most cost intensive legislation in the environmental sector, due to its rigorous requirements for providing wastewater treatment infrastructure in urban areas. The European Commission estimates that about €152 billion was invested in wastewater infrastructure over the period 1990–2010 (EC, 2000). Figure 7.7 shows the evolution of wastewater treatment since the application of the 91/217/EEC Directive in different regions of Europe.

The improvement of the nitrogen removal requires optimisation of existing WWTPs. Conventional biological extension of these WWTPs requires additional volume for aeration tanks and consequently a substantial investment. In order to avoid unnecessary capital expenditure and minimize investment costs, it has been shown to be more advantageous in many cases to increase nutrient removal capacity by making optimal use of the existing process units and by implementing complementary techniques. A promising complementary technical option for WWTPs with sludge digestion is the treatment of the liquor from sludge dewatering and sludge drying. This internal process stream, called reject water, contains up to 20–25% of the total influent nitrogen, but contributes only a minor 2% of the total influent flow. Normally, the reject water is recycled to the activated sludge reactors but some studies in the Netherlands and Germany have shown that the separated treatment of the reject water might be more economical compared to conventional extension of the WWTPs. One of the most interesting technologies for the reject water treatment is the biological nitrogen removal (BNR) via nitrite. This technology could suppose a 25% reduction of the total oxygen requirements as well as a reduction in the amount of sludge produced compared with the classical BNR via nitrate.

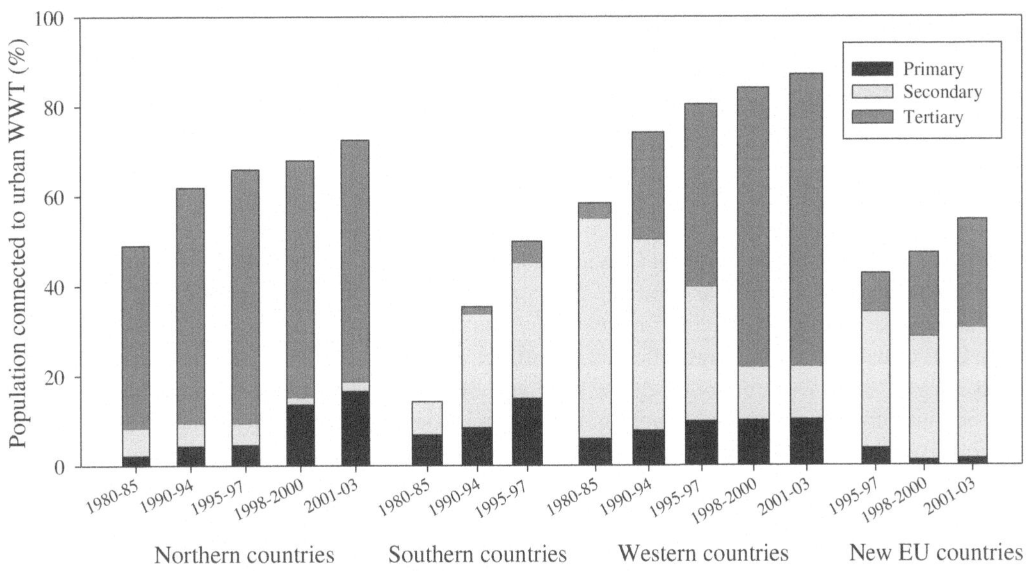

Figure 7.7 Evolution of wastewater treatment in different regions of Europe. Sources: (Eurostat, 2007; Fribourg and Boschet, 2004). Northern countries: Norway, Sweden, Finland and Iceland. Southern countries: Greece and Spain. Western countries: Austria, Netherlands, Germany and Ireland. New EU countries: Bulgaria, Poland, Estonia and Hungary. Primary treatment: physical and/or chemical processes. Secondary treatment: removal of at least 70% of BOD and at least 75% of COD. Tertiary treatment: removal of at least 70% nitrogen and at least 80% phosphorous

Economical viability of the partial nitrification technologies developed in project

Anaerobic digestion followed by dewatering is the main process in sewage sludge treatment for reducing and stabilising organic solids. This process results in a dewatered sludge and a high-strength ammonium wastewater (reject water), which is normally recycled back to the influent of the WWTP. This internal process stream contains up to 25% of the total nitrogen load of the WWTP but it only contributes to 2% of the total WWTP influent flowrate. The average ammonium concentration in the European reject waters is 1 g N-NH_4^+/L. This economical viability study includes the comparison of seven alternatives (four of them derived from this project). Other two alternatives (SHARON® plus heterotrophic denitrification and SHARON® plus Anammox®) are currently considered in the literature as the cheapest technology for treating reject water.

Proposed alternatives

Classic alternative (Case I)

Usually, the reject water is recycled back to the headwork of the WWTP. The classic alternative (case I) consists of upgrading the secondary treatment with nutrient removal of the WWTP. This upgrade requires two additional volumes: an extra aerobic volume for ammonium oxidation to nitrate (nitrification) and an extra anoxic volume for nitrate reduction to nitrogen gas (denitrification).

Advanced alternatives (Case II to VII)

A promising alternative for treating the reject water consists of a specific treatment of this stream based on the biological nitrogen removal via nitrite which consists of ammonium oxidation to nitrite (partial nitrification). This first stage has to be followed by denitrification stage that can be: (i) heterotrophic (from nitrite to nitrogen gas with an external organic carbon source) and (ii) autrotophic, called Anammox® process (from ammonium and nitrite to nitrogen gas without organic matter consumption). The combination of these processes results in other six alternatives: (case II) Nitrifying activated sludge with partial nitrification (Jubany *et al.*, 2009b) plus heterotrophic denitrification; (case III) Nitrifying airlift reactor with partial nitrification (Bartrolí *et al.*, 2010b) plus heterotrophic denitrification; (case IV) SHARON® plus heterotrophic denitrification; (case V) Nitrifying activated sludge with partial nitrification (Jubany *et al.*, 2009b) plus Anammox® process; (case VI) Nitrifying airlift reactor with partial nitrification (Bartrolí *et al.*, 2010b) plus Anammox® process; (case VII) SHARON® plus Anammox® process.

Economical evaluation of the alternatives

In this issue the investment and operational costs for the alternatives as explained before are estimated and analyzed. Spain was selected as an example to analyze the calculations involved in applying each technology. Spain has an organic load of 70 million population equivalent (p.e.), of which more than 34 million are related to stationary population, 10.5 million to seasonal population and 25 million to industrial and livestock load. Regarding the 91/217/EEC Directive, Spain had a conformity load of 59% in 2001, a non-conformity load of 22% and non-conformity under construction of 19%. The number of WWTPs was 1,326, of which 9% were primary treatment plants, 78% secondary treatment and 13% secondary treatment with nutrient removal (Environmental Ministry of Spain, 2001).

In order to estimate the cost of each alternative, the total reject water flow for each WWTP must first be defined. For this reason an approach based on the ratio between the total flow-rate of the WWTP and the reject water flowrate is applied (Q_{RW}/Q_T). Usually, the reject water flow is quantified within a range of 1–3% of the total flow treated by the WWTP. The assessment starts with the total organic load and by using data on medium load wastewater, the Q_{RW}/Q_T and the average ammonium concentration, the nitrogen load can be defined for the reject water flow. Thereafter, in order to define the requirements for each WWTP, the total reject water flow is divided by the total number of WWTPs. As mentioned before, these alternatives must only be applied to WWTPs with sludge digestion, as normally these facilities are related to WWTPs that treat significant loads. Therefore the ratio of the load treated by those WWTPs with more than 15,000 p.e. has been applied.

Figure 7.8 demonstrates the calculation sequence concerning the definition of the N-load of reject water in each WWTP with anaerobic digestion. In Table 7.2 the data obtained in each step of the calculation sequence is shown. Once the N-load of the reject water per WWTP of more than 15,000 p.e. is defined, the costs of the facilities required can be estimated. Nevertheless, some considerations in the calculations cost must be carried out:

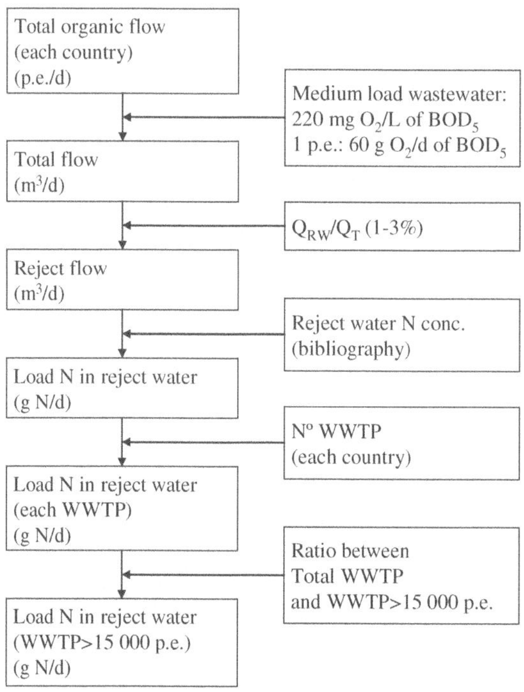

Figure 7.8 Diagram of the approach for defining the N-load by reject water flow for each WWTP bigger than 15,000 p.e.

Investment costs

In regard to the tanks:

a) Case I requires an aeration reactor for nitrification and a non-aerated reactor for denitrification. The design criteria are as follows: (i) nitrifying capacity of 0.03 kg N/kg VSS · d, (ii) biomass concentration of 2.0 kg VSS/m^3 in the aerobic reactor, (iii) denitrifying capacity of 0.06 kg N/kg VSS · d, and (iv) biomass concentration of 3.0 kg VSS/m^3 in the anoxic reactor (Carrera *et al.*, 2004b).

b) Case II requires a nitrifying volume divided in three parts and an anoxic reactor for denitrification. Regarding the volume for the nitrification process, a design N-load value of 3 kg N/m^3 · d has been selected from the results

Table 7.2 Total reject water flow and total N-load from reject water in Spain and per WWTP larger than 15,000 p.e.*

Total organic flow (p.e./d)	69,735,402
Total flow (m^3/d)	19,018,746
Q_{RW}/Q_T (%)	2
Reject flow (m^3/d)	380,375
Load N Reject flow (kg N/d)	389,366
Load N Reject flow per WWTP (kg N/d)	294
Load N Reject flow per WWTP > 15,000 (kg N/d)	250

**Data required: Total organic flow (Environment Ministry of Spain, 2001); BOD_5 for medium load wastewater; WWTP of Spain (Environmental Ministry of Spain, 2001); Ratio of load treated by WWTPs with more than 15,000 p.e. in Spain; 1 p.e. means the organic biodegradable load with a BOD_5 of 60gO_2/d according to the 91/217/EEC Directive.*

obtained in our study. The anoxic reactor volume for reduction of nitrite to nitrogen gas is determined by the value of denitrifying capacity from nitrite of 0.170 kg N/kg VSS·d and a biomass concentration of 3.0 kg VSS/m^3 (Carrera *et al.*, 2004b). The post-anoxic degasification tank is designed with 1 hour of hydraulic retention time and the settler with an overflow rate of 0.5 m^3/m^2 h.

c) Case III requires a nitrifying airlift reactor, an anoxic reactor for denitrification, a degasification tank and a settler. Regarding the airlift reactor for the nitrification process, a design N-load value of 5 kg N/m^3 · d has been selected from the results obtained in our study. The anoxic reactor volume is determined by the value of partial denitrifying capacity of 0.170 kg N/kg VSS · d and a biomass concentration of 3.0 kg VSS/m^3 (Carrera *et al.*, 2004b). The post-anoxic degasification tank is designed with 1 hour of HRT and the settler with an overflow rate of 0.5 m^3/m^2 h.
d) Case IV requires a sole reactor without settler. A design N-load value of 0.3 N/kg VSS·d has been selected from literature (Van Kempen *et al.*, 2001).
e) Case V requires a nitrifying volume divided in three parts and an Anammox® reactor for denitrification. It must be pointed out that only 50% of the total reject water has to be treated in this alternative. The rest of the inflow is by-passed to achieve the required ammonium/nitrite ratio for Anammox® process. The by-pass results in a design N-load value of 6 N/kg VSS · d for the nitrification process from the results obtained in this project. On the other hand, a design N-load value of 10 kg N/m^3 · d has been selected for the Anammox® process from the literature (Van der Star *et al.*, 2007).
f) Case VI requires a nitrifying airlift reactor, an anoxic reactor for denitrification, a degasification tank and a settler. It must be pointed out that only 50% of the total reject water has to be treated in this alternative. The rest of the inflow is by-passed to achieve the required ammonium/nitrite ratio for Anammox® process. The by-pass results in a design N-load value of 10 kg N/m^3 · d for the airlift reactor from the results obtained in our research study. On the other hand, a design N-load value of 10 kg N/m^3 · d has been selected for the Anammox® process from the literature (Van der Star *et al.*, 2007).
g) Case VII requires two reactors for SHARON® and Anammox® processes. Regarding the volume, a design N-load value of 1.2 kg N/m^3 · d has been selected for SHARON® process from literature (Van Dongen *et al.*, 2001). On the other hand, a design N-load value of 10 kg N/m^3 · d has been selected for the Anammox® process from the literature (Van der Star *et al.*, 2007).

Tanks and civil works items include the total costs of the construction of tanks and settlers. The facilities required for both biological processes are aeration for nitrification and the mixing equipment for anoxic and Anammox® reactors. Equipment item includes the costs of aeration equipment and mixers. The aeration equipment was chosen based on the stoichiometry amount of oxygen required for ammonium oxidation, its transfer efficiency and the height of the reactor. The biocarrier for the nitrifying airlift system will be sludge based activated carbon (SBAC). Its cost was estimated to be around 314 €/t with a biocarrier concentration of 40 kg SBAC/m^3 reactor.

Operational costs

Electrical consumption costs are mainly accounted by aeration. To estimate the electricity consumption, the aeration equipment and the mixers are taken into account at a cost of 0.09 €/kWh. For sludge treatment a cost of 0.076 €/kg was estimated, taking into account the fact that the sludge would need to be dried, incinerated and land filled (Bresters *et al.*, 1999). Regarding external carbon source for denitrification in cases I to IV, methanol is preferred based on a low sludge yield and low cost at 0.295 €/kg. The requirement of alkalinity is taken into account as follows: (i) due to the existing alkalinity of the dewatering liquids and the alkalinity formed through denitrification, a sufficient buffering capacity should maintain a suitable pH level for the overall process in alternatives with SHARON® process (cases IV and VII). However, a pump is required in other alternatives (cases II, III, V and VI) to recycle part of the effluent of the second stage to the first stage for using the alkalinity generated in the heterotrophic denitrification or the alkalinity by-passed to the Anammox® reactor. Nevertheless, an external addition of 10% of the required alkalinity (in form of calcium hydroxide) was considered in these alternatives. Maintenance costs are accounted for about 30% of the total costs (Fux and Siegrist, 2004b).

The following items; tanks, civil works, equipment, sludge treatment, electrical consumption, activated carbon and external carbon source addition were taken into account in order to estimate the total cost (euros per $kgN_{removed}$) for each alternative (Table 7.3). Other items such as electrical lead, equipment installation, maintenance, repair, analysis and technical staff are not taken into consideration here because they are common to all alternatives. Consequently, the final treatment cost for each alternative will be higher than the presented in table 7.3. However, the results of this work allow for economic comparison of these alternatives.

Table 7.3 Items considered, considerations applied and treatment costs per alternative (euros per kg N removed)[a]. Case I: classic alternative (no nitrite shortcut); Case II: activated sludge with biomass retention (Jubany *et al.*, 2009b) + heterotrophic denitrification; Case III: biofilm airlift reactor (Bartrolí *et al.*, 2010b) + heterotrophic denitrification; Case IV: SHARON® + heterotrophic denitrification; Case V: activated sludge with biomass retention (Jubany *et al.*, 2009b) + autotrophic denitrification (Anammox®); Case VI: biofilm airlift reactor (Bartrolí *et al.*, 2010b) + autotrophic denitrification (Anammox®); Case VII: SHARON® + autotrophic denitrification (Anammox®)

Data for a ratio Inflow/reject water flow = 2%							
Item	**Case I**	**Case II**	**Case III**	**Case IV**	**Case V**	**Case VI**	**Case VII**
				(€/kg $N_{removed}$)			
Tanks[b]	0.205	0.045	0.043	0.042	0.015	0.016	0.025
Civil works[b]	0.046	0.008	0.009	0.010	0.008	0.009	0.008
Equipment[b]	0.317	0.204	0.100	0.213	0.174	0.073	0.171
Sludge treatment	0.287	0.179	0.179	0.179	0.008	0.008	0.008
Maintenance/Repair	0.705	0.405	0.346	0.414	0.103	0.082	0.104
Electrical consumption	0.747	0.406	0.312	0.472	0.096	0.122	0.135
Activated Carbon	–	–	0.001	–	–	0.001	–
Products (Methanol)	0.747	0.463	0.463	0.463	–	–	–
Alkalinity	–	0.045	0.045	–	0.040	0.040	–
Total	3.05	1.76	1.50	1.79	0.44	0.35	0.45

[a] *All values are estimated on the basis of Spanish prices dating from 2007.*

[b] *In investment costs, the equipment has been annualised by a straight line depreciation of 5 years and a residual value of 10%, whereas tanks and civil work have been annualised by a straight line depreciation of 15 years and a residual value of 10%.*

Economical comparison between alternatives

The cost of each alternative regarding the most significant items for both investment and operation are compared in Table 7.3. As a result of the comparison between the treatment costs of all alternatives, we can conclude that:

(i) All the advanced alternatives with partial nitrification (cases II to VII) are clearly cheaper than the classic alternative: upgrading of a municipal WWTP for removing the nitrogen of reject water (case I).

(ii) The alternatives with autotrophic denitrification (Anammox® process) (cases V to VII) are noticeably cheaper than the alternatives with heterotrophic denitrification (cases II to IV) due to the lower costs of products (methanol), electrical consumption and sludge treatment. However, the main problem for implementing technologies with Anammox® is the very little experience with industrial applications of this technology and the long-term start-up required (Van der Star *et al.*, 2007).

(iii) The alternatives developed in this research study (cases II, III, V and VI) are economically comparable to the alternatives with SHARON® process (cases IV and VII). Consequently, the election of one of the advanced alternatives (cases II to VII) can not be only influenced by economical features.

REFERENCES

Anthonisen A., Loehr R., Prakasam T. and Srinath E. (1976). Inhibition of nitrification by ammonia and nitrous acid. *J. Water Pollut. Control Fed.*, **48**, 835–852.

Antileo C., Werner A., Ciudad G., Muñoz C., Bornhardt C., Jeison D. and Urrutia H. (2006). Novel operational strategy for partial nitrification to nitrite in a sequencing batch rating disk reactor. *Biochem. Eng. J.*, **32**, 69–78.

APHA. (1995). Standard methods for the examination of water and wastewater. 19th Ed. Washington DC, USA: American Publishers Health Association.

Bartrolí A., Pérez J. and Carrera J. (2010a). Automatic control for partial nitrification to nitrite with a biofilm airlift reactor. Conference proceedings. European Meeting on Chemical Industry and Environment (EMChIE 2010). Mechelen (Belgium), 17–19, May 2010.

Bartrolí A., Pérez J. and Carrera J. (2010b). Applying ratio control in a continuous granular reactor to achieve full nitritation under stable operating conditions. *Environ. Sci. Technol.*, **44**, 8930–8935.

Bernet N., Sanchez O., Cesbron D. and Steyer J. P. (2005). Modeling and control of nitrite accumulation in a nitrifying biofilm reactor. *Biochem. Eng. J.*, **24**, 173–183.

Bresters A. R., Coulomb A. R., Deak B., Matter B., Saabye A., Spinosa L. and Utvik A. (1999). Sludge Treatment and Disposal. Environmental Issues Series no.7. ISWA's working Group on Sewage and Waterworks Sludge.

Carrera J., Jubany I., Carvallo L., Chamy R. and Lafuente J. (2004a). Kinetic models for nitrification inhibition by ammonium and nitrite in a suspended and an immobilised biomass systems. *Process Biochem.*, **39**, 1159–1165.

Carrera J., Lafuente F. J. and Vicent T. (2004b). Biological treatment of high-strength ammonium industrial wastewaters. Design criteria (in Spanish). *Ingeniería Química*, **409**, 130–143.

Chandran K. and Smets B. F. (2000). Applicability of two-step models in estimating nitrification kinetics from batch respirograms under different relative dynamics of ammonia nitrite oxidation. *Biotechnol. Bioeng.*, **70**, 54–64.

Chung J., Bae W., Lee Y-W. and Rittmann B. E. (2007). Shortcut biological nitrogen removal in hybrid biofilm/suspended growth reactors. *Process Biochem.*, **42**, 320–328.

Coelho M. A. Z., Russo C. and Araújo O. Q. F. (2000). Optimization of a sequencing batch reactor for biological nitrogen removal. *Water Res.*, **34**, 2809–2817.

Dosta J., Galí A., Benabdallah E., Macé S. and Mata-Álvarez J. (2007). Operation and model description of a sequencing batch reactor treating reject water for biological nitrogen removal via nitrite, *Bioresour. Technol.*, **98**, 2065–2075.

EC (European Comission). (2000). Investment and employment related to EU policy on air, water and waste. http://ec.europa.eu/environment/enveco/industry_employment/inv_and_empl_exsum.pdf.

EEA (European Environmental Agency). (2002). Environmental signals. http://reports.eea.europa.eu/environmental_assessment_report_2002_9/en.

Environment Ministry of Spain. (2001). Informe de Coyuntura: Estado de la depuración de las aguas residuales urbanas. http://www.mma.es/portal/secciones/info_estadistica_ambiental/estadisticas_info/informes_coyuntura/informe_anual/pdf/pto32_01.pdf.

Eurostat (Statistical office of the European commission). (12 March of 2007). Population connected to urban waste water treatment: primary treatment, secondary treatment and tertiary treatment. http://epp.eurostat.ec.europa.eu/portal/page?_pageid=1996,45323734&_dad=portal&_schema=PORTAL&screen=welcomeref&open=/&product=Yearlies_new_environment_energy&depth=4.

Fribourg B. and Boschet A. (2004). Indicator Fact Sheet: Urban waste water treatment. European Environmental Agency. http://themes.eea.europa.eu/Specific_media/water/indicators/WEU16%2C2004.05/WEU16_UWWT_180504.pdf.

Frijters C. T. M. J., Silvius M., Fischer J., Haarhuis R. and Mulder R. (2007). Full-scale applications for both COD and nutrient removal in a CIRCOX® airlift reactor. *Water Sci. Technol.*, **55**(8–9), 107–114.

Fux C., Boehler M., Huber P., Brunner I. and Siegrist H. (2002). Biological treatment of ammonium-rich wastewater by partial nitritation and subsequent anaerobic ammonium oxidation (anammox) in a pilot plant. *J. Biotechnol.*, **99**, 295–306.

Fux C., Huang D., Monti A. and Siegrist H. (2004a). Difficulties in maintaining long-term partial nitrification of ammonium-rich sludge digester liquids in a moving-bed biofilm reactor. *Water Sci. Technol.*, **50**(10), 19–26.

Fux C. and Siegrist H. (2004b). Nitrogen removal from sludge digester liquids by nitrification/denitrification or partial nitritation/anammox: enviromental and economical considerations. *Water Sci. Technol.*, **49**(11–12), 53–60.

Fux C., Velten S., Carozzi V., Solley D. and Keller J. (2006). Efficient and stable nitritation and denitritation of ammonium-rich sludge dewatering liquor using an SBR with continuous loading. *Water Res.*, **40**, 2765–2775.

Ganigué R., Gabarró J., Sànchez-Melsió A., Ruscalleda M., López H., Vila X., Colprim J. and Balaguer M.D. (2009). Long-term operation of a partial nitrification pilot plant treating leachate with extremely high ammonium concentration prior to an anammox process. *Bioresour. Technol.*, **100**, 5624–5632.

Garrido J. M., van Benthum W. A. J., van Loosdrecht M. C. M. and Heijnen J. J. (1997). Influence of dissolved oxygen concentration on nitrite accumulation in a biofilm airlift suspension reactor. *Biotechnol. Bioeng.*, **53**, 168–178.

Hao X., Heijnen J. J. and Van Loosdrecht M. C. M. (2002). Sensitivity analysis of a biofilm model describing a one-stage completely autotrophic nitrogen removal (CANON) process. *Biotechnol. Bioeng.*, **77**, 266–277.

Hellinga C., Schellen A. A. J. C., Mulder J. W., van Loosdrecht M. C. M. and Heijnen J. J. (1998). The SHARON process: An innovative method for nitrogen removal from ammonium-rich wastewater. *Water Sci. Technol.*, **37**(9), 135–142.

Henze M., Gujer W., Mino T. and Van Loosdrecht M. (2000). Activated sludge models ASM1, ASM2, ASM2D and ASM3. Scientific and technical report no. 9. London: IWA Publishing.

Isanta E., Bartrolí A., Carrera J. and Pérez J. (2010). Assessment of the minimal oxygen limiting conditions required for full partial nitrification to nitrite in a biofilm airlift reactor. Conference proceedings. Sustainable Solutions for Small Water and Wastewater Treatment Systems (S_2Small2010). Girona (Spain), 19–22, April 2010.

Joo S. H., Kim D. J., Yoo I. K., Park K. and Cha G. C. (2000). Partial nitrification in an upflow biological aerated filter by O2 limitation. *Biotechnol. Lett.*, **22**, 937–940.

Jubany I. (2007). Operation, modelling and automatic control of complete and partial nitrification of highly concentrated ammonium wastewater, Ph.D. Thesis, Universitat Autònoma de Barcelona, Barcelona, Spain. ISBN: 978-84-690-7378-0 (http://www.tesisenxarxa.net/TESIS UAB/AVAILABLE/TDX-1005107-170622//ijg1de1.pdf).

Jubany I., Carrera J., Lafuente J. and Baeza J. A. (2008). Start-up of a nitrification system with automatic control to treat highly concentrated ammonium wastewater: Experimental results and modelling. *Chem. Eng. J.*, **144**, 407–419.

Jubany I., Lafuente J., Carrera J. and Baeza J. A. (2009a). Automated thresholding method (ATM) for biomass fraction determination using FISH and confocal microscopy. *J. Chem. Technol. Biotechnol.*, **84**, 1140–1145.

Jubany I., Lafuente J., Baeza J. A. and Carrera J. (2009b). Total and stable washout of nitrite oxidizing bacteria from a nitrifying continuous activated sludge system using automatic control based on Oxygen Uptake Rate measurements. *Water Res.*, **43**, 2761–2772.

Jubany I., Carrera J., Lafuente J. and Baeza J. A. (2009c). Expert control for a stable operation of a partial nitrification system to treat highly concentrated ammonium wastewater. *Water Sci. Technol.*, **60**(5), 1191–1199.

Jubany I., Baeza J. A., Lafuente J. and Carrera J. (2009d). Model-based study of nitrite accumulation with OUR control in two continuous nitrifying activated sludge configurations. *Water Sci. Technol.*, **60**(10), 2685–2693.

Kim D. J. and Seo D. (2006). Selective enrichment and granulation of ammonia oxidizers in a sequencing batch airlift reactor. *Process Biochem.*, **41**, 1055–1062.

Lemaire R., Marcelino M. and Yuan Z. (2008). Achieving the nitrite pathway using aeration phase length control and step-feed in an SBR removing nutrients from abattoir wastewater. *Biotechnol. Bioeng.*, **100**, 1228–1236.

Li Y. Z., Peng C. Y., Peng Y. Z. and Wang P. (2004). Nitrogen removal from pharmaceutical manufacturing wastewater via nitrite and the process optimization with on-line control. *Water Sci. Technol.*, **50**(6), 25–30.

Mobarry B. K., Wagner M., Urbain V., Rittmann B. E. and Stahl D. A. (1996). Phylogenetic probes for analyzing abundance and spatial organization of nitrifying bacteria. *Appl. Environ. Microbiol.*, **62**, 2156–2162.

Mosquera-Corral A., Gónzalez F., Campos J. L. and Méndez R. (2005). Partial nitrification in a SHARON reactor in the presence of salts and organic carbon compounds. *Process Biochem.*, **40**, 3109–3118.

Moussa M. S., Hooijmans C. M., Lubberding H. J., Gijzen H. J. and Van Loosdrecht M. C. M. (2005). Modelling nitrification, heterotrophic growth and predation in activated sludge. *Water Res.*, **39**, 5080–5098.

Mulder J. W., van Loosdrecht M. C. M., Hellinga C. and van Kempen R. (2001). Full-scale application of the SHARON process for treatment of rejection water of digested sludge dewatering. *Water Sci. Technol.*, **43**(11), 127–134.

Muñoz C., Rojas D., Candi O., Azocar L., Bornhardt C. and Antileo C. (2009). Supervisory control system to enhance partial nitrification in an activated sludge reactor. *Chem. Eng. J.*, **145**, 453–460.

Noto K., Ogasawara T., Suwa Y. and Sumino T. (1998). Complete oxidation of high concentration of ammonia by retaining incompatible nitrification activities in three-vessel system. *Water Res.*, **32**(3), 769–773.

Pambrun V., Paul E. and Spérandio M. (2006). Modeling the partial nitrification in sequencing batch reactor for biomass adapted to high ammonia concentrations. *Biotechnol. Bioeng.*, **95**, 120–131.

Pambrun V., Paul E. and Spérandio M. (2007). Control and modelling of partial nitrification of effluents with high ammonia concentrations in Sequencing Batch Reactor. *Chem. Eng. Processing*, **47**, 323–329.

Peng Y. and Zhu G. (2006). Biological nitrogen removal with nitrification and denitrification via nitrite pathway. *Appl. Microbiol. Biotechnol.*, **73**, 15–26.

Pérez J., Costa E. and Kreft J. U. (2009). Conditions for partial nitrification in biofilm reactors and a kinetic explanation. *Biotechnol. Bioeng.*, **103**, 282–295.

Picioreanu C., van Loosdrecht M. C. M. and Heijnen J. J. (1997). Modelling the effect of oxygen concentration on nitrite accumulation in a biofilm airlift suspension reactor. *Water Sci. Technol.*, **36**(1), 147–156.

Qiao S., Matsumoto N., Shinohara T., Nishiyama T., Fujii T., Bhatii Z. and Furukawa K. (2010). High-rate partial nitrification performance of high ammonium containing wastewater under low temperatures. *Bioresour. Technol.*, **101**, 111–117.

Tokutomi T. (2004). Operation of a nitrite-type airlift reactor at low DO. *Water Sci. Technol.*, **49**(5–6), 81–88.

Tokutomi T., Kiyowaka T., Shibayama C., Harada H. and Ohashi A. (2006). Effect of inorganic carbon on nitrite accumulation in an aerobic granule reactor. *Water Sci. Technol.*, **53**(12), 285–294.

Turk O. and Mavinic D. S. (1987). Benefits of using selective inhibition to remove nitrogen from highly nitrogenous wastes. *Environ. Technol. Lett.*, **8**, 419–426.

Vadivelu V. M., Keller J. and Yuan Z. (2006). Stoichiometric and kinetic characterisation of Nitrosomonas sp. in mixed culture by decoupling the growth and energy generation processes. *J. Biotechnol.*, **126**, 342–356.

Van der Star W., Abma W., Blommers D., Mulder J., Tokutomi T., Strous M., Picioreanu C. and van Loosdrecht M. (2007). Startup of reactors for anoxic ammonium oxidation: Experiences from the first full-scale Anammox reactor in Rotterdam. *Water Res.*, **41**, 4149–4163.

Van Dongen U., Jetten M. and Van Loosdrecht M. (2001). The SHARON®-Anammox® process for treatment of ammonium rich wastewater. *Water Sci. Technol.*, **44**(1), 153–160.

Van Hulle S., van Den Broeck S., Maertens J., Villez K., Donckels B., Schelstraete G., Volcke E. and Vanrolleghem P. (2005). Construction, start-up and operation of a continuously aerated laboratory-scale SHARON reactor in view of coupling with an Anammox reactor. *Water SA*, **31**, 327–334

Van Hulle S. W. H., Volcke E., López Teruel J., Donckels B., van Loosdrecht M. C. M. and Vanrolleghem P. (2007). Influence of temperature and pH on the kinetics of the Sharon nitritation process. *J. Chem. Technol. Biotechnol.*, **82**, 471–480.

Van Kempen R., Mulder J. W., Uijterlinde C. A. and van Loosdrecht M. C. M. (2001). Overview: full scale experience of the SHARON® process for treatment of rejection water of digested sludge dewatering. *Water Sci. Technol.*, **44**(1), 145–152.

Van Loosdrecht M. C. M. and Salem S. (2006). Biological treatment of sludge digester liquids. *Water Sci. Technol.*, **53**(12), 11–20.

Vázquez-Padín J. R., Pozo J. M., Jarpa M., Figueroa M., Franco A., Mosquera-Corral A., Campos J. L. and Méndez R. (2009). Treatment of anaerobic sludge digester effluents by the CANON process in an air pulsing SBR. *J. Hazard. Mat.*, **166**, 336–341.

Wagner M., Rath G., Koops H. P., Flood J. and Amann R. (1996). In situ analysis of nitrifying bacteria in sewage treatment plants. *Water Sci. Technol.*, **34**(1), 237–244.

Wett B. and Rauch W. (2003). The role of inorganic carbon limitation in biological nitrogen removal of extremely ammonia concentrated wastewater. *Water Res.*, **37**, 1100–1110.

Wyffels S., van Hulle S., Boeckx P., Volcke E., van Cleemput O., Vanrolleghem P. and Verstraete W. (2004). Modeling and simulation of oxygen-limited partial nitritation in a membrane-assisted bioreactor (MBR). *Biotechnol. Bioeng.*, **86**, 531–542.

Chapter 8

Minimisation of sludge production by utilisation of biological potential in membrane bioreactors

Matthias Kraume, Anja Drews and Jana Schaller

SUMMARY

Aerobic biological wastewater treatment produces a significant amount of excess sludge, which is expected to rise due to the increasing number of wastewater treatment plants and more stringent effluent criteria. Furthermore, sludge management accounts for about half of the overall wastewater treatment costs. In this context, the reduction of excess sludge production by utilisation of maintenance metabolism in membrane bioreactors (MBR) is an interesting alternative. Therefore the interest in high SRT operation is sparked again as an option for sludge minimisation, but has not been applied in full scale plants yet.

A cost model for submerged membrane bioreactors treating municipal wastewater incorporating energy demand for aeration and fouling prevention as well as sludge handling costs subject to local conditions was developed. The cost model is based on easily accessible input parameters like operational and biokinetic parameters and feed conditions. Based on the developed model simulation studies were performed to determine the optimal operational conditions by varying sludge treatment and electricity costs. First results showed that under certain conditions the optimum SRT is higher than currently applied, i.e. minimising sludge reduces the overall cost despite higher aeration demand.

METHODOLOGIES AND APPROACHES EMPLOYED

Two bench-scale MBRs were continuously operated at different sludge ages varying between 15–100 days to obtain various sludge characteristics like dewaterability, filterability, viscosity etc. and biokinetic parameters, which can be used for the cost model besides existing literature data.

Plant configuration

Two submerged MBRs for C- and N-removal equipped with an anoxic and an aerobic tank and an internal recirculation of 400% of the inflow were used to study the sludge characteristics at different operating conditions (see Table 8.1). They were of different size (10 L and 40 L, respectively) and equipped with different membrane modules (plate and frame and hollow fibre, respectively) and were fed with a complex synthetic wastewater representing a municipal wastewater with a targeted COD of 600 mg/L (see Table 8.2, composition adapted from Bracklow *et al.*, 2007). A nutrients solution adapted according to Smolders *et al.* (1994) was added to the synthetic wastewater: 450 µg/L $FeSO_4{\cdot}7H_2O$, 45 µg/L H_3BO_4,

 Reduction, Modification and Valorisation of Sludge (REMOVALS). Edited by Christophe Bengoa, Azael Fabregat, Josep Font and Frank Stüber. ISBN: 9781843393450. Published by IWA Publishing, London, UK.

6 µg/L $CuCl_2 \cdot 2H_2O$, 54 µg/L KJ, 30 µg/L $MnSO_4 \cdot H2O$, 18 µg/L $NaMoO_4 \cdot 2H_2O$, 18 µg/L $ZnCl_2 \cdot 2H_2O$, 45 µg/L $CoCl_2 \cdot 6H_2O$. Both MBRs were operated at room temperature (18–23°C). The plants were inoculated with activated sludge from an MBR treating municipal wastewater. Due to the diverging behaviour of sludge characteristics of both MBRs, they were reinoculated at the beginning of period 2 with activated sludge from an MBR treating municipal wastewater.

Table 8.1 Operational parameters of the investigated MBRs

	MBR I				MBR II		
	Period 1	Period 2	Period 3	Period 4	Period 1	Period 2	Period 4
Time of operation [d]	106	229	104	174	106	333	174
Volume [L]	10	10	10	10	40	40	40
Solid Retention Time [h]	140	140	140	15	140	140	75
Hydraulic Retention Time [h]	10	10	5	10	10	10	10
Steady state sludge loading*	0.130	0.094	0.094	0.217	0.084	0.059	0.090

[kg COD/(kgTS d)].

Table 8.2 Composition of the synthetic wastewater

Ingredient	(mg/L)	Ingredient	(mg/L)	Ingredient	(mg/L)
Peptone	25	Starch	200	$MgSO_4 \cdot 7H_2O$	4.2
Yeast extract	80	Sodium acetate	150	K_2HPO_4	29
Milk powder	180	NH_4Cl	150	Urea	40

Sampling and analysis

The influent synthetic wastewater and the permeate were sampled once a week and analysed for total and volatile suspended solids (TS, VSS), total COD and total N concentration. Dr. Lange cuvette test kits LCK 338 and LCK 238 were used for total nitrogen determination, LCK 114 and LCK 414 were used for total COD determination.

Sludge samples were taken once a week from the aeration tank of both plants and were analysed immediately for total and volatile suspended solids (TS, VSS) concentration, dewaterability (CST), filterability (TTF), rheology and oxygen transfer coefficient (k_La).

TS and VSS concentrations were measured according to German Standard DIN 38414 part 2 and 3. Capillary Suction Time (CST) as a method to characterise sludge dewaterability was measured in duplicate with a standard CST Device (Model 200, Triton Electronics). The measurement of the CST is very inaccurate due to variations in the structure of the filter paper, therefore the producer of the CST device specifies an accuracy of 20%. The filterability was measured by the filterability index time to filter (TTF) to filter a certain volume. Due to the availability of just small sludge samples, TTF was determined by filtering 3 mL of permeate from a 25 mL sludge sample. Sampling larger sludge volumes would have disturbed targeted SRT.

For the measurement of rheological properties a rotational viscometer type RheolabQC (Anton Paar Germany GmbH) was used. In order to include low viscosities, the viscometer was supplied with a double gap measurement device DG42. Measurements were carried out as double readings at a shear rate between 5 and 3000 1/s with undiluted samples. Furthermore the viscosity is strongly dependent on temperature; therefore all results were corrected to a temperature of 20°C.

Respirometric measurements and oxygen transfer measurements were carried out in a 1 L batch (Rosenberger, 2003). One litre of sludge was withdrawn once per week for respirometric tests and returned to the bioreactor afterwards. To provide sufficient mixing and oxygen transfer, the 1 L batch was equipped with a Rushton turbine and baffles and was kept at a constant temperature of 20° ± 2°C. Fine bubble aeration was employed through a porous tube. Aeration was set to 40 L/h; in order to avoid evaporation, the air was humidified prior to the reactor inlet. Dissolved oxygen (DO) and temperature were measured with an oxygen probe, type CellOx 325 (WTW GmbH), which was calibrated before each measurement and placed close to the stirrer.

ACHIEVEMENTS RELATED TO THE STATE OF THE ART

Plant performance

Both bench-scale MBRs showed good results in terms of COD elimination efficiency, which was mostly above 95%. The TN elimination efficiency in MBR II is mainly between 80–85%, whereas in MBR I the TN elimination varied between 70–85% due to stronger fluctuations in oxygen concentration.

Biomass accumulation

In the first period, the plants were initially inoculated with sludge from a decentralised MBR plant treating municipal wastewater. Both plants were operated with a HRT of 10 h at a nearly infinite sludge age. Sludge was only withdrawn for sampling which resulted in an actual sludge age of 140 days. As can be seen, the initial sludge concentration was around 9 g/L (see Figure 8.1). At the beginning a quick sludge growth could be observed in both plants, but then suddenly the sludge growth in MBR I stopped and the biomass concentration kept nearly constant. This occurred simultaneously with the appearance of *tubifex tubifex* worms in MBR I, which were eliminated by dosing of 4.2 mg/L $CuCl_2$ over a couple of days.

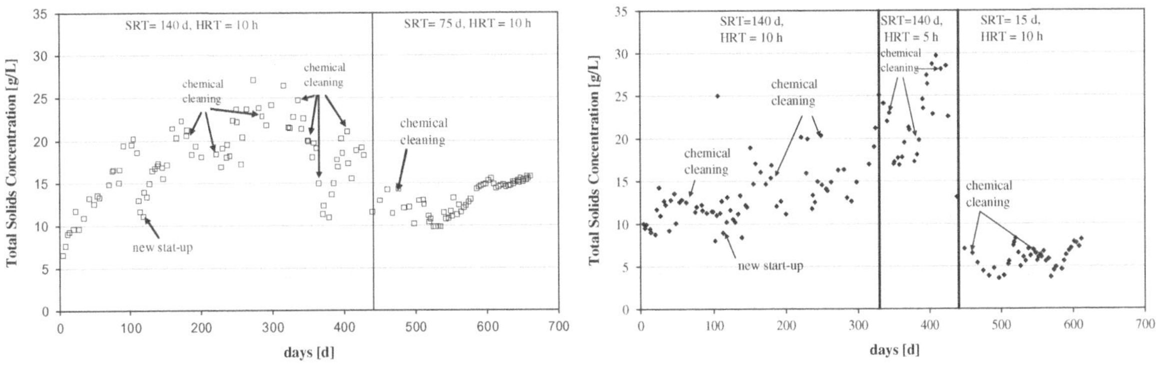

Figure 8.1 Accumulation in MBR I (left) and MBR II (right)

Due to problems with the high viscosity of the sludge and the oxygen transport into the sludge in MBR II both plants were reinocculated with sludge from an MBR treating municipal wastewater on day 106, when the second period started which lasted up to day 306 for MBR I and day 428 for MBR II, respectively. Both plants were operated at the same conditions as in the period before, i.e. a very high sludge age of 140 days with a HRT of 10 h. Membrane cleanings were performed after a breakdown of the aeration system or too high TMP.

The third period with changed conditions for MBR I started on day 335 whereas MBR II stayed at the same conditions as in period 2. In this period, the SRT has been retained unchanged, but the hydraulic retention time (HRT) of MBR I was dropped from 10 h to just 5 h to see the influence on sludge production. Due to several breakdowns of the aeration system and strong fouling events in MBR I, membrane cleaning was frequently performed in both plants, which resulted in partly heavy sludge losses of up to a tenth of the reactor volume, but still the SRT was larger than 100°d.

The forth period for both MBR started on day 439. Here, the SRT of both plants was changed to 15 and 75 days, respectively, while the HRT was kept constant at 10 h. A stable biomass concentration of 6.2 ± 0.7 g/L was reached after 90 days in MBR I and of 14.6 ± 0.2 g/L after two sludge ages in MBR II.

In Table 8.3, the obtained yield coefficients Y for the different periods are shown. The obtained values for high sludge ages (>75 d) are comparable to those of other MBRs operated at nearly complete sludge retention (Khongnakorn *et al.*, 2007 and Pollice *et al.*, 2007). They obtained values in the range of 0.06–0.12 g VSS/g COD.

Table 8.3 Yield coefficients for the different investigated periods

	MBR I				MBR II		
	Period 1	**Period 2**	**Period 3**	**Period 4**	**Period 1**	**Period 2**	**Period 4**
Yield coefficient [g VSS/g COD]	0.055	0.079	0.061	0.328	0.066	0.095	0.120

Sludge characteristics

Activated sludge is known to be a non-Newtonian fluid (Laera *et al.*, 2007). The apparent viscosity μ_∞ as a function of TS concentration is shown in Figure 8.2, where μ_∞ is the shear rate independent viscosity at high shear rate. In this study μ_∞ was evaluated at a shear rate of 2300 1/s. Furthermore apparent viscosity is strongly dependent on temperature; therefore all results were corrected to a temperature of 20°C. In agreement with literature (e.g. Pollice *et al.*, 2007 and Tixier *et al.*, 2003) apparent viscosity increases with increasing TS concentration which influences fluid dynamics in the plant.

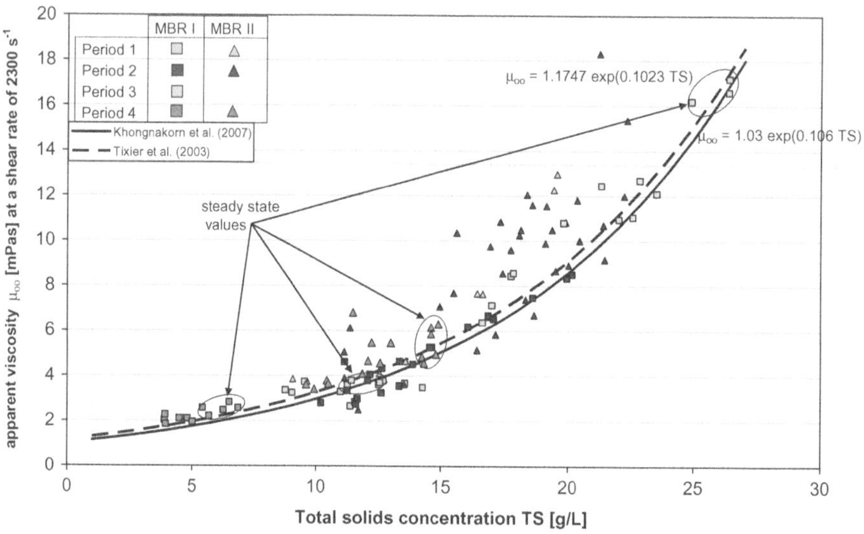

Figure 8.2 Apparent viscosity μ_∞ versus Total Solids concentration. The highlighted values are obtained for steady state conditions

The apparent viscosities of MBR II are slightly higher than those of MBR I at the same TS concentration, although both sludges are fed with the same synthetic wastewater. This might be due to fluctuations in the influent COD concentrations, which results in differently loaded wastewater and variable specific power input due to different MBR configuration. Tixier *et al.* (2003) also observed different viscosities at a defined TS concentration for activated sludges with different influent wastewater. The steady state viscosities for the different periods are showing a certain fluctuation, especially for higher TS values.

In comparison to the literature correlations, the obtained apparent viscosities are up to two times higher, especially for higher TS concentrations between 15 and 20 g/L. In literature correlations established by Khongnakorn *et al.* (2007) and Tixier *et al.* (2003) just a TS range from 3–17 g/L was used thus extrapolating results obtained for a lower TS range to higher TS values can result in an underestimation of up to 50% of the investigated parameter. This stresses the need for a larger database as a reliable basis for process optimisation and also shows that with the current correlations, e.g., mixing and pumping power requirements would have been underestimated. In this study more than half of the data obtained are valid for TS concentrations higher than 15 g/L.

In Table 8.4, some literature correlations for the viscosity as a function of TS are given and compared with the correlations found in this study. In agreement with most literature also an exponential relationship is found for the investigated MBRs. Especially the correlation including samples from both MBRs (MBR I and MBR II) is quite similar to the correlations found by Khongnakorn *et al.* (2007) and Tixier *et al.* (2003), just the coefficient is higher with 1.22 compared to 1.17 and 1.03, respectively.

Table 8.4 Viscosity correlations as a function of TS concentration found by different authors

	Correlation	
MBR I only	$\mu_\infty = 1.24 \cdot e^{0.098 \cdot TS}$	Eq. 1
MBR II only	$\mu_\infty = 1.23 \cdot e^{0.109 \cdot TS}$	Eq. 2
MBR I + MBR II	$\mu_\infty = 1.22 \cdot e^{0.105 \cdot TS}$	Eq. 3
Khongnakorn *et al.* (2007)	$\mu_\infty = 1.03 \cdot e^{0.106 \cdot TS}$	Eq. 4
Tixier *et al.* (2003)*	$\mu_\infty = 1.17 \cdot e^{0.102 \cdot TS}$	Eq. 5

**Correlation valid for biomass fed on domestic wastewater.*

Nevertheless, even the type of correlation between viscosity and TS concentration is contradictory in literature. Tixier *et al.* (2003) found an exponential relationship between viscosity μ_∞ and TS, whereas Pollice *et al.* (2007) found a correlation where a downward concavity appeared, indicating a decreasing influence of higher solid concentrations on the viscosity μ_∞. Obviously, the viscosity of an activated sludge is not only a function of TS but also a function of other sludge characteristics which have to be identified, and of operating conditions, specific power input and thus plant size.

The dewaterability of an activated sludge, an important factor influencing sludge disposal costs, can be characterized by the CST. Figure 8.3 shows the CST for both plants as a function of TS concentration. The absolute value of CST differs by up to a factor of 5 for both sludges at the same TS concentration. It is evident that below a critical TS concentration of approx. 15 g/L, the biomass concentration has nearly no influence on the dewaterability, as the CST values fluctuate between 10 s and 100 s. Above 15 g/L a dramatic increase in CST and a broad distribution can be observed. Furthermore it can be seen in Figure 8.3, that the obtained steady state CST values are normally lower or in the same range as the CST values obtained during the unsteady state conditions except for period 3 in MBR I. Therefore above 15 g/L higher TS values will result in a strongly lower sludge dewatering potential, however at steady state conditions the dewaterability has slightly improved compared to the unsteady state conditions. Knowledge of the sludge properties during the unsteady state period is especially at elevated SRT important, because it comprises quite a long period.

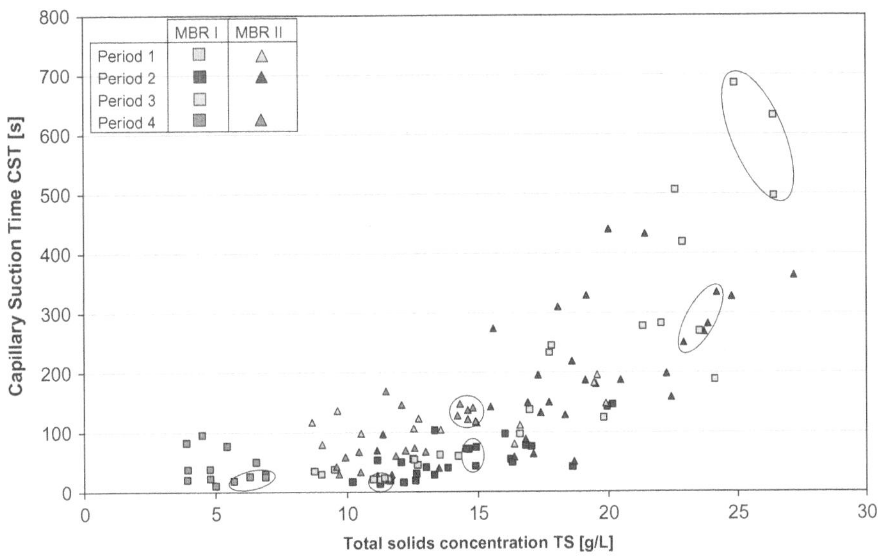

Figure 8.3 Capillary suction time versus Total Solids concentration. The highlighted values are obtained for steady state conditions

In comparison to literature where the CST was determined to be in the range of 5–20 s (Khongnakorn *et al.*, 2007) the obtained CST values in this study are very high, maybe due to higher viscosities μ_∞ in this study, thus indicating a poor dewaterability for both sludges. Therefore the potential to dewater these sludges is low and sludge treatment cost would increase dramatically. The strong differences may also result in the measurement method, because CST is strongly dependent on the filter paper used and the set-up of the device.

The mass transfer efficiency is important for the aeration requirements in a plant. However, the measurement of this parameter is quite complex, because it is dependent on the experimental setup. Therefore it is important to keep the same conditions (experimental set-up, air flow rate, stirrer speed etc.) for each measurement. In Figure 8.4, the relative mass transfer coefficient against TS concentration can be seen. The evolution of α versus TS is in agreement with literature (Rosenberger, 2003 and Germain *et al.*, 2007), where a decrease of the mass transfer coefficient with increasing TS concentration is shown thus raising the aeration requirements at higher TS concentrations. The α-correlations found in literature show an enormous variation of oxygen transfer efficiency at a given TS (see Figure 8.4), which can result in huge inaccuracy when optimising the total process. Therefore it is essential to identify the additional parameters that influence mass transfer efficiency like aeration device for a reliable optimisation. In this study the obtained α values fluctuate very strongly but considering just the steady state α values they are pretty well reproducible and can be regarded as reliable values for the optimisation.

In Table 8.5 the steady state sludge characteristics arising from the different operating conditions are given with the standard deviation. The steady state infinite viscosity for period 2 is not available due to the breakdown of the measurement device. No steady state conditions for MBR II were reached in period 1 due to problems with the sludge characteristics. In general, increasing SRT and decreasing HRT will result in a higher steady state biomass concentration. Along with the TS increase, CST as well as μ_∞ increases, whereas the oxygen transfer decreases. Just for MBR I the CST values in period 1 and 4 are quite similar, although the SRT were 15 d and 140 d, respectively. But looking at the other sludge characteristics a strong difference between these two periods is noticeable, which meets the expectations.

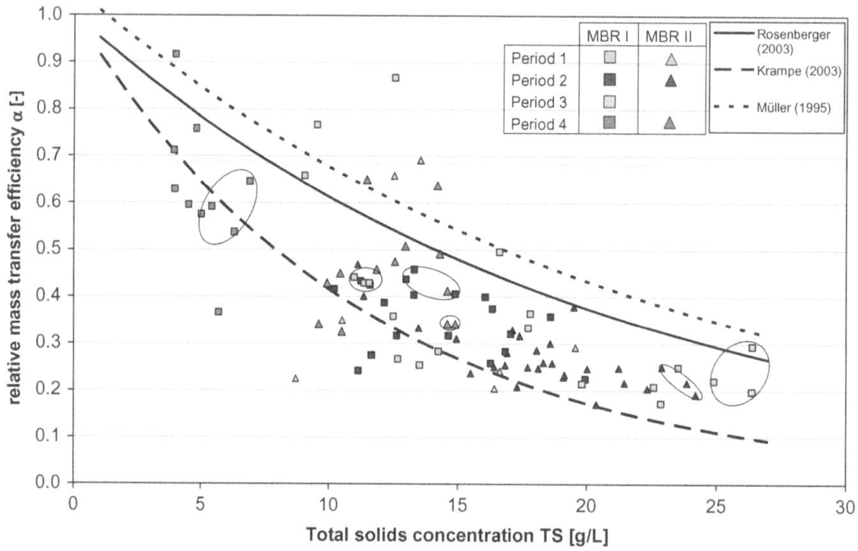

Figure 8.4 Relative mass transfer coefficient α versus TS. The highlighted values are obtained for steady state conditions

Table 8.5 Steady state sludge characteristics

	MBR I				MBR II			
	TS	μ_∞	CST	α	TS	μ_∞	CST	α
	[g/L]	[m Pas]	[s]	[-]	[g/L]	[m Pas]	[s]	[−]
Period 1	11.6 ± 0.4	3.7 ± 0.1	22 ± 1	0.43 ± 0.01				
Period 2	14.0 ± 2.2	n.a.	66 ± 10	0.44 ± 0.02	23.4 ± 0.6	n.a.	284 ± 30	0.22 ± 0.03
Period 3	25 ± 1.3	16.5 ± 0.3	600 ± 80	0.22 ± 0.03				
Period 4	6.2 ± 0.7	2.53 ± 0.3	28 ± 3	0.59 ± 0.04	14.6 ± 0.2	5.45 ± 0.8	132 ± 15	0.34 ± 0.01

Model development

The idea of this study is to set up a cost model (see Figure 8.5) for MBR users which is based on a few easily accessible input parameters like operational (HRT, SRT) and biokinetic parameters (yield, decay coefficient) and feed conditions (Schaller *et al.*, 2009).

Most of the cost models found in literature just consider the energy demand for aeration of the biology, fouling prevention, pumping etc. as the main factor influencing the operational costs. Just the model by Yoon *et al.* (2004) include the sludge handling costs using a fixed value for the sludge treatment, however, this model does not include a sub-model for fouling prevention. To set-up a model for the whole MBR process the following submodels are needed:

(i) model for sludge growth
(ii) aeration model for biological processes
(iii) aeration model for fouling prevention
(iv) sludge treatment model including excess sludge properties models and disposal models.

Sludge production model

The steady state biomass concentration in the membrane bioreactor is calculated from a mass balance for sludge and substrate in the reactor (ASM1, Xing *et al.*, 2003 and Yoon *et al.*, 2004).

$$V\frac{dX}{dt} = Y\frac{\mu_{max} \cdot S}{K_s + S} \cdot X \cdot V - b \cdot X \cdot V - Q_w \cdot X \qquad \text{Eq. 6}$$

$$V\frac{dS}{dt} = Y\frac{\mu_{max} \cdot S}{K_s + S} \cdot X \cdot V - Q \cdot S_0 - Q \cdot S \qquad \text{Eq. 7}$$

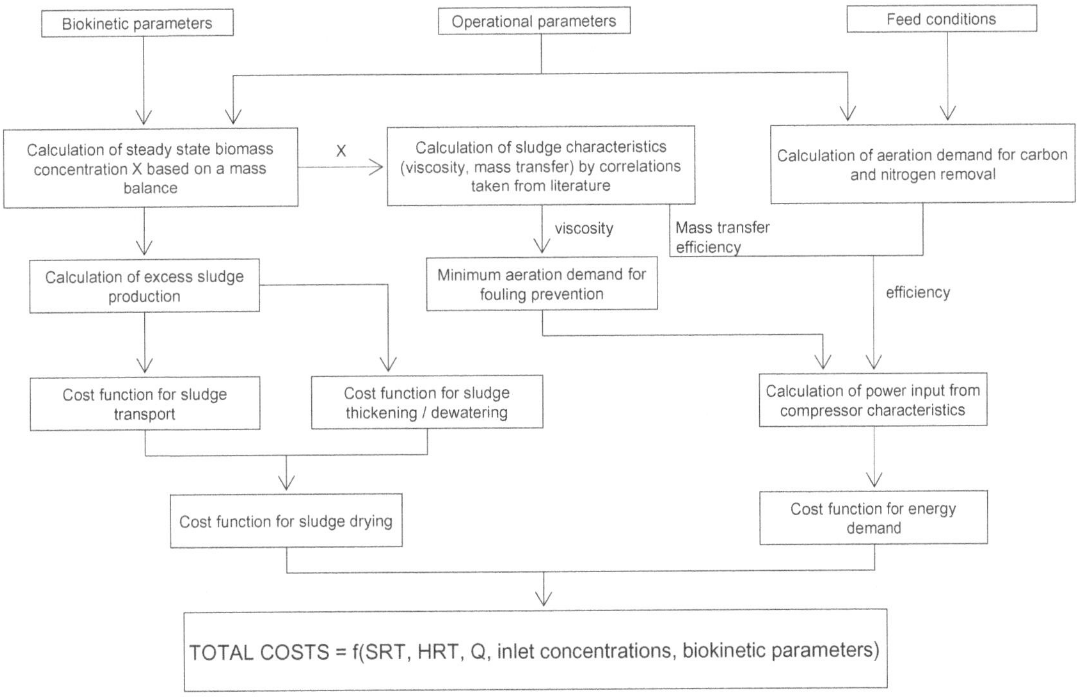

Figure 8.5 Scheme of the cost model

Assuming that the bioreactor is completely mixed, the biomass concentration at steady state conditions can be derived by rearranging Eq. (6) and (7) and solving for X.

$$X = \frac{1}{HRT} \cdot (S_0 - S) \cdot \frac{Y \cdot SRT}{1 + b \cdot SRT} \qquad \text{Eq. 8}$$

I.e., biomass concentration in the MBR is a function of operating parameters (SRT, HRT), inlet conditions (S_0) and biokinetic parameters (Y, b). Operating parameters and inlet conditions are well known, so only the biokinetic parameters have to be determined experimentally or taken from literature.

Knowing the sludge age SRT and the biomass concentration X, the excess sludge production can be calculated with Eq. (9):

$$ESP = \frac{V \cdot X}{SRT} = \frac{Y \cdot Q}{1 + b \cdot SRT} \cdot (S_0 - S) \qquad \text{Eq. 9}$$

Aeration model for the biological process and fouling prevention

The theoretical aeration demand to maintain the biological process is calculated according to ATV-DVWK-A 131 considering the oxygen requirement for carbon and nitrogen removal. To keep a constant oxygen concentration C_{set} of 2 mg/L in the reactor, the real aeration demand is then calculated considering the oxygen transfer efficiency α and the blower efficiency η.

$$Q_{air,real} = \frac{C^*}{C^* - C_{set}} \cdot Q_{air,theoret.} \cdot \frac{1}{\alpha \cdot \eta} \qquad \text{Eq. 10}$$

The oxygen transfer efficiency α is a function of MLSS concentration and can be calculated according to different correlations found in literature (Krampe and Krauth 2003; Muller *et al.*, 1995 and Rosenberger, 2003). The blower efficiency can be taken from data sheets. The power demand is then calculated using the correlation established by Krause and Cornel (2004) with $E_{pot} = 5.44\,Wh/(m^3 \cdot m)$ and the depth h of the aeration device.

$$P_{biology} = E_{pot} \cdot Q_{air,real} \cdot h \qquad \text{Eq. 11}$$

According to Cornel and Krause (2006) the aeration demand for fouling prevention varies between 0.4 and 1 kWh/m for immersed membranes. In a first approach a linear relationship between power demand for fouling prevention and the viscosity μ_∞ at a nearly infinite shear rate was assumed in this study.

$$P_{fouling} = 0.0375 \cdot \mu_\infty + 0.325 \qquad \text{Eq. 12}$$

Furthermore, the coarse bubble aeration already provides a portion (15%) of the dissolved oxygen for the biotreatment (Cornel and Krause, 2006).

Sludge handling model

The cost models for sludge thickening and transport are specific for each plant, due to different on-site sludge handling availabilities and thus they must be adapted for each case. In a first approach constant sludge treatment cost of 200–800 €/t_{DS} are assumed (Hall, 1999). This sub-model has to be refined in a next step.

Sensitivity analysis

In order to evaluate the derived model, simulation studies and sensitivity analyses were carried out. In a first step a plausibility check with the different sub-models was realised. According to Eq. 8 biomass concentration in the MBR can be calculated. Information about biokinetic parameters are less precisely known because they are strongly dependent on operating conditions and have to be determined experimentally or taken from literature.

Yield coefficients obtained in literature (Spérandio and Espinosa, 2008 and Wen *et al.*, 1999) vary between 0.38–0.13 mg VSS/mg COD for sludge ages from 5–110 days. Own measurements for a sludge age of 75 d resulted in a yield coefficient of 0.228 mg MLSS/mg COD. It is obvious that an inaccurate yield coefficient can result in an error of up to 50% in the determination of MLSS. Therefore a correlation for biokinetic parameters as a function of operating conditions needs to be implemented in the model.

According to Eq. 8 MLSS concentration changes linearly with yield coefficient and influent substrate concentration ($S_0 >> S$) and it is inversely proportional to HRT.

Further simulation studies for the different sub-models were performed with the values given in Table 8.6. The temperature T as well as C_{set} are estimated values. The influent concentration S_0 and TN are average values from the plant and the effluent concentration was calculated assuming 95% removal efficiency.

Table 8.6 Parameters of a bench-scale MBR used for the simulation

Operational parameters		Influent conditions	
Q [m^3/d]	0.096	S_0 [mg/L]	600
HRT [h]	10	S [mg/L]	30
F/M [kgCOD/d]	0.0576	TN [mg/L]	80
T [°C]	14		
Cset [mg/L]	2	**Biokinetic parameters**	
ρ_{air} [kg/m^3]	1.225	Yield	0.228
		b [1/d]	0.009

In Figure 8.6 the real aeration demand to maintain the biological process were determined using different α correlations taken from literature (Krampe and Krauth, 2003; Muller *et al.*, 1995 and Rosenberger, 2003) and assuming a blower efficiency of 50%. With increasing SRT the difference between the aeration demands increases up to 60%. It is obvious that the big differences in the calculated aeration demand especially for high sludge ages results in a significant difference of aeration costs. Therefore oxygen transfer for higher MLSS concentrations should be investigated more intensively to get an improved knowledge of the parameters influencing oxygen transfer.

Cost considerations

The costs of an MBR process vary strongly depending on local conditions due to differences in the electricity tariff and sludge treatment depending on the method used, on local conditions, like plant size, plant equipment, local regulations, transportation costs and sludge characteristics.

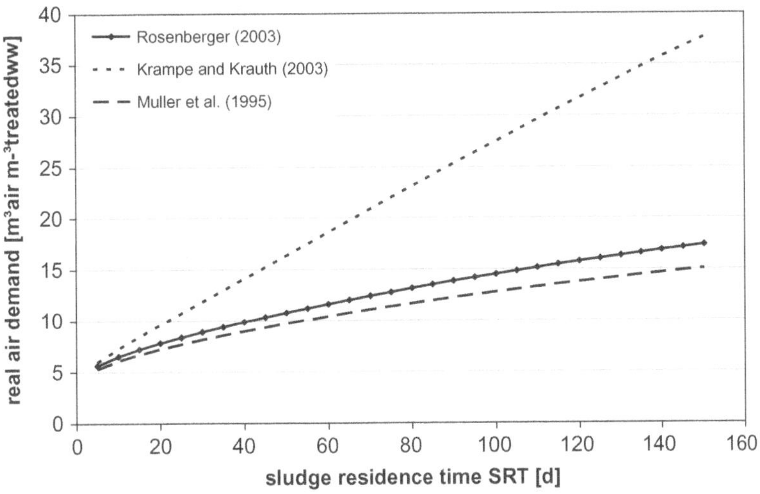

Figure 8.6 Calculated aeration demand to maintain the biological process with different α correlation assuming a blower efficiency of 50%

In Figure 8.7 the total aeration costs for the biology and fouling prevention as well as the sludge treatment costs per m^3 treated wastewater versus sludge age are shown, assuming different electricity tariffs and sludge treatment cost, respectively. The total aeration costs were calculated with the parameters given in Table 8.1, the α correlation found by Rosenberger (2003), a blower efficiency of 50% and the assumption that the coarse bubble aeration for fouling prevention provides 15% of the aeration demand to maintain the biological process. Total costs in the range of 10 c/m^3 have been reported by Verrecht *et al.* (2009) which shows that the model yields realistic values.

Increasing the sludge treatment cost by 40% from 500 €/t_{DS} to 700 €/t_{DS} at constant electricity costs of 10 €cents kW/h will result in an increase of the total operation cost of 10–25% depending on sludge age. The strongest increase was observed for decreasing sludge ages. An increase in the electricity tariff by 25% at constant sludge treatment cost results in an increase of 10–18% of total operating costs. The increase was more pronounced for higher sludge ages. At a given combination of sludge treatment costs and energy costs a minimum of total costs can be observed at different sludge ages depending on local costs. In the 3 combinations shown in Figure 8.7, the optimal SRT varies between 50 and 100 days. Therefore the model can be used to optimise the sludge age for a given MBR plant.

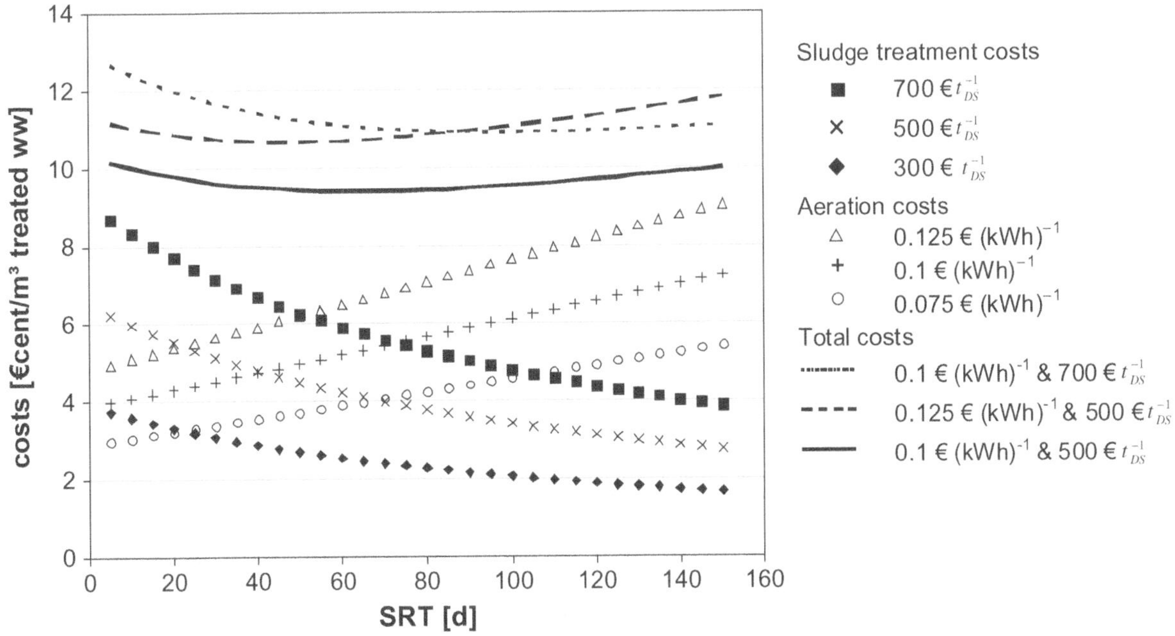

Figure 8.7 Calculated costs per m^3 treated wastewater for aeration and sludge treatment versus sludge age assuming different electricity tariffs and sludge treatment cost, respectively

Conclusion

With increasing biomass concentration, apparent viscosity increases, while the viscosity for a given MLSS concentration is slightly different in both plants. The available literature correlations were fitted to a small TS range from 3–17 g/L. Especially for higher TS concentrations between 15 and 20 g/L the obtained apparent viscosities in this study are up to two times higher in comparison to literature correlations. For the first time a correlation for the viscosity as a function of TS using a large TS range from 4 g/L to 27 g/L was established. In agreement with literature also an exponential relationship is found for the investigated MBRs.

A comparative study of two different MBRs operated under various conditions (SRT: 15–140 d, HRT: 5–10 h) fed with a complex synthetic wastewater to eliminate inlet fluctuations showed that elevated SRT operation is feasible and therefore excess sludge reduction is in a considerable degree attainable. For the first time sludge characteristics like viscosity, dewaterability and oxygen transfer were investigated for a very large TS range and correlations including the whole range could be established. It was shown that extrapolating results obtained for a lower TS range to higher TS values can result in an underestimation of up to 50% of the investigated parameter, e.g. viscosity value. The process demonstrates the same removal efficiency as MBR processes with lower SRT values.

A cost model for immersed membrane bioreactors treating municipal wastewater incorporating energy demand for aeration and fouling prevention as well as the related sludge handling costs subject to local conditions is presented. The model presented is suitable for MBR users and is based on a few easily accessible input parameters like operational (HRT, SRT) and biokinetic parameters (yield, decay coefficient) and feed conditions. Only the information for biokinetic parameters and oxygen transfer efficiency vary strongly in literature, therefore the correct choice of these parameters is essential for an applicable model, otherwise the impact of aeration demand is over- or underestimated.

In first simulation studies the model was found to be appropriate to predict the total costs of an immersed MBR, nevertheless some improvements need to be made to be more universally valid. Especially the information of biokinetic parameters as a function of sludge age is needed to have a reliable prediction of the biomass concentration. Furthermore the sludge treatment costs have to be modelled more in detail splitting them into costs for transport, sludge thickening and dewatering and maybe also other sludge treatment routes like anaerobic digestion.

Nomenclature

b	decay coefficient [1/d]
C^*	oxygen saturation concentration [mg/L]
C_{set}	Oxygen conc. in aerated tank [mg/L]
E_{pot}	Potential engergy [Wh]
ESP	Excess sludge production [g h-1]
h	Depth of aeration device [m]
HRT	Hydraulic retention time [h]
K_s	Half saturation constant [mg/L]
P	Power demand [kW]
Q	Influent flow rate [m^3/h]
Q_w	sludge flow rate [m^3/h]
Q_{air}	Air flow rate [m^3/h]
S	COD concentration influent [mg/L]
S_0	COD concentration effluent [mg/L]
SRT	Sludge retention time [d]
t	Time [d]
V	Volume [m^3]
X	Biomass concentration [mg/L]
Y	Yield coefficient [mg MLSS/mg COD]
α	Dimensionless oxygen transfer coefficient [-]
η	Blower efficiency [-]
μ_{max}	Maximum specific growth rate [1/d]
μ_∞	Shear independent viscosity at high shear rates [mPas]

REFERENCES

ATV-DVWK. (2000). Arbeitsblatt ATV – DVWK – A 131. GFA.

Bracklow U., Drews A., Vocks M. and Kraume M. (2007). Comparison of nutrients degradation in small scale membrane bioreactors fed with synthetic/domestic wastewater. *Journal of Hazardous Materials*, **144** (3), 620–626.

Cornel P. and Krause S. (2006). Membrane bioreactors in industrial wastewater treatment – European experiences, examples and trends. *Wat. Sci. Technol.*, **53**(3), 37–44.

Germain E., Nelles F., Drews A., Pearce P., Kraume M., Reid E., Judd S. J. and Stephenson T. (2007). Biomass effects on oxygen transfer in membrane bioreactors. *Wat. Res.*, **41**(5), 1038–1044.

Hall J. (1999). Ecological and economical balance for different sludge management options. In *Proc. of the workshop on 'Problems around sludge'*, Stresa (Italy) 18–19 Nov.

Khongnakorn W., Wisniewski C., Pottier L. and Vachoud L. (2007). Physical properties of activated sludge in a submerged membrane bioreactor and relation with membrane fouling. *Sep. and Purif. Technol.*, **55**, 125–131.

Krampe J. and Krauth K. (2003). Oxygen transfer into activated sludge with high MLSS concentrations. *Water Sci. Technol.*, **47**(11), 297–303.

Laera G., Giordano C., Pollice A., Saturno D. and Mininni G. (2007). Membrane bioreactor sludge rheology at different solid retention times. *Wat. Res.*, **41**(18), 4197–4203.

Muller E. B., Stouthamer A. H., Vanverseveld H. W. and Eikelboom D. H. (1995). Aerobic domestic waste-water treatment in a pilotplant with complete sludge retention by cross-flow filtration. *Wat Res.*, **29**(4), 1179–1189.

Pollice A., Giordano C., Laera G., Saturno D. and Mininni G. (2007). Physical characteristics of the sludge in a complete retention membrane bioreactor. *Wat. Res.*, **41**(8), 1832–1840.

Rosenberger S. (2003). Charakterisierung von belebtem Schlamm in Membranbelebungsreaktoren zur Abwasserreinigung. PhD thesis, Department of Chemical Engineering, Technische Universität Berlin.

Schaller J., Camacho M., Drews A. and Kraume M. (2008). Operation of Different Membrane Bioreactors at High Sludge Ages for Minimising Excess Sludge Production. In *Book of Abstracts of SIDISA08*, Florence 24–27 June.

Schaller J., Drews A. and Kraume M. (2009). Development of a Cost Model for Membrane Bioreactors Including Sludge Handling Costs. In: *Proc. of 5th IWA Specialised Membrane Technology Conference for Water and Wastewater Treatment*, Beijing 1st–3rd Sept. (submitted to *Desalination and Water Treatment*, January 2010).

Smolders G. J. F., van der Meij J., van Loosdrecht M. C. M and Heijnen J. J. (1994). Model of the anaerobic metabolism of the biological phosphorus removal process: stoichiometry and pH Influence. *Biotechnol. and Bioeng.*, **43**(6), 461–470.

Spérandio M. and Espinosa M. C. (2008). Modelling an aerobic submerged membrane bioreactor with ASM models on a large range of sludge retention time. *Desalination*, **231**(1–3), 82–90.

Tixier N., Guibaud G. and Baudu M. (2003). Determination of some rheological parameters for characterization of activated sludge. *Bioresource Technol.*, **90**(2), 215–220.

Verrecht B., Gugliemi G., Mulder J. W., Brepols C. and Judd S. (2009). An aeration energy model for an immersed membrane bioreactor. In *Proc. of Final MBR-Network Workshop*, Berlin 31 March–1st April.

Wen X., Xing C. and Qian Y. (1999). A kinetic model for the prediction of sludge formation in a membrane bioreactor. *Process Biochemistry*, **35**(3–4), 249–254.

Wu Z., Wang Z., Zhou Z., Yu G. and Gu G. (2007). Sludge rheological and physiological characteristics in a pilot-scale submerged membrane bioreactor. *Desalination*, **212**(1–3), 152–164.

Xing C.-H., Wu W.-Z., Qian T. and Tardieu E. (2003). Excess sludge production in membrane bioreactors: a theoretical investigation. *J. of Environ. Eng.*, **129**(4), 291–297.

Yoon S. H., Kim H. S. and Yeom I. T. (2004). The optimum operational condition of membrane bioreactor (MBR): cost estimation of aeration and sludge treatment. *Water Res.*, **38**(1), 37–46.

Chapter 9

Destruction of organic compounds in sewage sludge suspensions by ultrasound and catalytic wet air oxidation

Michèle Besson, Claude Descorme, François Di Gregorio, Anne-Marie Wilhelm, Henri Delmas and Rana Kidak

SUMMARY

Project objectives

The objectives of this work were to efficiently degrade organic compounds in sewage sludge suspensions by catalytic wet air oxidation (CWAO) using heterogeneous catalysts, after a pre-treatment by power ultrasound and (or) thermal oxidation in order to break down macromolecules and solubilize the organic solids.

Work performed

Municipal and industrial sewage sludges containing ca. 3 g/L TOC were subjected to different treatment sequences:

- Ultrasonication (20 kHz, power 0–210 W).
- Ultrasonication + batch Wet Air Oxidation (WAO) or Catalytic Wet Air Oxidation (CWAO) at 190–210°C, 37–51 bar air, 1–8 h.
- Batch WAO (210°C, 51 bar air, 4 h) + batch CWAO of the supernatant (190°C, 37 bar air).
- Batch WAO + continuous CWAO of the supernatant in a trickle-bed reactor.

Different supported metallic catalysts were prepared, fully characterized and used in the CWAO. Platinum and ruthenium were chosen as the active phase. Titanium dioxide, zirconium dioxide and ZrO_2-CeO_2 mixed oxides were selected supports because they were shown to be stable under the oxidation process operating conditions.

The final suspensions were analysed in terms of residual TOC concentration in the solid, TOC concentration in the liquid phase and overall TOC abatement.

End results

Ultrasonic pre-treatments, using a 20 kHz system led to an increase in the solubilized COD and the solubilized TOC, a very severe decrease of the average diameter of the sludge flocs, a disruption of the bacterial cells with release of proteins into the liquid phase, but no clear reduction of the overall organic content. The effect of sonication strongly

 Reduction, Modification and Valorisation of Sludge (REMOVALS). Edited by Christophe Bengoa, Azael Fabregat, Josep Font and Frank Stüeber. ISBN: 9781843393450. Published by IWA Publishing, London, UK.

depends on the nature of the sludge. The municipal sludge was more sensitive to ultrasound than the industrial one, since it is clearly heterogeneous (fibers) and flocs have higher average size.

WAO experiments carried out with or without a sonication pre-treatment also revealed difference between the municipal and industrial sludge. In the case of the municipal sludge only, the transfer of the solid organic compounds slightly increased with the ultrasonic power applied. As expected, WAO largely increased the solubilization of the solid materials. The effect of the pre-treatment on the overall TOC abatement was limited (32–37% for the municipal sludge, 43–47% for the industrial sludge).

A comparison of the results of the CWAO of the sludge in the presence of different noble metal supported catalysts, without or with pre-sonication, revealed that the effect of ultrasounds was a disintegration of the flocs and the formation of finer solid particles with even higher ability to adsorb on the catalyst surface. The beneficial effect of the catalyst on the mineralization of the organic compounds in the liquid phase was then annealed. Finally, the overall conversion of TOC was in the same range with or without pre-treatment.

The final strategy we tested was a combination of a thermal oxidation of the sludge, to ensure sufficient solubilization of the organic material followed by a CWAO of the supernatant. Upon batch experiments, this sequence led a final TOC abatement of the industrial sludge of 80–90% when a Ru catalyst was used. The remaining solid phase contained only ca.3 % TOC after the WAO treatment at 210ºC and 70 bar total pressure.

CWAO of aqueous supernatants in a continuous trickle-bed reactor confirmed the high activity and stability of the catalyst for TOC abatement. The final effluent contained only acetic acid. In parallel, nitrogen containing compounds were selectively transformed to dinitrogen and nitrates.

METHODOLOGIES AND APPROACHES EMPLOYED

Methodologies

Wet air oxidation (WAO), based on the oxidation of pollutants in the liquid phase at high temperature and pressure, in the presence of an oxidizing agent (oxygen), is currently gaining interest, in particular for sewage sludges. The main purpose is the destruction of the particulate organic materials and their conversion into biodegradable compounds and inorganic carbon dioxide. Catalysts are often used in Wet Air Oxidation to improve the process, by enhancing the reaction rate and allowing the use of milder operating conditions. Soluble transition metal salts (such as copper and iron salts) have been found efficient. However they require a post-treatment to be separated from the effluent and recycled. With this respect, heterogeneous catalysts are preferred. Catalysts based on noble metals deposited on stable supports under WAO operating conditions are active and stable catalysts even in the treatment of complex industrial effluents. However, the deposition of solid organic matters present in the sludge at the catalyst surface may rapidly deactivate the catalyst.

To overcome this problem, ultrasonication or (and) thermal oxidation were considered as pre-treatment methods to improve the solubilization of the solid fraction before the catalytic wet air oxidation of the liquid in the presence of metallic solid catalysts. High-power ultrasounds at low frequencies generate intensive cavitations and produce hydromechanical shear forces. As a result an effective disintegration of the insoluble particulate matter and cells is observed, leading soluble molecules and a substantial release of the cell intra-cellular material into the liquid. Studies were conducted with Pt and Ru metal catalysts, since they are resistant towards leaching. TiO_2, ZrO_2 and mixed CeO_2-ZrO_2 oxides were chosen as supports because they remain stable whatever the reaction conditions encountered in CWAO.

Approaches

Approaches followed during the project are presented on Figure 9.1.

At the beginning of the project, batches of industrial and municipal sludges were collected. Since sewage sludges are fermentable and to be sure to perform the experiments on the same initial effluent, the particulate materials were separated from the suspension by centrifugation. The received 'pasty sludge' was then transferred in pillboxes and stored in a freezer. For each experiment, a given amount of sludge was suspended in water to recreate a suspension containing an initial TOC value of ca. 3 g/L.

The ultrasonic apparatus was an ultrasonic homogenizer Autotune (Vibracell) with a horn diameter of 13 mm. This device operated at 20 kHz and the maximum applied power was 200 W. Batch experiments were performed using 300 mL suspension at 25ºC. The ultrasonic power input was determined by calorimetric measurements.

Batch WAO and CWAO studies were performed in a 300 mL Hastelloy autoclave equipped with a magnetically driven stirrer and a heating jacket. After a given reaction time, the whole suspension was recovered and filtered. The solid and liquid phases were subsequently analysed separately.

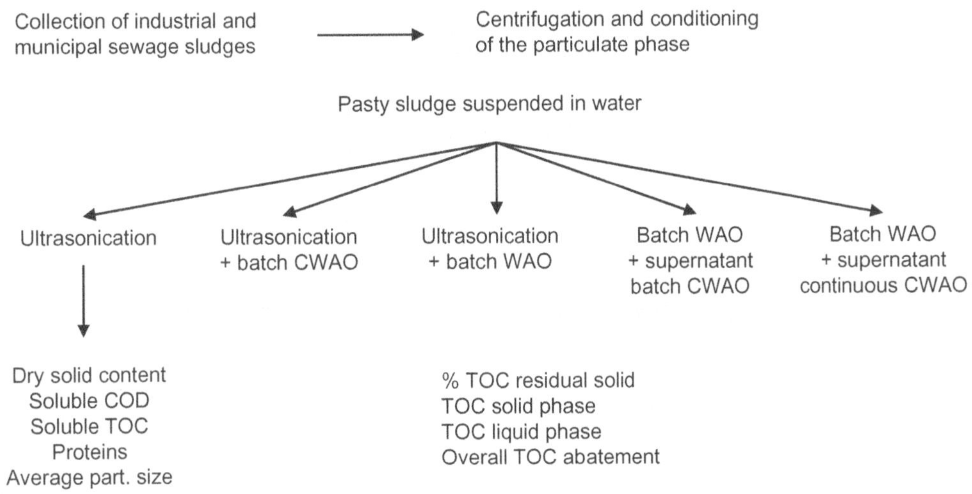

Figure 9.1 Approaches employed

TOC (Total Organic Carbon) analyses on the solid and aqueous phases were performed using a Shimadzu TOC analyser. The liquid phase was also analysed for solubilized COD, using colorimetric tubes and protein content. A few organic acids in the oxidized liquors were analysed by HPLC, in particular acetic acid. The average particle size of the sewage sludge suspension after sonication was measured using a Malvern Laser Diffraction particle Size Analyzer, which measures materials from 0.02 microns to 2000 microns. The protein content was measured using the so-called modified Lowry method.

ACHIEVEMENTS RELATED TO THE STATE OF THE ART

Preliminar

Upon sewage sludge Catalytic Wet Air Oxidation, the presence of particulates has been identified as an inhibiting parameter (Moulut-Godard). Therefore the pre-treatment of sewage sludge via mechanical or thermal disintegration shall improve the subsequent CWAO. However, no information was available on the impact of different sludge disintegration degrees on further catalytic processes.

Three different sludges have been used for this study. Two were industrial sludges (STVULBAS and SOTREFI), while the third one was from a municipal wastewater treatment plant (TOULOUSE).

Influence of power ultrasound as pre-treatment process

The ultrasonic pre-treatment of TOULOUSE and STVULBAS sludges was studied as a function of the ultrasonic power (50 W, 100 W, 200 W) and exposure time (Figure 9.2). The parameters analysed were the dry solid content, the solubilized COD, the solubilized and non-solubilized TOC, the dp50 and the total protein content.

Table 9.1 Analysis of the sludges used in this study (na: not analyzed)

Sludge	STVULBAS	SOTREFI	TOULOUSE
C (%)	35.4	32.2	41.1
H (%)	5.2	4.6	6.15
N (%)	5.6	4.9	4.2
S (%)	1	0.9	0.6
Fe (g/L)	7.5	4.4	0.6
Ca (mg/l)	3.5	15.6	2.5
P (%)	2.9	3.9	2.0
Al (%)	0.2	3.8	2.9
Na (g/l)	0.6	0.2	0.1
Si (%)	1.0	na	2.9
Cu (%)	<0.1	0.1	na
K (%)	0.2	0.9	0.3
Mg (%)	0.1	0.9	0.3
Ti (%)	0.7	0.1	0.1

Figure 9.2 Apparatus used for lab-scale ultrasonic disintegration

The nature of sludge has an important effect, because of their physical differences. The municipal sludge was constituted of sediments from the primary settling tank together with the waste activated sludge. It contained a large quantity of fiber particles and the range of particle size is wide with a bimodal distribution at 79 and 1,096 microns. The STVULBAS sludge had a more homogeneous structure with mean particle size of 245 microns. After sonication, the particle size gradually changed and decreased to 70 microns and 4.5 microns for the two sludges.

To compare results with those in the literature, the DDCOD (degree of disintegration based on COD) was calculated according to the formula:

$$DD_{COD} = \frac{COD_t - COD_0}{COD_{NaOH} - COD_0} \cdot 100\% \qquad \text{Eq. 1}$$

where: DD_{COD}: degree of disintegration based on COD, COD_t: COD soluble at time t (g O_2/L), COD_0: COD soluble at time zero (g O_2/L), COD_{NaOH}: COD soluble in NaOH solution: total estimated disintegration (g O_2/L)

The COD_{NaOH} is accepted as the maximum possible soluble COD concentration (Neis *et al.*, 2001). The evolution of the DD_{COD} as a function of time for different powers is given in Figure 9.3 for TOULOUSE sludge. In 60 min, 56%, 64% and 80% of DD_{COD} have been reached, applying 50, 100 and 200 W, respectively.

An energy parameter has been defined which is obtained by dividing the energy given to the heterogeneous system by the mass of total dry solids present (Bougrier *et al.*, 2005).

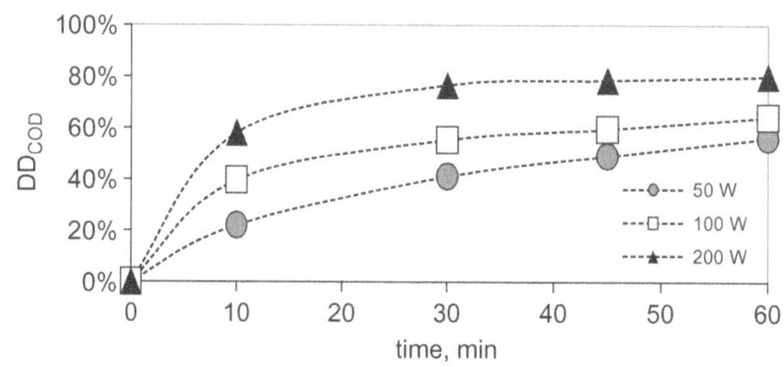

Figure 9.3 Evolution of DD_{COD} (degree of disintegration on the basis of COD) in liquid phase of municipal sludge TOULOUSE liquid phase. The concentration was 8 g/L at 20°C

$$E_s(\text{kJ/kg}) = \frac{\text{Power(W)} \cdot \text{time(sec)}}{1000 \cdot \text{mass}_{\text{dry solids}}(\text{kg})} \qquad \text{Eq. 2}$$

The power, the time of reaction and the mass of dry solid were varied, keeping the two other parameters constant. Figure 9.4 shows the degree of disintegration as a function of Es applied.

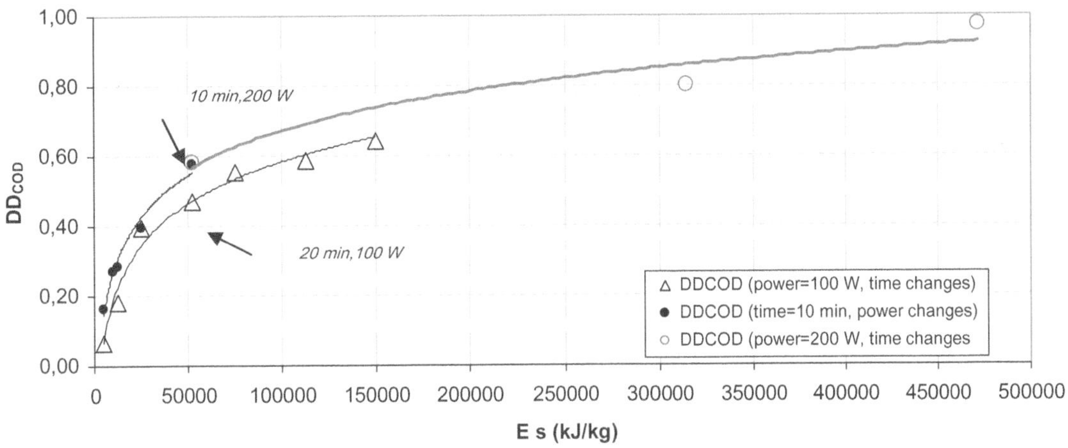

Figure 9.4 Evolution of DD_{COD} in liquid phase of municipal sludge TOULOUSE solution, with respect to specific energy

The main conclusions can be summarized as 'high power-short reaction time' is more effective than 'low power-long retention time', while keeping the volume small and the concentration below a certain level.

The changes in TOC repartition between the solid and liquid phases were analysed after 90 min sonication. Figure 9.5 shows these results. It was shown that ultrasound did not induce any significant mineralization. Soluble matter concentration increased slightly, whereas particulate matter concentration decreased.

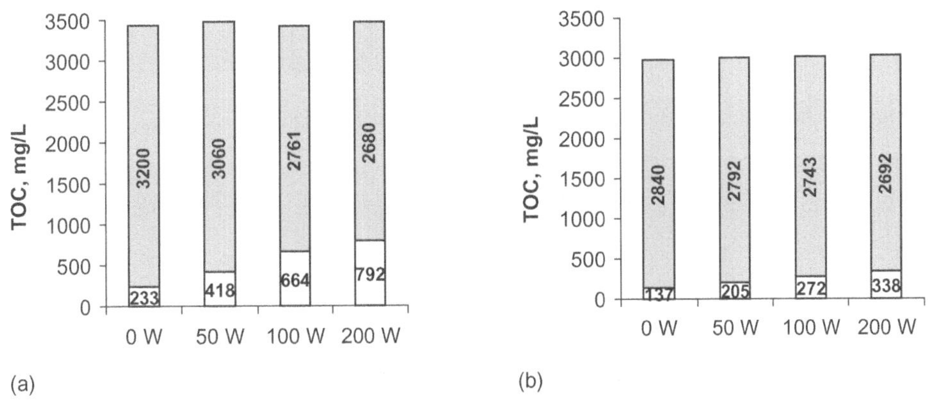

Figure 9.5 Change of TOC in (a) TOULOUSE municipal sludge (TOC_0 = 3433 mg/L) and (b) ST VULBAS industrial sludge (TOC_0 = 2977 mg/L). (■: TOC in the solid phase, □: TOC in the liquid phase) versus ultrasonic power. The process time was 90 min

The increase of applied power from 50 W to 200 W had a positive effect on the solubilization of the organic particles. At 200 W, while 19% of the solid in the municipal sludge was solubilized, only 7% of the solid was solubilized in the industrial sludge. The evolution of COD in liquid phase followed the same tendency.

Synthesis and characterization of catalysts

Ultrasonication and subsequent Catalytic Wet Air Oxidation

Different Pt and Ru catalysts containing 3 wt.% active phase were prepared on commercial oxide supports (TiO_2, ZrO_2 and $Ce_xZr_{1-x}O_2$) via a classical incipient-wetness and liquid-phase impregnation methods using aqueous solutions of metallic salts. Catalysts were reduced before use. They were characterized for metallic dispersion. Most catalysts were very well dispersed with average metal particle sizes less than 4 nm (Figure 9.6), as observed by X Ray Diffraction and Transmission Electron Microscopy (TEM).

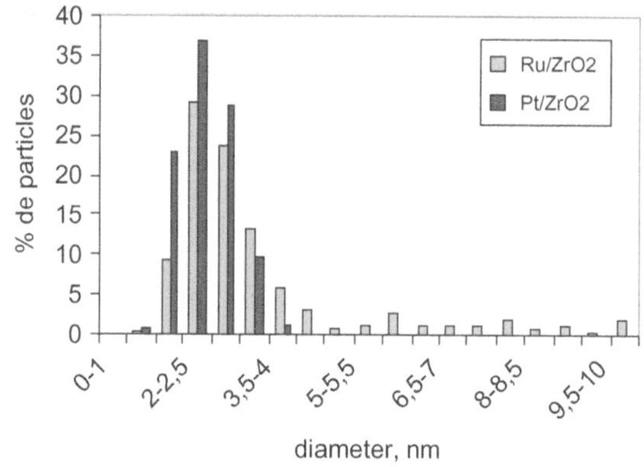

Figure 9.6 Particle size distributions obtained from TEM observations of Pt and Ru catalysts

Sewage sludges (initial TOC content ca. 3000 mg/L) were treated in a batch reactor (Figure 9.7) without or with pre-sonication at different power in the presence or in the absence of a catalyst. Based on previous works, the operating conditions were chosen as 190°C and 37 bar air pressure with different reaction times.

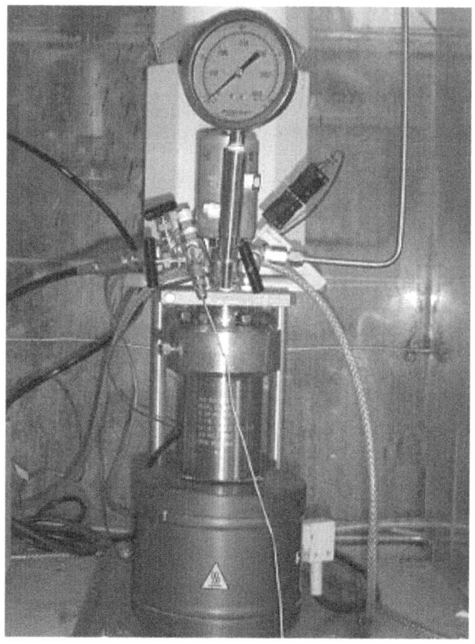

Figure 9.7 Lab-scale batch reactor

Table 9.2 compares the TOC concentrations in the liquid phase, the solid phase and 'overall' suspension and the overall TOC conversion, for some experiments performed on the untreated STVULBAS sludge. Figure 9.8 illustrates these results.

Heating the suspension to the reaction temperature under argon did not change the overall TOC content in the suspension (3,000 mg/L). The result was a transfer of the organic matter from the solid phase to the liquid phase (from 2,754 mg/L to 1,773 mg/L). Then, after the introduction of air, the TOC content in the solid phase decreased rapidly, the TOC content in the liquid phase continued to increase during the first hour, and significant mineralization occurred (25–32%).

Further mineralization occurred (40–48% after 4 h) and a plateau at 47–55% was reached after 8 h. Thus, it appears that the hydrothermal oxidation degrades the solid organic matter in two steps: the hydrolysis of the complex polymeric substances and the bacterial cells, and the subsequent oxidation (Shanableh, 2000).

Table 9.2 TOC concentration in the liquid phase, the solid phase and the 'overall' suspension as a function of reaction time upon CWAO of STVULBAS sludge. (TOC_0: 3000 mg/L, T: 190°C, 37 bar air pressure, catalyst concentration: 6.67 g/L)

Time (h)	TOC concentration (mg/L)															
	Without catalyst				3% Ru/ZrO_2				3% Pt/ZrO_2				3% Pt/CeO_2-ZrO_2			
	liq	sol	susp	conv	liq	sol	susp	conv	liq	sol	susp	conv	liq	sol	susp	conv
<0[a]	246	2754	3000	–	246	2754	3000	–	246	2754	3000	–	246	2754	3000	–
0[b]	1199	1773	2972	0	1199	1773	2972	0	1199	1773	2972	0	1199	1773	2972	0
0.5	–	–	–	–	–	–	–	–	1553	762	2315	23	–	–	–	–
1	1532	682	2214	25	1527	640	2167	27	1451	578	2029	32	1407	671	2078	31
4	1417	269	1686	43	1492	295	1787	40	1290	281	1571	48	1364	397	1761	41
8	1396	179	1575	47	1388	189	1577	47	1200	138	1338	55	1292	239	1531	49

[a]*The TOC concentration in the liquid phase and in the suspended matter contents were determined in the initial suspension of 10 g of sludge into 150 ml of distilled water.*

[b]*TOC concentrations dxetermined after heating the suspension to the reaction temperature (190°C) under Ar under stirring. This point was taken as 'zero time'.*

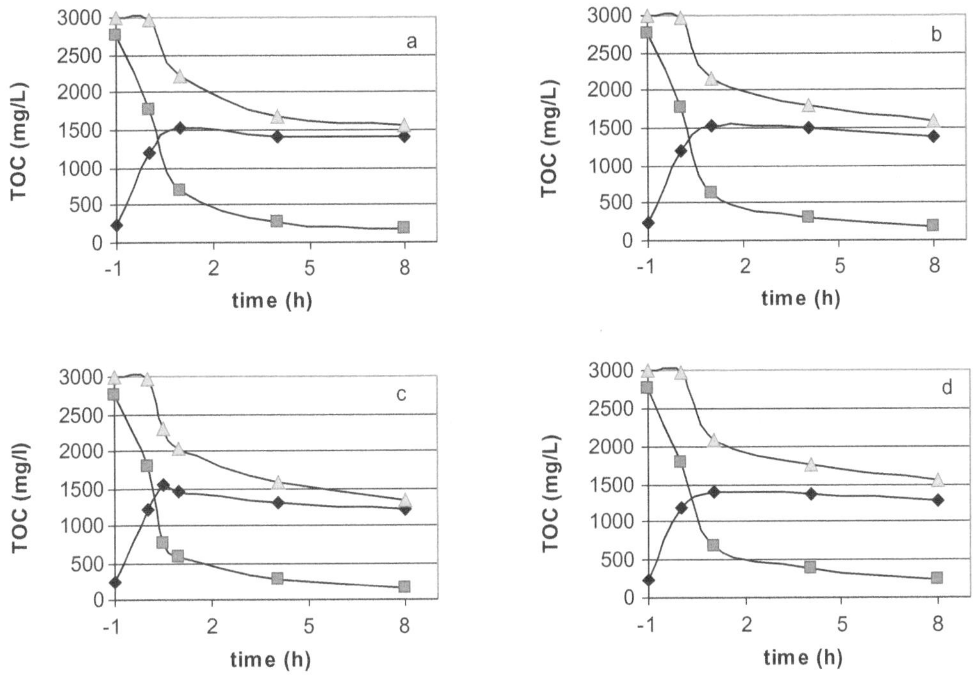

Figure 9.8 Evolution of TOC concentration in the liquid phase (◆), the solid phase(■), and the 'overall' suspension (▲) upon WAO or CWAO of non pre-treated STVULBAS sludge in the absence of catalyst (a), or in the presence of 3%Ru/ZrO_2 (b), 3%Pt/ZrO_2 (c), 3%Pt/CeO_2-ZrO_2 (d)

Finally, the addition of any solid catalyst prepared did not significantly improve the final TOC abatement. Moreover, the Pt catalysts supported on the CeO_2-ZrO_2 supports gave even higher amounts of residual TOC in the solid phase than in the blank experiment (397 and 387 mg/L for two supports with different Ce and Zr ratios compared to 281 mg/L). These mixed supports with high redox properties and high capacities for oxygen storage and mobility appeared to be attractive to promote the oxygen activation step upon oxidation. Indeed, $Ce_xZr_{1-x}O_2$ used as support for noble metal have shown higher activity in the oxidation of acetic acid (Mikulova *et al.*, 2007a, 2007b) or 2-chlorophenol (Li *et al.*, 2007a, 2007b) in the literature. This negative effect in our case on the transfer from the solid to the liquid phase suggested a strong adsorption of the organics on the solid surface, which, indeed, was higher in the case of $Ce_xZr_{1-x}O_2$ (150–170 m^2/g compared to 90–100 m^2/g). The retention of the organic material on the catalyst surface may also explain the absence of any significant increase in the overall TOC conversion.

The effect of sonication on the subsequent CWAO treatment was also examined. Three different sludges were sonicated at different power for one hour and then oxidized at 190°C, 37 bar air during 4 h. Selected results are shown in Table 9.3 for the STVULBAS sludge and Table 9.4 for the SOTREFI sludge.

Table 9.3 Effect of ultrasonication (200 W, 1 h) on the STVULBAS sludge subsequent CWAO (TOC_0 = 3000 mg/L, 190°C, 37 bar air, 6.67 g/L catalyst, 4 h)

Sonication power (W)	Catalyst	TOC liquid (mg/L)	TOC solid (mg/L)	TOC suspension (mg/L)	Overall TOC conversion (%)
No	No	1419	277	1696	43
No	3% Pt/ZrO_2	1290	281	1571	48
200	No	1331	254	1585	47
200	3% Pt/ZrO_2	1255	391	1646	45

After ultrasonication of the STVULBAS sludge and in the absence of a catalyst, the solid TOC concentration remained about constant and the liquid phase mineralization was only slightly decreased after sonication. As a consequence, no clear effect of ultrasonication on the overall TOC conversion was observed after WAO. On the other hand, surprisingly, no increase in the TOC conversion was achieved after CWAO of the pre-sonicated sludge, compared to the CWAO treatment alone. Instead, the use of a catalyst increased the TOC content in the solid. This result suggested again that heterogeneous catalysts behave as adsorbents for the sludge, as it was observed previously with the high surface area CeO_2-ZrO_2-supported catalysts.

This effect was also observed upon CWAO of the SOTREFI sludge, performed in the same conditions. The use of a catalyst did not lead any significant increase in the overall TOC abatement. However, it seems that the ability of the catalyst to adsorb the organic matter is more important. Indeed, the concentration of TOC in the solid was 66–98 mg/L after WAO, while this value reached 241–270 mg/L after CWAO in the presence of platinum catalysts.

This was further observed in the experiments performed after mechanical disintegration of the SOTREFI using an Ultra-turrax apparatus. The results shown in Table 9.4 as a function of the rotation speed, clearly show that the mechanical pre-treatment did not significantly modify the TOC concentrations in the liquid and solid phases after WAO. Whatever the rotation speed, no improvement in the overall conversion was noted. However, again, in the presence of any heterogeneous catalyst, the TOC content in the solid was significantly higher.

Table 9.4 Effect of ultrasonication (200 W, 1 h) or Ultra-turrax desintegration on the SOTREFI sludge subsequent CWAO (TOC_0 = 2490 mg/L, 190°C, 37 bar air, 6.67 g/L catalyst)

	Catalyst	TOC liquid (mg/L)	TOC solid (mg/L)	TOC suspension (mg/L)	Overall TOC conversion (%)
No pre-treatment					
	No catalyst	1393	66	1399	44
	3% Pt/ZrO_2	1225	241	1467	41
	3% Ru/ZrO_2	1133	202	1335	46
Ultrasonication pre-treatment at different power					
200 W	No catalyst	1125	98	1223	51
200 W	3% Pt/ZrO_2	1018	270	1288	48
Ultra-turrax pre-treatment at different rotation speed					
3600 rpm	No catalyst	1445	70	1515	39
3600 rpm	3% Pt/ZrO_2	1180	180	1360	45
3600 rpm	3% Ru/ZrO_2	1243	166	1409	43
11200 rpm	No catalyst	1330	70	1400	44
11200 rpm	3% Pt/ZrO_2	1207	180	1387	44
11200 rpm	3% Ru/ZrO_2	1138	185	1323	47
20000 rpm	No catalyst	1307	81	1388	44
20000 rpm	3% Pt/ZrO_2	1242	170	1412	43
20000 rpm	3% Ru/ZrO_2	1184	208	1392	44

Both after ultrasonication or Ultra-turrax disintegration and even though an efficient disintegration of the sludge was observed, the organic solids were only partly solubilized and the subsequent WAO or CWAO treatments did not give higher TOC abatements. Particulate organic compounds were not sufficiently converted into soluble compounds to avoid the deposition of finely divided solids onto the heterogeneous catalyst surface and CWAO was not improved. The benefits of the catalyst are significantly annealed by the adsorption of solids.

Ultrasonication and subsequent Wet Air Oxidation (WAO)

The effect of the ultrasonication power on the subsequent WAO process was studied in more details, using the STVULBAS and TOULOUSE sludges.

Different behaviours were evidenced for the two sludges. After 90 min, sonication at different powers did not induce mineralization. However, the carbon partition between the liquid and solid phases varied with the specific supplied energy. Solubilization occurred and reached 7% maximum for the STVULBAS sludge and 19% maximum for the TOULOUSE sludge. These differences were attributed to the heterogeneous nature of the municipal sludge (fibers) and the higher average size of the flocs.

WAO of the non-sonicated and sonicated sludges also revealed differences between the two sludges (Figure 9.9). While the solid TOC was reduced in the municipal sludge as the sonication power increased, sonication had no effect in the case of the industrial sludge.

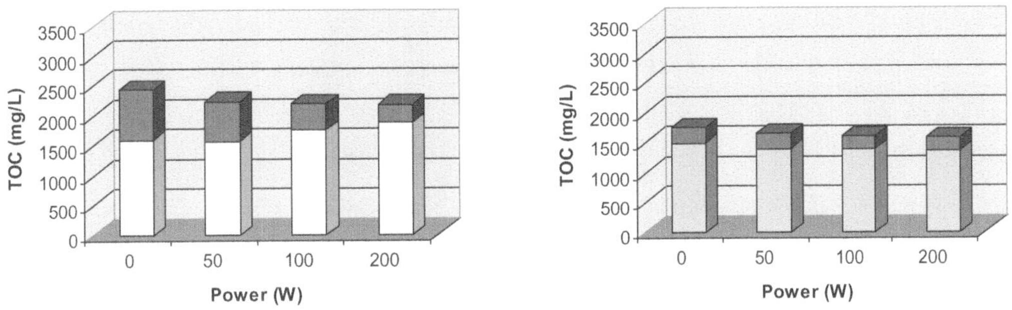

Figure 9.9 Effect of pre-sonication on the subsequent WAO treatment (190°C, 37 bar air, 4 h) of the municipal TOULOUSE sludge (left) and the industrial STVULBAS sludge (right)

Influence of a thermal WAO pre-treatment on the subsequent CWAO of the supernatant, in a batch reactor

To try to solve the inherent difficulties of treating sludges containing solid organic particles, a two step process was tested. The sludge was first pre-oxidized in the absence of any catalyst and the oxidized liquour was then further treated in the presence of a catalyst.

The WAO pre-treatment was performed at two different temperatures 190°C, 37 bar air and 210°C, 51 bar air for 4 h. While after WAO treatment at 190°C the residual solid still contained 15% TOC, more severe conditions (210°C) yielded a solid containing less than 5% TOC. CWAO results of the oxidized liquor are given in Figure 9.10 for TOC and TN (Total Nitrogen) abatements.

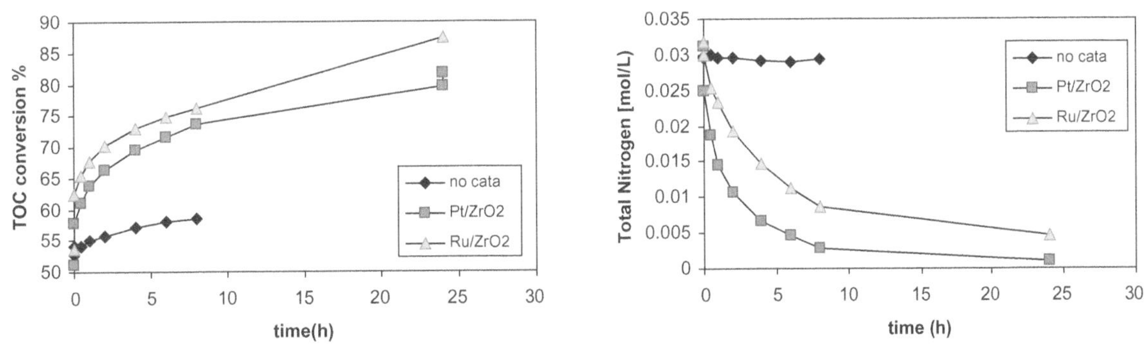

Figure 9.10 TOC and TN abatements upon catalytic oxidation of the supernatant received from a first thermal pre-oxidation of the STVULBAS sludge (210°C, 51 bar air, 4 h)

While in the absence of catalyst, no further TOC and TN abatement was observed, the addition of 3% Pt/ZrO_2 or Ru/ZrO_2 led a final TOC conversion higher than 80% and nearly complete TN abatement. These two catalysts were not only able to mineralize the organic carbon, but also to transform the remaining nitrogen selectively into dinitrogen.

The capacity for supported noble metal catalysts to convert nitrogen-containing compounds was previously mentioned in the literature. However, studies usually dealt with model compounds (ammonium salts, aniline…) (Oliviero *et al.*, 2003 and Cybulski, 2007). The use of these catalysts for complex effluents was never described. These new results further demonstrate their very promising properties.

Our results showed that with a two step treatment, combining a first WAO of the suspension at T > 210°C (to ensure a sufficient transfer of the solid organics to the liquid phase so that the final solid contains less than 5% TOC) with a CWAO of the supernatant, a 85–90% overall TOC conversion could be attained.

Operation in a continuous trickle bed reactor

Stability in the continuous CWAO, following WAO of the sludge suspension

Further work was performed to test the catalyst stability upon CWAO in a continuous reactor (trickle-bed reactor, 15 cm length, 1 cm inner diameter), (Figure 9.11). A batch of catalyst was specifically prepared using a TiO_2 shaped support (St Gobain 90 m^2/g) according to procedures developed for powder catalysts.

Figure 9.11 Trickle-bed reactor used for the catalytic continuous experiment

All liquids from WAO batch experiments on the TOULOUSE and STVULBAS sludges were collected separately (2 L each, TOC_0 ca. 1 g/L) and subsequently treated in a continuous reactor at 190°C, 50 bar air over 1.5 g of 3% Ru/TiO_2. The initial conditions were a liquid flow rate of 0.5 mL/min, a gas flow rate of 5 NL/h. Contact time was increased by recirculation of the treated effluent. The results are shown on Figure 9.12 for the TOULOUSE sludge, and in Figure 9.13 for the STVULBAS sludge.

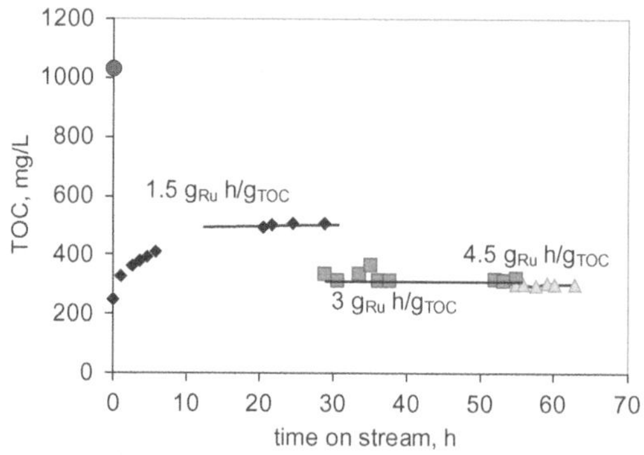

Figure 9.12 CWAO in a continuous reactor of the supernatant recovered from the WAO of the municipal sludge (TOULOUSE). TOC_0 = 1030 mg/L, F_{liq} = 30 mL/h, F_{gas} = 5 NL/h, 190°C, 50 bar air, 1.5 g 3% Ru/TiO_2 catalyst

CWAO of both supernatants led to a large TOC abatement which increased with increasing contact times. The final effluent only contained acetic acid, and nitrogen was transformed to N_2 and nitrates. The catalyst demonstrated a remarkable stability.

These experiments demonstrate that WAO (after pre-sonication or not) can be used as a pre-treatment of the sludge. The residual solid contains less than 5% TOC, which means that the filtered cake from the process could be directly disposed of at landfill sites. The supernatant can then be treated catalytically with high mineralization degrees. TOC in the final effluent mainly consists of acetic acid and the organic and ammonium nitrogen are converted into dinitrogen and nitrates.

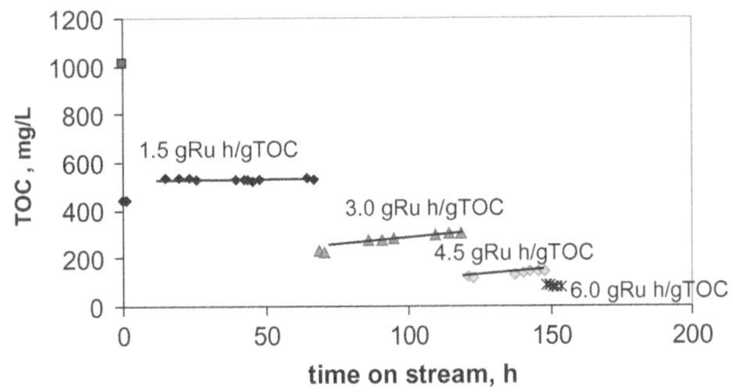

Figure 9.13 CWAO in a continuous reactor of the supernatant recovered from the WAO of the industrial sludge (STVULBAS). TOC_0 = 1013 mg/L, F_{liq} = 30 mL/h, F_{gas} = 5 NL/h, 190°C, 50 bar air, 1.5 g 3% Ru/TiO_2 catalyst

To summarize, the work performed provided new and valuable scientific data on noble metal catalysts to be used in the CWAO of complex effluents and demonstrated the efficiency of such catalysts in the CWAO of supernatants obtained after a first WAO pre-treatment.

IMPACT OF THE TECHNOLOGY ON ITS INDUSTRY OR RESEARCH SECTOR

Sludge thermal processing attracts more and more attention because of on-going discussions on safe sludge disposal methods according to legal requirements. In this context, Wet Air Oxidation (WAO) has become an economically viable alternative. The process can be defined as the oxidation of substances, which are present either in the liquid or in the suspension, with the use of an oxidant (oxygen or air) at elevated pressure and temperature. The insoluble organic matter is converted into simpler soluble organic compounds that are further oxidized. Currently there are more than 200 plants operated around the world, the majority being dedicated to the treatment of sewage sludge (Patria *et al.*, 2004). Some industrial applications were reviewed by Luck (1995) and Kolaczkowski *et al.* (1999).

To achieve optimum results, a catalyst may be added. The objective is then to reduce the operating temperature and pressure and/or to treat pollutants that cannot be destroyed during non-catalytic WAO. Despite, successful laboratory applications, the absence of stable catalysts has prevented Catalytic WAO (CWAO) from being widely employed at industrial wastewater treatment plants. The main reasons are that soluble catalysts have to be removed from the effluent in a subsequent step, while the heterogeneous catalysts have to maintain their activity for sufficiently long periods. Soluble catalysts such as copper salts were very effective on several organics (Mantzavinos *et al.*, 1996; Hao *et al.*, 1993; Shende and Mahajani, 1994 and Lin and Ho, 1996). Such catalysts were used in the Ciba-Geigy (Cu^{2+} separated as copper sulphide) and Loprox (Fe^{2+}) processes for industrial effluents. They are presently used in the MINERALIS® and ATHOS® processes dedicated to sewage sludges (www.veoliawaterst.com/athos/en/). Mixtures of metal oxides (Cu, Zn, Co, Mn, Bi) are reported to exhibit good activity, but leaching of these catalysts was detected (Mantzavinos *et al.*, 1996; Pintar and Levec, 1992 and Fortuny *et al.*, 1998). On the contrary, heterogeneous catalysts based on noble metals deposited on stable supports are much less sensitive to leaching (Imamura *et al.*, 1988; Duprez *et al.*, 1996; Béziat *et al.*, 1999; Pintar *et al.*, 2001 and Pham Minh *et al.*, 2008). Considerable efforts were focused on reducing the unwanted catalyst deactivation. Stable supported catalysts with high activity towards model pollutants as well as industrial effluents (Kraft paper pulp bleaching effluent, olive mill wastewater) have been successfully synthesized and tested at the laboratory scale. Pd, Pt and Ru have received most attention. The Nippon Shokubai-LC and Osaka Gas processes in Japan use monolith reactors with $Pt\text{-}Pd/TiO_2\text{-}ZrO_2$ catalyst or mixtures of titania or zirconia-titania supported metals (honeycomb or spheres) (Patria *et al.*, 2004 and Luck, 1995). These catalysts remain, however, relatively expensive materials and to be used, they must demonstrate high stability. In the case of sewage sludges, the presence of solid organic materials has been identified as an inhibiting parameter.

Ultrasonic irradiation might be used to increase solubilization. In the last years, ultrasounds have also gained attention in environmental science and technology as a non conventional method for improving wastewater and sludge treatment processes. Ultrasounds appeared as a possible technology for the treatment of sludges in the 90's. Ultrasounds have been studied at the laboratory, the pilot and more recently at the industrial scales. Many studies concerning the application of ultrasounds focused on the anaerobic digestion process at biological wastewater treatment plants. Due to micro-scale bubble explosions caused by ultrasonic waves, the structure of sludge flocs was effectively destroyed, cells underwent lysis and, in a subsequent stage, part of the insoluble organic particulates could be transferred to the liquid phase (Tiehm *et al.*, 1997; Thiem *et al.*, 2001 and Lehne *et al.*, 2001). The use of pre-treated sludge significantly improved the COD removal efficiency and the biogas production in anaerobic digesters. Recent developments of high power transducers allowed the treatment of large amounts of sewage sludge. Since a few years, a few firms are commercializing ultrasound homogenizers for the pre-treatment of sewage sludges before anaerobic digestion (Hielsher GmbH, Sonix, Sonotronic GmbH. . .).

The use of sludge disintegration as a pre-treatment technique may however not be limited to sludge digestion. Sludge disintegration may also be a valuable option before CWAO. Thermal lysis may be another option. Up to now, no reports have been published on possible various pre-treatments before CWAO. Accordingly, the objective of this study was to investigate the effects of various pre-treatments (ultrasonic and/or thermal) on sludge solubilization, the mineralization enhancement upon CWAO and the catalyst stability. This work elucidated experimentally the effects of ultrasonic treatments on the physical and chemical properties of two waste-activated sludges before further catalytic wet air oxidation (CWAO) in the presence of heterogeneous platinum and ruthenium catalysts.

The industry and research sector dealing with the catalytic wet air oxidation of complex effluents will benefit from the knowledge developed in this project. It is likely that these data will be important to decide, whether soluble or solid catalysts shall be preferred. Some drawbacks of the CWAO have been outlined: organic particles in the sludge get deposited on the catalyst surface. The problem of solid deposition is even more stronger after disintegration of the sludge since finer particles are produced. However, it has been shown that CWAO may be a promising technology for abatement of organic pollutants when dissolved. An alternative may be recommended in the case of sewage sludges: the liquid phase might be enriched by performing a thermal WAO treatment before the oxidized liquor will is submitted to CWAO to transform the organic compounds into biodegradable molecules. The use of a monolithic reactor could also allow to overcome this problem. Luck *et al.* (1995) reported for instance the use of a 5% CuO catalyst deposited on a titania monolith with honeycomb structures containing square channels with a width of 3.56 mm and a wall thickness of 0.65 mm for the oxidation of waste streams containing sludges at 235°C, 40 bar. COD removal reached up to 74%, while in the blank non-catalytic tests it was 58%. However, copper was leached from the catalyst. The Nippon Shokubai process (NS-LC) involves a Pt-Pd/TiO_2-ZrO_2 honeycomb catalyst which is claimed not to be sensitive to the deposition of solids on the catalyst surface (Luck, 1995 and Cybulski, 2007).

REFERENCES

Béziat J. C., Besson M., Gallezot P. and Durécu S. (1999). Catalytic wet air oxidation on a Ru/TiO_2 catalyst in trickle-bed reactor. *J. Catal.*, **182**, 129–135.

Bougrier C., Carrère H. and Delgenès J. P. (2005). Solubilization of waste-activated sludge by ultrasonic treatment. *Chem. Eng. J.*, **106**, 163–169.

Cybulski A. (2007). Catalytic Wet Air Oxidation: are monolithic catalysts and reactors feasible? *Ind. Eng. Chem. Res.*, **46**, 4007–4033.

Duprez D., Delanoë F., Isnard J. P., Blanchard G. and Barbier Jr J. (1996). Catalytic oxidation of organic compounds in aqueous media. *Catal. Today*, **29**, 317–322.

Fortuny A., Font J. and Fabregat A. (1998). Wet air oxidation of phenol using active carbon as catalyst. *Appl. Catal. B: Environmental*, **19**, 165–173.

Hao O. J., Phull K. K., Chen J. M., Davis A. P. and Maloney S. W. (1993). Factors affecting wet air oxidation of TNT: rate studies. *J. Haz. Mater.*, **34**, 51–68.

Imamura S., Fukuda I. and Ishida S. (1988). Wet oxidation by ruthenium supported on cerium (IV) oxides. *Ind. Eng. Chem. Res.*, **27**, 718–721.

Kolaczkowski S. T., Plucinski P., Beltran F. J., Rivas F. J. and McLurgh D. B. (1999). Wet air oxidation: a review of process technologies and aspects in reactor design. *Chem. Eng. J.*, **73**, 143–160.

Lehne G., Müller A. and Schwedes J. (2001). Mechanical disintegration of sewage sludge. *Water Sci. Technol.*, **43**, 19–26.

Li N., Descorme C. and Besson M. (2007a). Catalytic wet air oxidation of aqueous solution of 2-chlorophenol over Ru/zirconia catalysts. *Appl. Catal. B: Environmental*, **71**, 262–270.

Li N., Descorme C. and Besson M. (2007b). Catalytic wet air oxidation of 2-chlorophenol over Ru loaded CexZr1-xO2 solid solutions. *Appl. Catal. B: Environmental*, **76**, 92–100.

Lin S. H. and Ho S. J. (1996). Catalytic wet-air oxidation of high-strength industrial wastewater. *Appl. Catal. B: Environmental*, **9**, 133–147.

Luck F. (1995). A review of industrial catalytic wet air oxidation processes. *Catal. Today*, **24**, 195–202.

Luck F., Djafer M. and Bourbigot M. (1995). Catalytic wet air oxidation of biosolids in a monolithic reactor. *Catal. Today*, **24**, 73–78.

Mantzavinos D., Hellenbrand R., Livingston A. G. and Metcalfe I. S. (1996). Catalytic wet air oxidation of p-coumaric acid: partial oxidation intermediates, reaction pathways and catalyst leaching. *Appl. Catal. B: Environmental*, **7**, 379–396.

Mikulova J., Barbier Jr J., Rossignol S., Mesnard D., Duprez D. and Kappenstein C. (2007a). Wet air oxidation of acetic acid over platinum catalysts supported on cerium-based materials: influence of metal and oxide crystallite size. *J. Catal.*, **251**, 172–181.

Mikulova J., Rossignol S., Barbier Jr J., Mesnard D., Kappenstein C. and Duprez D. (2007b). Ruthenium and platinum catalysts supported on Ce, Zr, Pr-O mixed oxides prepared by soft chemistry for acetic acid wet air oxidation. *Appl. Catal. B: Environmental*, **72**, 1–10.

Moulut-Godard M. L. (2004). Etude du traitement par oxydation en voie humide catalytique de boues d'épuration. PhD thesis, University Lyon 1, N° 211–2004.

Neis U., Nickel K. and Tiehm A. (2001). Ultrasonic disintegration of sewage sludge for enhanced anaerobic biodegaradation. *Adv. Sonochem.*, **6**, 59–90.

Oliviero L., Barbier Jr J. and Duprez D. (2003). Review: Wet air oxidation of nitrogen-containing organic compounds and ammonia in aqueous media. *Appl. Catal. B: Environmental*, **40**, 163–184.

Patria L., Maugans C., Ellis C., Belkhodja M., Cretenot D., Luck F. and Copa B. (2004). IWA Publishing Advanced Oxidation Processes for Water and Wastewater Treatment, Ed. S. Pearsons, 247–274.

Pham Minh D., Gallezot P., Azabou S., Sayadi S. and Besson M. (2008). Catalytic wet air oxidation of olive oil mill effluents: 4. Treatment and detoxification of real effluents. *Appl. Catal. B:Environmental*, **84**, 749–757.

Pintar A. and Levec J. (1992). Catalytic oxidation of organics in aqueous solutions: I. Kinetics of phenol oxidation. *J. Catal.*, **135**, 345–357.

Pintar A., Besson M. and Gallezot P. (2001). Catalytic wet air oxidation of Kraft bleach plant effluents in a trickle-bed reactor over Ru/TiO_2 catalysts. *Appl. Catal. B: Environmental*, **31**, 275–290.

Shanableh A. (2000). Treatment of sewage sludge using hydrothermal oxidation –technology application challenges. *Water Sci. Technol.*, **41**, 85–92.

Shende R. V. and Mahajani V. V. (1994). Kinetics of wet air oxidation of glyoxalic acid and oxalic acid. *Ind. Chem. Eng. Res.*, **33**, 3125–3130.

Thiem A., Nickel K. and Neis U. (2001). Ultrasonic waste activated sludge disintegration for improving anaerobic stabilisation. *Water Res.*, **35**, 2003–2009.

Tiehm A., Nickel K. and Neis U. (1997). The use of ultrasound to accelerate the anaerobic digestion of sewage sludge. *Water Sci. Technol.*, **36**, 121–128.

Veolia Water Solutions & Technologies. www.veoliawaterst.com/athos/en/

Chapter 10

Production of activated carbon from sludge

Karl M. Smith, Geoffrey D. Fowler, Suangusa Pullket and Nigel J. D. Graham (Imperial College London – ICL)

SUMMARY

The aim of this study was to establish whether carbon adsorbents that performed competitively in wastewater treatment applications could be produced from sewage sludge. To this end, a range of sludge types and adsorbent production and characterisation techniques were employed. In order to assess the extant research on sewage sludge derived adsorbents, a major literature review of the topic was also conducted.

This chapter is derived from a report produced for the project's sponsor: The European Union. At the time of writing the research presented by the lead authors of this chapter has resulted in two journal publications (Pullet *et al.*, 2009 and Smith *et al.*, 2009) and more papers are in preparation. If more technical information is sought, then the consultation of these publications is strongly recommended.

Project objectives

1) Produce and characterise sludge-based adsorbents (SBAs)/catalysts under different experimental conditions.
2) Establish cause-effect relationships between the characteristics of sludge-based adsorbents and the performance in different wastewater treatment applications.

Work performed

Literature review

The literature review of SBAs and their use in aqueous phase applications was published in the journal *Water Research* in 2009: Smith *et al.* (2009).

Sludge sample collection

Sludge samples from a total of eight different wastewater treatment plants (WWTPS) and five different countries (England, Wales, Germany, France and Poland) were collected.

 Reduction, Modification and Valorisation of Sludge (REMOVALS). Edited by Christophe Bengoa, Azael Fabregat, Josep Font and Frank Stüber. ISBN: 9781843393450. Published by IWA Publishing, London, UK.

Adsorbent production and characterisation

Three different techniques were employed to prepare adsorbents from the sludge samples: carbonisation; physical activation; chemical activation. A range of different characterisation techniques (Fourier Transform Infrared Spectroscopy (FTIR), X-Ray Diffraction crystallography (XRD), ash content, CHN content, Inductively Coupled Plasma – Atomic Emission Spectrophotometry (ICP-AES), peroxide number (catalytic activity), gaseous CCl_4 uptake, pH assessment, ball pan hardness) were employed, but the fundamental parameter used to evaluate the SBAs was the BET surface area. This is generally regarded as being the principal measure of an activated carbon's utility. For the chemical and physical activation procedures, the Response Surface Methodology (RSM) technique was employed to optimise the processes. The physical activation work was reported in a paper published in the *Journal of Residuals Science & Technology*: Pullket *et al.* (2009).

Research was also conducted on the use of different binding techniques to produce granular activated carbons (GACs) that exhibited an enhanced attrition resistance.

Appraisal of SBAs' performance in wastewater treatment applications

Tests conducted by ICL established the SBAs' adsorption capacity for phenol, methylene blue and bisphenol A (BP-A – a common endocrine disrupting chemical (EDC)). Tests conducted by Institut National Polytechnique de Tolouse and Universitat Rovira i Virgili ascertained the SBAs' performance in, respectively, Adsorption-Oxidation (AD-OX) and Catalytic Wet Air Oxidation (CWAO) applications. Tests were also undertaken by Universitat Autònoma de Barcelona and the Université de Nantes to appraise, respectively, the use of the SBAs as biofilters for the elimination of industrial waste gas and as a general use adsorbent for the treatment of water contaminated with water metals and organic dyes.

End results

Adsorbent production and characterisation

For all of the production techniques, the sludge type was found to exert a strong influence over the surface areas obtained. Generally, it was observed that the higher the organic content, the higher the surface area attained. Accordingly, the sludge samples that had not been subjected to digestion procedures constituted the best feedstocks.

For carbonisation, it was determined that the optimum carbonisation temperature for maximising the BET surface area is 900–1,000°C. The high variation reflects the fact that this figure is sludge dependent. The maximum BET surface area obtained was ca. 155 m^2/g. Carbonisation was found to be the most effective technique for producing highly alkaline SBAs.

The maximum BET surface area obtained by physical activation, using steam as the activant, was ca. 270 m^2/g. The optimum activation temperature for preparation of SBAs was determined to be 838°C. The optimum dwell time was found to be sludge dependent, but typically was 70–80 mins (Pullket *et al.*, 2009). The high ash content of the sludge samples was the key impediment to the attainment of higher surface areas.

The maximum BET surface area obtained by chemical activation was 1,979 m^2/g and the results suggest that surface areas in excess of 2,000 m^2/g are achievable. In addition to the sludge type, the key factors that influenced the properties of the SBAs produced by chemical activation were: the reagent to sludge ratio; the activation reagent, the reagent impregnation method (wet vs. dry) and the sludge preparation/pre-treatment method. Further particulars regarding the chemical activation procedure can be found in the publications arising from this research.

SBA assessment in wastewater treatment applications

It was found that for applications involving the adsorption of phenol, BP-A and by inference, most common phenolic compounds, SBAs with a high surface area (i.e., $>500\,m^2/g$) were most appropriate; i.e., SBAs produced by chemical activation. However, SBAs produced by physical activation were also suitable if their ash (inorganic) content was lowered through the use of appropriate acid washing techniques. The porosity is primarily associated with the organic content. Hence, raising the organic to inorganic ratio resulted in a rise in the surface area.

For the catalytic wet air oxidation (CWAO) treatment of phenolic compounds the surface area was found by Universitat Rovira i Virgili to be less influential. Thus, the performance of untreated SBAs produced by the physical activation of dewatered, raw (DRAW) sludge was found to be acceptable (Marques *et al.*, 2011). However, these SBAs exhibited a low attrition resistance and a high propensity towards leaching. Thus, prior to use, it was necessary to subject these SBAs to a combined hardening and leaching mitigation process.

For the adsorption of organic dyes, Université de Nantes ascertained that the mesoporosity was the critical factor. A key finding was that, due to their high mesoporosity, the SBAs produced by the physical activation of the DRAW sludge exhibited a higher adsorption capacity than the reference carbon.

For the use of SBAs as biofilters for the elimination of industrial waste gas, Universitat Autònoma de Barcelona found that only the external surface was of importance: the internal surfaces were not accessible to the micro-organisms.

METHODOLOGIES AND APPROACHES EMPLOYED

Sludge preparation

As described by Pullet *et al.* (2009), the two UK sludge samples employed were: Dewatered, RAW filter cake (DRAW), sourced from Little Marlow WWTP, UK and Dewatered, Mesophilic Anaerobically Digested (DMAD) sludge, which was obtained from Ashford WWTP, UK. Both samples were sterilized at 121°C for 15 minutes (Priorclave direct steam autoclave, London, UK). They were crushed to below 10_mm using a cross beater mill (Glen Creston, London, UK) and stored in air tight containers prior to further use. The non-UK sludge samples employed are as follows: dewatered, aerobically digested, secondary biological sludge with added organic polymer coagulant (DSBS), from Nantes, France; a mixture of primary sludge stabilised by anaerobic fermentation and excess sludge stabilised by aerobic digestion (KAN), from Kalisz, Poland; the limed form of KAN sludge (KAW), mesophilic anaerobically digested sludge (GOS), from Łódź, Poland; thickened, primary, municipal sewage sludge seeded with a mesophilic anaerobic digested sewage sludge and then digested in a laboratory scale, fermentative bio-reactor for hydrogen production (GLAM) (Massanet-Nicolau *et al.*, 2008), from Glamorgan, Wales; sludge obtained from a lab-scale membrane bio-reactor with enhanced phosphorous removal (BERLIN), from Berlin, Germany. All sludge samples were dried to constant weight at 105–110°C before utilisation. After drying, the non-UK sludge samples were crushed below 9.5 mm in size using a pestle and mortar prior to use: the relatively small quantities in which these samples were provided discounted the use of the cross beater mill.

Sludge carbonisation/preparation

Carbonisation/physical activation

The carbonised and physically activated SBAs were prepared using a Carbolite 11/150 laboratory scale rotary furnace (Carbolite, UK), which is shown in Figure 10.1. For each run, 210 g of prepared sludge was loaded into a cylindrical quartz reactor, 4.5 litres in capacity, which was set to rotate at 10 rpm. This was heated at a rate of either 5 or 10°C/min to the desired temperature, under a nitrogen flow of 500 ml/min. For steam activation, once the desired temperature was reached, water was directly injected into the reactor using a peristaltic pump at a flow rate of 0.7 g/min and isothermal conditions were maintained for the specified activation period. Subsequently, the furnace was automatically cooled down to room temperature.

Figure 10.1 The rotary furnace employed for the carbonisation and physical activation experiments

Chemical activation

The sludge samples were utilised in three forms: untreated (i.e., wet), dried (105°C to constant mass) and carbonised. Two different reagent impregnation methods were employed: wet impregnation and dry impregnation. During dry impregnation, the sludge was mixed directly with the reagent: no water was added. Under wet impregnation, the sludge was mixed with the reagent and water was added to the mixture to aid the dissolution of the reagent. The ratio of reagent to sludge was 1:1 (by dry basis).

Before being activated, all of the impregnated samples were dried at 105°C for a minimum of 12 hrs (i.e., overnight). After drying the untreated sludge samples were crushed below 2 mm and again dried overnight before being activated: this was necessary because after drying the sludge-reagent mixture was highly agglomerated. The quantity of dried material loaded into the crucible was 30 g: for each activation experiment, one crucible was loaded into the reactor. The crucible and reactor were both made from Alsint (Multi-lab, Newcastle (UK)). The reactor was heated in a Carbolite CTF 12/65 horizontal tube furnace (Carbolite, UK). During activation a constant flow of 500 ml/min of N_2 into the reactor was maintained.

Sample characterisation

Prior to characterisation the SBAs were crushed and sieved to a particle size below 150 μm. Their surface area and porosity was characterised using a Coulter Omnisorp® 100 (Beckman-Coulter, UK). Prior to analysis, the SBAs were degassed at 150°C (250°C was employed for the chemically activated SBAs) for 7–8 hours until a pressure below 10^{-5} Torr was reached. Their N_2 adsorption and desorption isotherms were measured at liquid nitrogen temperature. From the adsorption isotherms the BET surface area (Brunauer *et al.*, 1938) and t-plot surface area (Lippens and deBoer, 1965), which indicates the mesoporosity and macroporosity of the activated carbon, were determined. The micropore surface area was determined by subtracting the t-plot surface area from the BET surface area.

The ash content of the carbons was measured using the method prescribed by ASTM standard D 2866-94. The pH of the SBACs was determined using a method derived from Bagreev *et al.* (2001a), whereby, after being ground and dried for 3 hrs at 150°C, 0.4 g of the sample was added to 20 ml of deionised water. The mixture was mechanically agitated for 24 hrs and after the removal of the SBACs by filtration, the pH of the recovered solution was measured using a pH meter (Fisherbrand Hydrus 500 [Fisher Scientific, UK]) equipped with a VWR, epoxy, gel, pH electrode (VWR, Lutterworth, Leicestershire, UK).

The phenol, bis-phenol A and methylene blue uptakes of the SBAs were all assessed by placing a specified mass of SBA into a fixed volume of a solution of the compound. After a fixed period of time, the UV absorbance of each sample-solution mixture was measured at a wavelength that corresponded to the compound's peak absorbance: a UV spectrophotometer (Shimadzu, Japan) was utilised for this analysis.

The Ball Pan Hardness method was based upon the ASTM method D 3802-79 (ASTM, 1994) which enables the assessment of a sample's resistance to particle size degradation under service conditions. The 0.5–2_mm size fraction of the test carbon was used in each test.

The carbon, hydrogen and nitrogen content was determined using a Carlo-Erba EA1108-CHNS analyser by MEDAC Ltd, UK.

Hardened carbon preparation

The hardened carbons were prepared by milling the carbonised (1 hr at 700°C under N_2) or steam activated DRAW sludge to below 45 μm. This material was then mixed with a binder and an appropriate solvent. Once the mixture exhibited sufficient signs of agglomeration it was then pelletised by means of either a manual extruder or a uni-axial hydraulic press. The pellets were dried for 4 hrs at 105°C. Subsequent to drying, some of the pellets were steam activated as described previously.

ACHIEVEMENTS RELATED TO THE STATE OF THE ART

Production of high surface area SBAs

It has been established that it is possible to produce, by chemical activation, SBAs with BET surface areas in excess of 2,000 m²/g. The highest BET surface area achieved thus far was 1,979 m²/g and the results indicate that higher surface areas are readily achievable. This figure was particularly significant because it was obtained using a reagent:sludge ratio of 1:1: some other studies on the chemical activation of sewage sludge have resorted to ratios of 3:1 or greater (Bagreev *et al.*, 2001b and Ros *et al.*, 2006). Low ratios are advantageous because they make the process more economically attractive through reducing the quantity of reagent required and hence the operating costs. The highest BET surface area reported in the literature to date for a SBA is 1,883 m^2/g (Lillo-Ródenas *et al.*, 2008). However, this figure was obtained using a synthetic sludge and the behaviour of a real sludge sample may be different.

A major element that contributed to the high degree of success of this achievement was the use of a range of reagents that had not previously been utilised to prepare SBAs. This served to maximise the probability that a breakthrough would be made in the development of a high surface area SBA. The novel reagents were: K_2CO_3; $FeCl_3$; $CaCl_2$; $MgCl_2$; KCl and $K_3PO_4{\cdot}H_2O$.

Development of a one-step production method for the production of SBAs

The physical activation work conducted for this WP constitutes the first major study undertaken on the production of SBAs by a single step physical activation. Previously, all of the large scale studies on steam activation had utilised a two-step method, whereby the sludge was carbonised prior to being activated (Mitchell *et al.*, 1978; Rio *et al.*, 2006a, 2006b and Fitzmorris *et al.*, 2006a, 2006b). The use of a two stage process is, from a commercial perspective, unattractive because of the high energy requirements associated with subjecting the sludge sample to two, distinct, high temperature heating processes and because of the longer production time. However, the single-step process is feasible because the process starts at room temperature and thus the sludge is essentially carbonised as it is heated up to the activation temperature. The single-step process has been used extensively at Imperial College London for the production of activated carbons from a variety of waste materials, including waste tyres (San Miguel et *al.*, 2003).

If the dried sludge was admitted directly to the furnace at temperature, i.e., when the furnace was already at the activation temperature, then the sudden and uncontrolled release of the volatile component of the sludge (i.e., the oils and gases) could adversely affect the operation of the process. Thus, it remains to be seen whether the single-step method could be adopted commercially. An alternative approach would be to improve the commercial viability of the two-step approach through utilising the off-gases arising from the carbonisation process to pre-heat the sludge.

Utilisation of the response surface methodology method to evaluate the optimum steam activation conditions

Optimisation of the activation conditions is essential to the realisation of SBA production as a competitive alternative to conventional sludge disposal routes. The response surface methodology (RSM) is a tool that can be used to reduce the number of experiments required to determine the optimum activation conditions, in relation to the classical, empirical approach (Wächter and Cordery, 1999 and Rio *et al.*, 2005). It uses statistical methods to mathematically express the relationship between two or more variables (Karacan *et al.*, 2007) and thus ascertain the interactions between the examined parameters.

Previously, the RSM method has been utilised to optimise the chemical activation (Rio *et al.*, 2005) and physical activation [by steam (Rio *et al.*, 2006a, 2006b)] of sewage sludge. However, these studies did not consider the effect that the sludge type would have on the process. Given the wide range of extant sludge types (primary, secondary, limed, anaerobically digested, etc.) this factor was considered to be too significant to ignore. An investigation was therefore conducted into the use of the RSM method to optimise two disparate sludge samples.

The two sludge types were: Dewatered, Mesophilic Anaerobically Digested (DMAD) sludge and Dewatered RAW (DRAW) sludge. A fixed steam flow rate, sample weight and heating rate were employed. The two control variables were the temperature and the dwell time.

From the RSM models developed, it was determined that the optimum activation conditions for the DMAD and DRAW sludge samples were 838°C and 73 mins and 838°C and 80 mins respectively. The shared optimum temperature indicated that the nature of the fixed (non-volatile) carbon component did not vary significantly between different sludge types. The disparate dwell times reflected the different fixed carbon content of the sludge samples, which necessitated different dwell times to achieve the optimum degree of carbon burn-off.

The relationship between the dwell time and temperature and the BET surface area of the SBAs, as determined from the RSM approach, is illustrated in Figure 10.2. Under the respective, optimised steam activation conditions for the two sludge samples, the DRAW derived SBA exhibited a BET surface area that was ca. 100 m^2/g greater than the DMAD derived SBA's surface area. This outcome was again ascribed to the DRAW sludge's higher organic content: a high ash content was detrimental to the surface area because of the negligible porosity of the ash fraction.

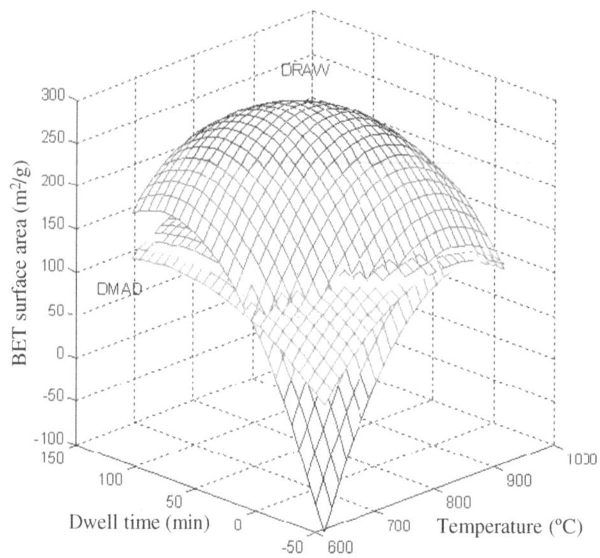

Figure 10.2 Relationship between the activation time, temperature and BET surface area of the SBAs produced by the steam activation of the DRAW and DMAD sludge samples

This work was presented both at a conference (Pullket *et al.*, 2008) and in a paper published in a peer reviewed journal (Pullket *et al.*, 2009).

Determination of the influence of the sludge type on the properties of the SBAs produced by physical activation

The work conducted to ascertain the optimum steam activation conditions was used as a foundation for undertaking an in-depth study into the effect of the sludge type on its response to steam activation. The optimum steam activation conditions described were applied to six more sludge samples. From this research, much of which was published in the journal paper by Pullket *et al.* (2009), it was established that the treatment/process that the sludge has been subjected to has a fundamental effect on its properties as a SBA feedstock.

The optimum sludge type was, due its high organic content, undigested sludge. Sludge that had been stabilised by liming, on the other hand, was observed to be a very poor quality activated carbon feedstock. This was predominantly due to the rise in the ash content of the sludge. The ash fraction is non-porous and therefore does not contribute to the surface area of the adsorbent. Moreover, the primary component of lime is calcium and this is a noted catalyst of the steam-carbon gasification/oxidation reaction (Cannon *et al.*, 1994). Its presence therefore accelerates the burn-off of the organic fraction during physical activation. This therefore makes it harder to control the effects of the activation process.

Another sludge found to be of very poor quality was obtained from the University of Glamorgan. This sludge was produced from an anaerobically digested, primary sludge that had been fermented in a bio-rector for the purposes of hydrogen production (Massanet-Nicolau *et al.*, 2008). The cause of its poor performance was not immediately clear, as prior to activation its ash content, at 25 wt.%, was not high relative to the other sludge samples. The DRAW and DMAD sludge samples, for example, exhibited ash contents of 20 and 40 wt.% respectively. The mass loss it exhibited during carbonisation/devolatilisation was not also not significantly greater than the mass loss exhibited by the other sludge samples. However, it was clear that after being activated, the sample exhibited a very high ash content (94.9 wt.%). This indicated either that the fixed carbon content was very reactive, or that the sludge contained a high proportion of catalytically active metals.

Production of high surface area SBAs by physical activation

The work conducted on the physical activation of sewage sludge corroborated the findings previously reported in the literature, namely, that because of its high ash content, it was not feasible to produce SBAs with a BET surface area in excess of 300 m^2/g from sewage sludge. However, it was determined that through inducing the partial dissolution and hence removal, of the inorganic fraction, acid washing the sludge-derived carbons was a highly effective technique for raising the surface area of the SBAs. The highest BET surface area achieved in this fashion was 622 m^2/g; prior to washing, the SBA exhibited a BET surface area of ca. 125 m^2/g. Although there is a precedent in the literature for the acid washing of SBAs, the highest BET surface area previously achieved in this fashion is 428 m^2/g (Ros *et al.*, 2007).

Even the use of a dilute acid solution at ambient conditions, using a residence time of 24 hrs, was observed to increase a SBA's BET surface area from 255 m^2/g to 497 m^2/g. Moreover, with respect to its performance in the CWAO process as a catalyst/adsorbent for the destruction and removal of phenolic compounds in water, this SBA performed on a par with the reference commercial carbon (Marques *et al.*, 2011).

Production of SBAs with a high EDC uptake

Endocrine disrupting chemicals (EDCs) are a class of micro pollutants that are a well documented cause of reproductive disturbance in wildlife. Indeed, EDCs are described by Liu *et al.* (2009) as a latent crisis to humans and the environment. Bisphenol-A (BP-A) is a common EDC with a relatively high solubility in water and consequently, it was selected as the model EDC compound that would be used for assessing the SBAs' EDC uptake.

The key finding was that the BP-A uptake of the SBAs was heavily influenced by their surface area. Thus, the SBA that exhibited the highest surface area – 1,979 m^2/g – also exhibited the highest BP-A uptake, which was ca. double that of the reference commercial activated carbon. This was consistent with the difference in scale between the two adsorbents' surface areas. Accordingly, the SBAs produced by physical activation, which exhibited BET surface areas under 280 m^2/g, did not attain high BP-A uptakes.

No other study has been conducted into the analysis of the EDC uptake of SBAs. The finding that it is possible, by chemical activation, to produce SBAs that can attain a BP-A uptake that is ca. double that of a commercial carbon is likely to have great significance. Chemical activation is generally regarded as being a specialist production process, the use of which is usually only justified when the application specifically demands the use of high grade adsorbents. The research finding reported here may be sufficient to render SBA production by chemical activation a commercially viable enterprise.

Development of post-activation techniques to mitigate the high leaching propensity of SBAs

A key problem associated with the use of SBAs is their high propensity towards leaching. This issue was studied by Fitzmorris *et al.* (2006a), who observed that when placed into an aqueous solution with a pH of 1, all of the metals under investigation (As, Cd, Cr, Cu, Ni, Pb, Se and Zn) leached into solution: Cu and Zn leached in the greatest quantities. At a pH of 5 only Zn was leached in significant quantities but prior to testing, the SBA had been washed in both 0.1 M HCl and distilled water. This treatment would therefore have been liable to remove much of the SBA's readily leachable component.

The SBAs prepared by physical activation were found, when their performance as adsorbents/catalysts in the catalytic wet air oxidation (CWAO) process was tested by Universitat Rovira i Virgili, to exhibit a high propensity towards leaching, most notably with regard to Fe (Marques *et al.*, 2011). A range of different post activation leaching mitigation techniques were tested and thermal treatment, which comprised heating the SBA at 950°C under nitrogen, was found to be the most effective. One side effect of this treatment was that the SBA exhibited a greater tendency towards leaching calcium. However, this was not considered to pose a significant problem.

The high propensity towards leaching exhibited by SBAs is well documented. The discovery that it is possible to mitigate this problem through the application of a single-step heating step could therefore provide the impetus necessary to make SBA production a commercial reality. The inhibition of leaching is also discussed in the next section.

Production of leaching resistant, hardened pellets

A key performance issue identified with the SBAs submitted to Universitat Rovira i Virgili for use in the CWAO process was that they exhibited a low attrition resistance. Under the high pressures and temperatures employed by the CWAO process the SBAs disintegrated. This problem became acute when the SBAs were produced from DRAW sludge. The DRAW-derived SBAs exhibited the best performance of the SBAs tested. If the successful utilisation of SBAs in the CWAO process was to be realised, then it was recognised that it would be necessary to devise a method of improving the SBAs' hardness.

Work undertaken to achieve this goal was successful. Lignosulphonite based binders have been found to be the most effective. Pellets prepared with these binders can not be prepared by manual extrusion: uni-axial compression was found to be an effective pelletisation method. Moreover, after drying the pellets must be heated to temperatures in excess of 500°C in order to attain the necessary hardness and water resistance. However, it was found that subjecting the pellets to the steam activation process was sufficient to cure the pellets. Thus, if the pellets were prepared with carbonised sludge, then the curing and activation steps can be combined in a single step.

No precedents exist in the literature for the production of attrition resistant, activated carbons through the application of a dedicated hardness enhancement procedure. A key factor in this is that conventional activated carbon feedstocks, e.g., coconut shells, readily yield carbons that exhibit a high attrition resistance. However, it seems likely that many potential feedstocks have traditionally been overlooked for the very reason that they would yield carbons with a low attrition resistance. The development of this method may therefore engender a reappraisal of the use of feedstocks that have, hitherto, been discounted.

A further discovery of this work concerns the hardened SBAs that were produced using the lignosulphonite based binders and were then steam activated after being pelletised. When utilised in the CWAO process they were observed to exhibit a propensity towards leaching that was lower than the reference commercial carbon. This represents a major advance in this research field and does much to improve the commercial viability of the hardening process. This development could result in the utilisation of feedstocks that, due to their high content of leachable metals, had previously not been considered for use in activated carbon production.

Appraisal of the effect of seasonal variation

It has been determined that sewage sludge does exhibit some seasonal variation in its fixed carbon and volatile content: hot summers, in particular, were observed to induce a drop in the volatile content of the sludge. However, it was found that the magnitude of the seasonal variation was too low to engender adverse effects on the sludge's properties as an activated carbon feedstock. This finding is highly significant because it will nullify any fears that potential producers of SBAs may have had over the variability of the feedstock.

IMPACT OF THE TECHNOLOGY ON INDUSTRY OR RESEARCH SECTOR

Research impacts

Review of the literature

One of the key outputs from this project has been the production and subsequent publication in the journal *Water Research* of a literature review on the production of SBAs and their utilisation in liquid phase applications (Smith *et al.*, 2009). This article has drawn attention to the fact that, despite exhibiting low BET surface areas, SBAs often exhibit a high uptake of numerous different contaminants. It has also identified areas of research with respect to SBAs that merit further investigation. It is therefore anticipated that this paper may stimulate a renewed interest in research into the production and utilisation of SBAs. Furthermore, for researchers already working with SBAs, the review should act as: firstly, a convenient tool for ascertaining what work has already been done and secondly, as an invaluable reference guide to the key works in the field.

This review paper has also highlighted the limitations of SBAs. However, it is hoped that because it has also suggested possible strategies for mitigating these drawbacks, the paper has made a convincing case for the investigation of SBAs and their prospective value. As described in the previous section it is, moreover, believed that some of the problems identified have been resolved as a direct consequence of the work undertaken for this study. One paper on the influence of the sludge type (Pullket *et al.*, 2009) has already been published and at the present time of writing, further papers are in preparation. The scope of these publications are:

1) The production of attrition resistant, granular, activated carbons from sewage-sludge derived materials
2) Activated carbons produced from sewage sludge by chemical activation and the study of the thermal behaviour of the activation process
3) The EDC uptake of the chemically activated sludge based adsorbents
4) The uptake of phenolic compounds by the physically activated sludge based adsorbents
5) The utilisation of the RSM approach to optimise the production of sludge based adsorbents by chemical activation

Advancement of knowledge of utilisation of sewage sludge for adsorbent production

This research has resulted in numerous scientific breakthroughs with respect to the development of SBAs. Some of the key findings include: raw/undigested sludge is the best sludge feedstock for SBA production; the use of chemical activation can result in SBAs with surface areas approaching 2,000 m^2/g; high surface area SBAs can exhibit a EDC uptake that is double that of commercial activated carbons; the treatment, by acid washing, of SBAs prepared by physical activation is sufficient to render these materials competitive with commercial carbons.

The highest BET surface area attained (1,979 m^2/g) exceeds all published values for the surface ares of SBAs. Thus, it is anticipated that this result will have a galvanising impact on research into SBA production. The finding that chemical activation can produce adsorbents that exhibit not only a high surface area but also a high EDC uptake should bolster research efforts into the use of chemical activation.

The discovery that lignosulphonate binders can both improve the attrition resistance and mitigate the propensity towards leaching of SBAs should have major implications not only for research into SBAs, but also for research on the use of waste materials for activated carbon production.

Social and environment impacts

Introduction

Sewage sludge production has reached such a scale that if it is not appropriately dealt with, then it will begin to exert a strongly negative effect on the environment and society. In the UK alone, over one million tonnes of sewage are produced every year (dry solids). Although 62% of this sludge is recycled to land (Water UK, 2006) this of course means that the remaining 38% must be disposed of. In 2004, 50% of the sludge not applied to agricultural land was disposed of either by incineration or landfill (Water UK, 2006). Nether approach can be regarded as being sustainable and indeed, both are associated with negative impacts on the environment and ergo, the populations local to the incineration/landfill facilities. Moreover, as the legislation governing the permissible levels of pathogens, heavy metals and other substances in sludge intended for application to agricultural land becomes increasingly stringent, the quantity of sludge requiring disposal is liable only to rise.

This study has demonstrated that the conversion of sewage sludge into an adsorbent is a highly viable alternative to the existing disposal and valorization routes. The primary advantages that it offers are described below.

Reduction in quantity of material

The reduction in weight typically incurred through the conversion, by physical activation, of dried sludge into activated carbon was 60–80 wt.%. In other words, up to four fifths of the mass of the dried sludge was lost. Much of this weight loss arises from the conversion of the sludge's volatile fraction into hydro-carbon rich gases and oils, which can potentially be used in energy recovery/production applications. The remainder of the weight loss is principally due to the partial gasification of the fixed carbon fraction, which occurs during activation. Thus, irrespective of the obvious benefits of producing a high value commodity, i.e., activated carbon, the physical activation process is highly advantageous because of the sharp reduction it induces in the quantity of sludge to be dealt with.

Abrogation of odours

A common problem associated with WTTPs is their impact on the surrounding area in terms of the highly unpleasant odours that they generate. Consequences include depressed house prices and moreover, harm to the psychological wellbeing of the local residents, which is induced by the discomfort of being perpetually exposed to malodorous smells and the attendant social opprobrium this entails. The odours stem from the release of gases which are produced during microbial decomposition processes. H_2S is one such gas and it is released during anaerobic decay (Ansari *et al.*, 2005). H_2S is of particular concern not only because of its strongly odoriferous nature but because it is, firstly, toxic and secondly, because in the atmosphere it is easily oxidised to SO_2. When SO_2 comes into contact with atmospheric moisture it can result in the formation of sulphurous (H_2SO_3) acid and hence, acid rain (Ansari and Bandosz, 2005).

Thus, any process which results in the conversion of the sludge into a decomposition-resistant and therefore, odour-less, form will clearly induce a swift reduction in the noxious and toxic gases associated with WWTPs. The SBA production methodology exhibits this potential in abundance. On the other hand, during the carbonisation process H_2S is generated. However, as this occurs under controlled conditions its release can be safely managed so as not to cause a public or an environmental nuisance.

Mineralisation/stabilisation of fixed carbon in sludge

Biochar is essentially the char arising from the pyrolysis (carbonisation) of biomass. Biochar has long been used as a soil amendment because of its noted effects on plant growth and crop yield (Glaser *et al.*, 2002 and Gaunt and Lehmann, 2008). It has recently been recognised that because it results in the biomass becoming degradation resistant, the production of biochar can be a highly effective means of sequestering much of the carbon in biomass. Research has indicated that the carbon consumed through the production of the energy necessary to drive the biochar production process is exceeded by the carbon sequestered (Gaunt and Lehmann, 2008). However, this research assumes that the oils and gases released are utilised for energy production, thus offsetting against the carbon footprint of using fossil fuels to produce the same quantity of energy.

The production of adsorbents from sludge by physical activation is tantamount to the conversion of the sludge into biochar. The issue of whether the SBAs produced are suitable for use as soil amendments remains to be resolved. However, the fact that the SBA production process results in the stabilisation of much of the fixed (i.e., non volatile) carbon is beyond doubt. Due to the fact that some of the fixed carbon is gasified through its reaction with the activant to form either CO or CO_2, it is currently not known whether the SBA production results in a net sequestration or release of carbon (assuming that the furnace is powered with fossil-fuels). However, it is quite feasible that the former scenario applies. Moreover, it is possible to produce SBAs by carbonisation/pyrolysis alone.

Researchers have shown that SBAs produced by carbonisation can attain high levels performance in gas phase contaminant capture/immobilisation applications (Bagreev *et al.*, 2001c). On the other hand, this study has shown that the adsorbents produced solely by carbonisation are not suitable for use in aqueous phase applications. However, these SBAs have been shown by Universitat Autònoma de Barcelona to be suitable for use as biofilter substrates. In this application the SBAs act as supports for microbial populations that are able to digest/consume hazardous gases. The activated carbon used in the activated sludge process fulfils a similar role in the aqueous phase, in that it essentially acts as a support for bacteria., It is therefore probable that SBAs produced by carbonisation could also be used in the activated sludge process.

Utilisation of adsorbents for decontamination of polluted water and gases

A key aim of this study was to produce adsorbents that could achieve high levels of performance in water treatment applications. Thus far, the SBAs produced have been shown to attain competitive levels of performance in the CWAO process (Marques *et al.*, 2011) and to exhibit high capacities for organic dyes and EDCs. The results indicate, moreover, that SBAs exhibit great potential for many other gaseous phase pollutants.

Hence, the study's aim is considered to have been comprehensively achieved. The ramifications of this are immense. In essence what has been developed is a low/zero cost feedstock for producing activated carbons. This therefore raises the possibility of the decontamination of polluted water courses and effluents that, previously, had lain untouched due to the prohibitive cost of their treatment.

Although, in the developed world, the effects of this discovery may not be on a significant scale, in the developing world they could be far reaching. There are already many waste streams that can be utilised to produce ACs (Dias *et al.*, 2007; Ioannidou and Zabaniotou, 2007 and Pollard *et al.*, 1992), but the case for using sewage sludge is particularly strong because: firstly, it is present wherever large urban populations are to be found and secondly, it is a waste that requires disposal. Many other waste materials can be stockpiled but sewage sludge, because of both its odorous properties and its high content of potentially hazardous pathogens, is not a waste that can be readily stored. Moreover, unlike many other activated carbon precursors, for example, agricultural by-products, the availability of sewage sludge is not affected by seasonal variation.

Development of a low-tech solution to the growing volumes of sludge being produced in the developing world

In the long term, the conversion of sewage sludge into adsorbents could represent a leading solution to the problem of how to dispose of the excess sewage sludge in the developing world. Moreover, demand for activated carbon in the developing world is certain to increase as industrialisation and awareness of the primacy of the environment and the importance of having access to clean water, grows apace.

The technology required to carbonise and physically activate sewage sludge is not regarded as being complicated: its use as a purifying agent dates back to 1500 BC, when carbonised wood was used by the Egyptians as a medicinal agent. Furthermore, the ancient Hindus in India utilised charcoal to filter their drinking water (Bansal *et al.*, 1988). The issue of where to derive the energy necessary to power the plant from could be resolved by making recourse to the carbonisation/pyrolysis hydrocarbon products of the conversion process. The drying stage can be highly energy intensive, but if utilised in countries with warm/hot climates then drying the sludge through prolonged exposure to the air should be feasible. The spent SBAs could be used as biochars or combusted to produce the energy required to drive the production process.

Industrial and economic impacts

Introduction

The approach taken to determining these impacts has been to ascertain whether a commercial plant for producing SBAs by physical activation could operate at a profit. This economic evaluation, which is presented in Table 10.1, was based upon the work conducted by Rio *et al.* (2006). However, it is important to stress that a detailed analysis of the economics of SBA production was beyond the scope of this study. The numbers presented, many of which were derived from the study by Rio *et al.* (2006), should be viewed merely as an indication of the potential financial costs and earnings of the proposed scheme. A description of how the different components of Table 10.1 were produced is given in the text below.

Table 10.1 Estimated annual costs and earnings from producing activated carbon from sewage sludge

	Costs		
SBA preparation step:	**Drying**	**Pyrolysis**	**Steam activation**
Investments (capital payment) (€)	2,300,000		600,000
Pyrolyser	NA	1,260,000	NA
HT Gas treatment		180,000	
Linear depreciation costs (€/year)	153,300	96,000	40,000
Maintenance costs (€/year)	23,000	14,400	6,000
Energy consumption (kWh/tonne)	50	630	700
Daily consumption of sludge (tonnes)	30	6	2.4
Annual electricity consumption	547,500	1,379,700	613,200
Annual electricity cost (€)	41,610	104,857	46,603
Steam consumption (kg/tonne char)	N/A	N/A	280
Annual steam consumption (tonne/year)			245.28
Cost of 1 tonne of steam (€/tonne)			17.80
Steam consumption costs (€/year)			4,365.98
General costs* (€/year)	11,260	23,860	32,300
Total cost (€/year)	229,170	239,117	89,269
Total cost (€/tonne)	21	109	102
Annual cost of running the SBA plant (€/year)		557,556	
Daily cost of running the SBA plant (€/day)		1,527	
Cost of SBA production per tonne of wet sludge (€)		51	
Revenue streams			
Income from landfill costs avoidance (€/tonne wet sludge)		83	
Daily income from landfill costs avoidance		2,490	
Estimated SBA retail price (€/tonne SBA)		360	
Daily income from SBA retail (€)		684	
Total daily income (€)		3,174	
Net profits			
Daily profit (€/)		1,647	
Profit per tonne of SBA produced (€)		867	

**20% of exploitation costs.*

Costs from the conversion of sludge into activated carbon (SBA preparation)

In keeping with the scenario chosen by Rio *et al.* (2006), the sludge has a solids content of 20 wt.% dry solids by weight and the unit processes 30 tonnes of wet sludge a day (approx 11,000 wet tonnes per annum).

The investment/capital costs refer to all those associated with the purchase of the plant (the dryer, activation furnace, pyrolyser, *etc.*). These costs would be incurred, firstly, in the initial start-up phase and subsequently, at the end of each unit's lifetime. The depreciation has been assumed to be over a 15 year period, which is the assumed lifetime of the equipment. However, a lifetime of 25 years is perhaps a more realistic figure for major plant such as furnaces/kilns and driers. The use of a shorter and hence more severe depreciation regime could strengthen the economic case for the SBA production facility, by demonstrating its financial robustness under a shorter period and hence more stringent operational criteria. Usually, it would be necessary to take a loan to pay for the capital costs. Clearly, the loan peiod will have an impact on the operational costs of the facility. However, present interest rates are currently at an historic low which would benefit this project if the right terms were negotiated. This study was undertaken to obtain an indication of the potential costs of creating and operating a SBA system. Some simplifications have been made in the costing of the proposed scheme, thus the data can only be regarded as indicative.

The drying costs are predicated on the assumption that the sludge is dried for 3 hrs at 105°C and that the sludge is carbonised at 600°C for 1 hr. Much of the activation work conducted for this project did not include a dedicated carbonisation step. However, if a singe step approach was adopted commercially, then the uncontrolled discharge of the volatile component of the sludge into the activation chamber could result in the fouling of the activation furnace walls and would certainly result in the incursion of significant quantities of water into the pyrolysis products.

It was assumed that the carbonisation step induced a 60% loss in mass of the dried sludge, which is broadly consistent with the loss in mass exhibited by the DRAW sludge at this carbonisation temperature. The DRAW sludge was selected as the feedstock because the SBAs produced from it exhibited the highest surface areas of those produced from this project.

Accordingly, the weight loss induced by the activation of the char was assumed to be 20 wt.%, which aligned with the weight loss exhibited by the DRAW-derived char under the optimised conditions for steam activation. Thus, 2.4 tonnes char would yield ca. 1.9 tonnes of activated carbon and 15.8 tonnes of wet sludge would be required to produce one tonne of activated carbon. The same steam flow rate that was employed for the steam activation experiments in the lab (0.7 g H_2O/200 g sludge char) was utilised. It was also assumed that the dwell time for the activation step was 80 minutes.

Revenue from abrogation of landfill costs

It was assumed that if the sludge was converted into activated carbon then the cost of disposing of the wet sludge would be obviated. At present, landfill tax in the UK is levied at £40/tonne waste and this is set to rise at a rate of at £8/year until at least 2013 (H M Revenue & Customs, 2009). A report by WRAP (2008) estimated that the 'gate fee' for disposing of one tonne of waste was £21/tonne. Thus, the total cost of disposing of one tonne of sewage sludge would be £61/tonne. However, when the cost of transporting the sludge and other secondary costs are factored in, it is likely that the cost of disposal would total £70/tonne. Using an exchange rate of £1 = €1.18 (the value of currencies change on a daily basis so these are only nominal values), this equates to €82.60/tonne. The spent SBAs may also have an associated disposal charge. However, it is anticipated that this could be circumvented through their use as biochars (see section 'Mineralisation/stabilisation of fixed carbon in sludge').

Revenue from retail of SBA

The 2008 edition of what is perhaps the definitive guide to the activated carbon markets, namely, the publication The economics of activated carbon (Roskill Information Services Ltd, 2008) stated that in 2007, coal-based steam-activated granular activated carbon for municipal water treatment retailed in the region of US$1,000/tonne. Using an exchange rate of $ 1 = €0.72 (again, these are only nominal values), this equates to €720/tonne. The SBAs produced by physical activation are expected to be primarily used in low grade applications. Consequently, it was therefore to reflect this by pricing the SBA at 50 % of the retail price of the commercial carbon: the price quoted in Table 10.1 is €360/tonne.

Calculation of net profit

As shown in Table 10.1, the average cost of converting one tonne of wet sludge into activated carbon is €51. The daily net profit is found by, firstly, summing the income received (*i.e.*, money saved) from diverting 30 tonnes of wet sludge away from landfill with the income received from selling the SBA produced by its conversion into activated carbon (1.9 tonnes). The cost of the conversion process is then subtracted from the income to yield the daily profit figure.

Other potential sources of revenue/cost offsetting

This study assumes that all the energy needed for the process is bought in. There is significant potential to recover energy from the pyrolysis stage of the process and also to utilise waste heat for sludge drying, which would greatly reduce the need to generate heat for both steps of the process. The calculation of the potential revenue that would be generated by such a scheme was beyond the scope of this project. However, for future work it would constitute an important line of research.

The implementation of a SBA production scheme would obviate the need to have any sludge stabilisation/digestion facilities. The costs saved by this should, at least in part, offset the costs of the SBA production unit.

A further source of income is carbon credits. As previously mentioned, the SBAs generated can potentially also be described as biochars. This therefore makes the process eligible for the accruement of carbon credits, which may then be sold/redeemed for the market value of these credits. The drive to decarbonise the economy is growing, as is evidenced by the heated negotiations that have surrounded the 2009 G8 summit held in L'Aquila (Italy), where there have been renewed efforts to agree on a workable cap on greenhouse gas emissions. The potential of the proposed SBA production scheme to sequester carbon can not be overstated and is likely to be a key factor when making the case for the implementation of this scheme.

The prospective costs of the other options for sludge disposal

This study has considered landfill as the only option to SBA production for the disposal/valorization of sewage sludge. Incineration is another alternative, but it is an expensive process, with costs estimated at between €140–300 per tonne of dry solids (Smith, 2004 and Rio *et al*., 2006). Its utilisation is compounded by the fact that it generates residual ash, which can comprise up to 40% of the original volume of the dry sludge, adding to the final disposal charge. Moreover, it could be classified as hazardous waste, this incurring further landfill taxes and disposal costs. The problems associated with incineration are illustrated by the decision, in 2008, of the UK water company Severn Trent Water to close their sludge incineration plant (the Coleshill incinerator). With regards to land application, it was assumed that the proposed system would utilise only sewage sludge that was either ineligible for this use or was surplus to demand.

Conclusions

The utilisation of sewage sludge as an activated carbon feedstock evidently has the potential to be commercially lucrative, with a prospective profit of €392/tonne. This figure is, of course, based on a number of assumptions and would be likely to drop if the application of performance enhancement treatments such as acid washing or hardening were required. However, despite this the business case for the development of a commercial scale SBA production unit remains strong and it is hoped that the realisation of this concept will constitute the ultimate result of this research. A schematic of how the proposed SBA production facility would be incorporated into a sludge management scheme is presented in Figure 10.3.

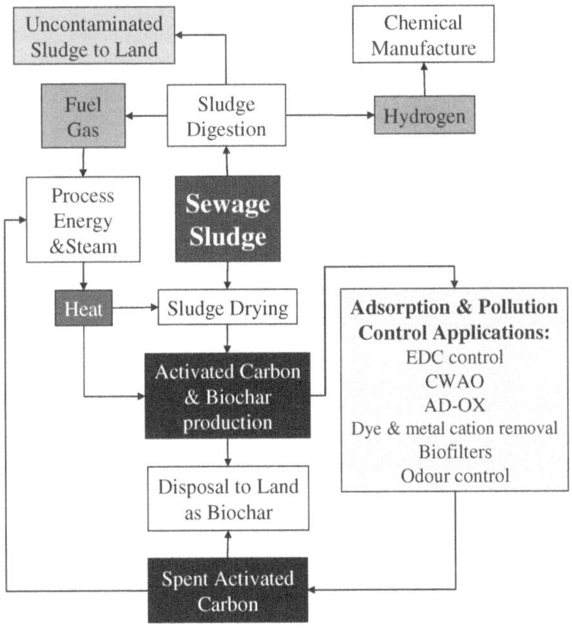

Figure 10.3 A schematic of how a SBA production facility would be incorporated into a sludge management solution

ACKNOWLEDGEMENTS

We wish to acknowledge the following project partners for their provision of technical advice and/or samples: Chemviron Carbon Ltd: Tony Ryan, Harry Sharrock; Université de Nantes: Laurence Le Coq, Rengaraj Selvaraj, Claire Gerente; Technical University of Lódź: Tomasz Bedyk, Stanislaw Ledakowicz, Lech Nowicki; Universitat Rovira i Virgili: Frank Stüber, Rita R.N. Marques, Maretva Baricot and Institut National Polytechnique de Tolouse: Carine Julcour, Henri Delmas.

REFERENCES

Ansari A. and Bandosz T. J. (2005). Inorganic-organic phase arrangement as a factor affecting gas-phase desulfurization on catalytic carbonaceous adsorbents. *Environmental Science & Technology*, **39**, 6217–6224.

Ansari A., Bagreev A. and Bandosz T. J. (2005). Effect of adsorbent composition on H2S removal on sewage sludge-based materials enriched with carbonaceous phase. *Carbon*, **43**, 1039–1048.

ASTM (American Society for Testing and Materials). (1994). Standard Test Method for Ball-Pan Hardness of Activated Carbon.: West Conshohocken, PA.

ASTM (American Society for Testing and Materials). (2002). Standard Test Method for Total Ash Content of Activated Carbon, No. D 2866-94 (Reapproved 1999), West Conshohocken, PA.

Bagreev A., Bandosz T. J. and Locke D. C. (2001a). Pore structure and surface chemistry of adsorbents obtained by pyrolysis of sewage sludge-derived fertilizer. *Carbon*, **39**, 1971–1979.

Bagreev A., Locke D. C. and Bandosz T. J. (2001b). H_2S adsorption/oxidation on adsorbents obtained from pyrolysis of sewage-sludge-derived fertilizer using zinc chloride activation. *Industrial & Engineering Chemistry Research*, **40**, 3502–3510.

Bagreev A., Bashkova S., Locke D. C. and Bandosz T. J. (2001c). Sewage sludge-derived materials as efficient adsorbents for removal of hydrogen sulfide. *Environmental Science & Technology*, **35**, 1537–1543.

Bansal R. C., Donnet J. B. and Stoeckli F. (1988). Active Carbons, New York, Marcel Dekker, p. 482.

Brunauer S., Emmett P. H. and Teller E. (1938). Adsorption of Gases in Multimolecular Layers. *Journal of the American Chemical Society*, **60**, 309–319.

Cannon F. S., Snoeyink V. L., Lee R. G. and Dagois G. (1994). Reaction-mechanism of calcium-catalyzed thermal regeneration of spent granular activated carbon. *Carbon*, **32**, 1285–1301.

Dias J. M., Alvim-Ferraz M. C. M., Almeida M. F., Rivera-Utrilla J. and Sánchez-Polo M. (2007). Waste materials for activated carbon preparation and its use in aqueous-phase treatment: A review. *Journal of Environmental Management*, **85**, 833–846.

Fitzmorris K. B., Lima I. M., Marshall W. E. and Reimers R. S. (2006a). Anion and cation leaching or desorption from activated carbons from municipal sludge and poultry manure as affected by pH. *Water Environment Research*, **78**, 2324–2329.

Fitzmorris K. B., Lima I. M., Marshall W. E. and Reimers R. S. (2006b). Anion and cation removal from solution using activated carbons from municipal sludge and poultry manure. *Journal of Residuals Science & Technology*, **3**, 161–167.

Gaunt J. L. and Lehmann J. (2008). Energy balance and emissions associated with biochar sequestration and pyrolysis bioenergy production. *Environmental Science & Technology*, **42**, 4152–4158.

Glaser B., Lehmann J. and Zech W. (2002). Ameliorating physical and chemical properties of highly weathered soils in the tropics with charcoal – a review. *Biology and Fertility of Soils*, **35**, 219–230.

H M Revenue & Customs. (2009). Notice LFT1: A general guide to landfill tax. London, p. 58.

Ioannidou O. and Zabaniotou A. (2007). Agricultural residues as precursors for activated carbon production – A review. *Renewable and Sustainable Energy Reviews*, **11**, 1966–2005.

Karacan F., Ozden U. and Karacan S. (2007). Optimization of manufacturing conditions for activated carbon from Turkish lignite by chemical activation using response surface methodology. *Applied Thermal Engineering*, **27**, 1212–1218.

Lillo-Rodenas M. A., Ros A., Fuente E., Montes-Moran M. A., Martin M. J. and Linares-Solano A. (2008). Further insights into the activation process of sewage sludge-based precursors by alkaline hydroxides. *Chemical Engineering Journal*, **142**, 168–174.

Lippens B. C. and deBoer J. H. (1965). Studies on Pore systems in Catalysts V. The t-plot method. *Anales de Quimica*, **4**, 319–323.

Liu Z.-H., Kanjo Y. and Mizutani S. (2009). Removal mechanisms for endocrine disrupting compounds (EDCs) in wastewater treatment – physical means, biodegradation, and chemical advanced oxidation: A review. *Science of The Total Environment*, **407**, 731–748.

Marques R. R. N., Stüber F., Smith K. M., Fabregat A., Bengoa C., Font J., Fortuny A., Pullket S., Fowler G. D. and Graham N. J. D. (2011). Sewage sludge based catalysts for catalytic wet air oxidation of phenol: Preparation, characterisation and catalytic performance. *Applied Catalysis B: Environmental*, **101**, 306–316.

Massanet-Nicolau J., Dinsdale R. and Guwy A. (2008). Hydrogen production from sewage sludge using mixed microflora inoculum: Effect of pH and enzymatic pretreatment. *Bioresource Technology*, **99**, 6325–6331.

Mitchell M., Ernst W., Rosmussen E., Bagherzadeh P. and Lightsey G. (1978). Adsorption of textile dyes by activated carbon produced from agricultural, municipal and industrial wastes. *Bulletin of Environmental Contamination and Toxicology*, **19**, 307–311.

Pollard S. J. T., Fowler G. D., Sollars C. J. and Perry R. (1992). Low-Cost Adsorbents for Waste and Waste-Water Treatment – a Review. *Science of the Total Environment*, **116**, 31–52.

Pullket S., Smith K. M., Fowler G. D. and Graham N. J. D. (2008). Influence of source and treatment method on the properties of activated carbons produced from sewage sludge, ECSM'08 – European Conference on Sludge Management, 1–2 September, Liege, Belgium.

Pullket S., Smith K. M., Fowler G. D. and Graham N. J. D. (2009). Influence of Source and Treatment Method on the Properties of Activated Carbons Produced from Sewage Sludge. *Journal of Residuals Science & Technology*, **6**, 43–49.

Rio S., Faur-Brasquet C., Le Coq L., Courcoux P. and Le Cloirec P. (2005). Experimental design methodology for the preparation of carbonaceous sorbents from sewage sludge by chemical activation – application to air and water treatments. *Chemosphere*, **58**, 423–437.

Rio S., Le Coq L., Faur C. and Le Cloirec P. (2006a). Production of porous carbonaceous adsorbent from physical activation of sewage sludge: application to wastewater treatment. *Water Science and Technology*, **53**, 237–244.

Rio S., Le Coq L., Faur C., Lecomte D. and Le Cloirec P. (2006b). Preparation of adsorbents from sewage sludge by steam activation for industrial emission treatment. *Process Safety and Environmental Protection*, **84**, 258–264.

Ros A., Lillo-Rodenas M. A., Canals-Batlle C., Fuente E., Montes-Moran M. A., Martin M. J. and Linares-Solano A. (2007). A new generation of sludge-based adsorbents for H_2S abatement at room temperature. *Environmental Science & Technology*, **41**, 4375–4381.

Ros A., Lillo-Rodenas M. A., Fuente E., Montes-Moran M. A., Martin M. J. and Linares-Solano A. (2006). High surface area materials prepared from sewage sludge-based precursors. *Chemosphere*, **65**, 132–140.

Roskill Information Services Ltd. (2008). The Economics of Activated Carbon, London, p. 322.

San Miguel G., Fowler G. D. and Sollars C. J. (2003). A study of the characteristics of activated carbons produced by steam and carbon dioxide activation of waste tyre rubber. *Carbon*, **41**, 1009–1016.

Smith S. R. (2004). Is Agricultural Recycling More Sustainable Than Incineration?, The Proceedings of the 9th European Biosolids and Biowastes Conference, 15–17 November, Wakefield (UK).

Smith K. M., Fowler G. D., Pullket S. and Graham N. J. D. (2009). Sewage sludge-based adsorbents: A review of their production, properties and use in water treatment applications. *Water Research*, **43**, 2569–2594.

Wächter R. and Cordery A. (1999). Response surface methodology modelling of diamond-like carbon film deposition. *Carbon*, **37**, 1529–1537.

Warhurst A. M. and Fowler G. D., Mcconnachie G. L. and Pollard S. J. T. (1997). Pore structure and adsorption characteristics of steam pyrolysis carbons from Moringa oleifera. *Carbon*, **35**, 1039–1045.

Wrap (2008). Wrap Gate Fees Report: Comparing the cost of alternative waste treatment options, Banbury, Oxfordshire (UK), WRAP, p. 3.

Chapter 11

Utilisation of activated carbon as catalyst in wet air oxidation

Lead authors: Frank Stüber, Azael Fabregat Llangostera, Christophe Bengoa, Josep Font Capafons, Agustí Fortuny (Universitat Rovira i Virgili), Rita R.N. Marques, Rosa M. Quinta-Ferreira and Rodrigo J. G. Lopes (Universidade de Coimbra)

Co-authors: Chemviron Carbon Ltd., Imperial College of London, Tratamientos y Recuperaciones Industriales S.A.

SUMMARY

Project objectives

The main goal of this work was to investigate the utilisation of Wet Oxidation technology for phenolic wastewater remediation employing activated carbons prepared from sewage sludge (SBAC = Sludge Based Activated Carbon) as the catalytic promoter of the reaction process. The study included experiments in laboratory and semi-industrial pilot trickle-bed reactors, and process analysis by means of kinetic modelling and Computational Fluid Dynamics (CFD) simulation.

Work performed

In order to select carbons suitable for their use in catalytic wet air oxidation (CWAO) of phenolic contaminants, more than 15 SBACs as well as two commercial carbons were tested in batch and trickle bed reactors. The sludge feedstock used for carbon preparation was two England sourced samples labelled as DRAW (Dewatered RAW) and DMAD (Dewatered Mesophilic Anaerobically Digested). The experiments were undertaken to ascertain the catalytic activity (compound and TOC removals) and stability (resistance to carbon burn off, metal leaching and mechanical attrition) of the prepared SBACs. Imperial College of London was responsible for the production of carbons; however the final optimisation of carbons was achieved by continuous feedback form Universitat Rovira i Virgili on obtained oxidation results. Due to its industrial relevance, phenol was the model contaminant subjected to CWAO, although other phenolic compounds such as o-cresol, o-chlorophenol and p-nitrophenol were also oxidised in batch and continuous reactors.

The work on CFD (Universidade de Coimbra) was started with the development of Euler-Euler k-fluid and Volume-of-Fluid CFD models to provide a behaviour analysis in transient conditions of a pilot scale TBR. CFD codes

 Reduction, Modification and Valorisation of Sludge (REMOVALS). Edited by Christophe Bengoa, Azael Fabregat, Josep Font and Frank Stüber. ISBN: 9781843393450. Published by IWA Publishing, London, UK.

were used for the purpose of multiphase reacting flow prediction, namely based on Eulerian and Volume-of-Fluid models. Key hydrodynamic parameters such as liquid holdup and pressure drop were previously validated with the exception of catalyst wetting efficiencies under trickle-flow regime. Also, the catalytic wet air oxidation based on manganese-cerium catalysts was modelled with Volume-of-Fluid model under transient conditions. The hydrodynamic prediction of catalyst wetting efficiency by means of interface tracking schemes will be accomplished using level set methods for gas-liquid-solid surface reconstruction. Finally, the integration of sludge based activated carbon kinetics in the computational model will be accomplished towards the successful implementation and deployment of CWAO technology.

End results

An important outcome of the completed experimental work is the knowledge on how carbon preparation affects the catalytic performance of SBACs. At the conclusion of the project several SBACs are available that demonstrated on laboratory scale to perform acceptable (carbonised and physically activated carbons) or very similar (chemically activated carbons plus acid washing) in CWAO of phenolic contaminants when compared to commercial carbons specially recommended for wastewater treatment. This knowledge should allow establishing guidelines for the production of SBACs with catalytic activity tailored to the specific application needs. Moreover, a range of temperature (140 to 160°C) and partial oxygen pressure (2 to 4 bar) ensuring both carbon activity and stability were assessed using a laboratory scale Trickle Bed Reactor as wastewater treatment on industrial scale has generally to be conducted in continuous mode.

A mayor result of the process analysis is that computational flow modelling techniques can successfully predict gas-liquid distribution and pressure gradient in TBR, mainly with Eulerian codes. Local temperature variation and Total Organic Carbon profiles were evaluated axially and radially providing a more rigorous physical description of the underlying complex two phase flow. Computational runs exhibited back mixing phenomena, poor radial mixing and revealed the existence of hot spots in the simulated trickle flow regime.

METHODOLOGIES AND APPROACHES EMPLOYED

Integrated approach employed

To achieve the main goal, that is to obtain stable and active SBACs for the continuous CWAO of phenolic wastewater, a combined experimental-theoretical approach was developed in the work (see Figure 11.1). The carbons initially prepared by ICL from recollected municipal wastewater excess sludge were tested by URV and further optimised by a means of continuous feedback between partners on their observed CWAO performance and suggestions for improvement of the preparation methodology. In a parallel design orientated study, Universidade de Coimbra analysed in detail the operation and scale up of larger Trickle Bed Reactor applied to CWAO by using advanced tools of CFD simulation.

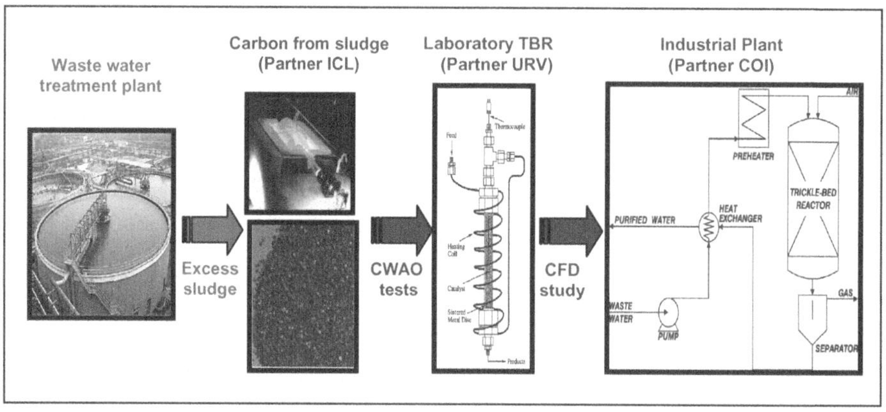

Figure 11.1 Schematic chart of integrated approach employed in work 'Utilisation of activated carbon as catalyst in wet air oxidation'

Experimental methodology

Stirred batch reactor

Figure 11.2 shows a detailed scheme of the batch reactor used for the oxidation tests. The main element consists in a 316 SS Autoclave (capacity of 300 ml) proved up to 450°C and 345 bar. A jacket-type furnace and a control device allow monitoring and controlling the oven temperature. To this end, a thermocouple is introduced into a thermo well to assess the reaction temperature and to actuate on the furnace power supply. For proper temperature control inside the reactor an internal coil where ambient water circulates at a convenient flow rate is required. Vigorous stirring is ensured by a motor impelling a straight blade turbine-type agitator. Agitation speed can be fixed by means of a speed controller device. The gas stream proceeds from an external high pressurised bottle and the reactor pressure can be adjusted using a pressure reducer placed upstream of the reactor gas inlet. The system can be depressurised with the help of a valve fixed on the vessel top. During experiments, the pressure inside the reactor is monitored through a pressure gauge, while liquid samples are regularly withdrawn through a sample valve assembly.

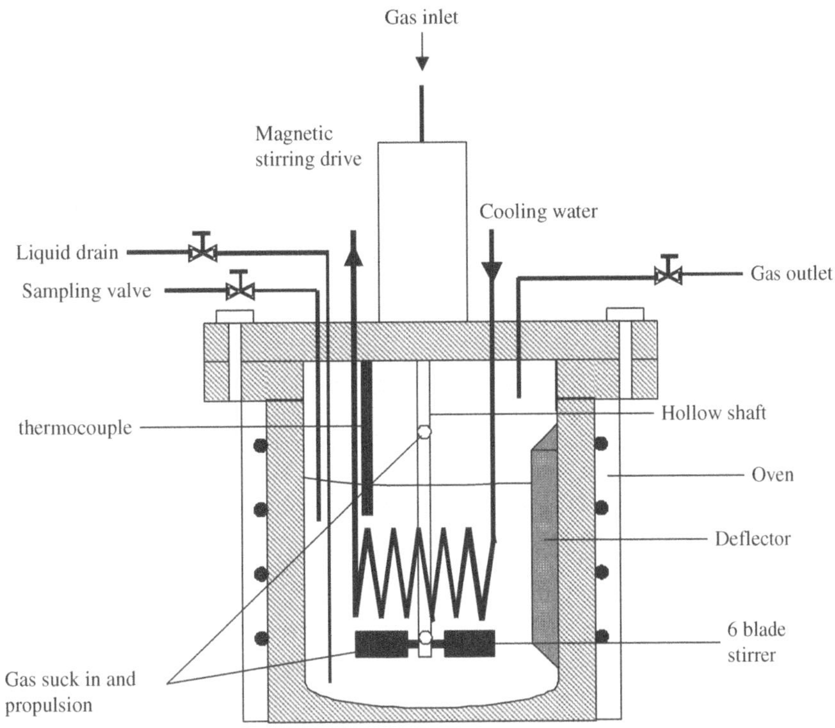

Figure 11.2 Batch reactor for CWAO tests of sludge derived carbons

Typical batch experiments with 1 or 5 g/L of contaminant solutions are conducted for 4 h at 160°C and 4 bar of oxygen partial pressure. To start the experiment, 100 mL of solution and 2 g of sludge based carbon are loaded into the reactor vessel, which is then closed. Then, the reactor is pressurised with nitrogen to check for possible leaks. If no leaks are detected, heating and stirring at low speed are started to establish the desired temperature inside the reactor. As soon as the temperature attains the set point, cooling water is fed to the coil at a fixed flow rate to dampen temperature fluctuations caused by the furnace off set. The system needed about 40–50 min to reach stable conditions of temperature. Prior to start the reaction, a first liquid sample is withdrawn to assess the amount of contaminant adsorbed on the carbon and synthetic air is allowed to the vessel to establish a system pressure of 25 bar. The reaction is finally initiated by a sudden increase of the stirrer speed to 1000 rpm. During the 4h lasting experiment liquid samples are withdrawn at regular time intervals to measure phenol and TOC concentration as well as pH of reaction solution. At the end of each experiment, the reactor is opened to recover both aqueous solution and catalyst for further analysis or reuse.

Continuous trickle bed reactor

A laboratory Trickle Bed Reactor (TBR) was employed for the continuous oxidation tests (Figure 11.3). It contains a 5 L stirred feed tank, which is connected to a high precision metering pump that can deliver up to 150 mL/h at a maximum pressure of 350 bar. After dampening the pump pulsation, the liquid flow rate is measured by means of a mass flow meter. The liquid and gas are fed in downflow through a SS tube (20 cm long, 1.1 cm i.d.) normally loaded with 7 g of carbon. A thermocouple is inserted axially to measure the mid temperature inside the bed. The carbon is retained by two sintered metal grids at the bottom and top of the reactor. The reactor with the preheating coil is placed in a temperature-controlled oven. The gas and the liquid streams are mixed and pre-heated before entering the reactor. Downstream of the reactor exit, there are 2 Gas-Liquid separators, a liquid sampling valve, a needle valve and a flow meter to adjust and measure the gas flow rate.

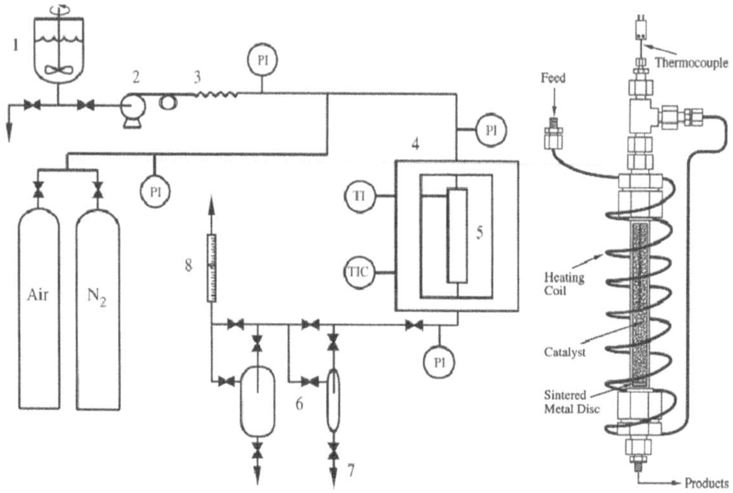

Figure 11.3 Experimental set-up for continuous CWAO

Standard 72 h experiments with 5 g/L of contaminant solutions were conducted in the TBR at a given liquid space time of 0.12 h and 120–160°C and 2–8 bar of oxygen partial pressure. Prior to start the experiment, the reactor was carefully loaded with carbon particles crushed down to a diameter of 0.3 to 0.7 mm to obtain a uniform catalyst bed. Then, the reactor was closed and pressurised first under N_2 to check for possible leaks in the system. If no leaks were detected, the oven was heated to the selected reaction temperature, and air pressure was readjusted. As soon as reactor temperature and air pressure reach its operating values, liquid flow is set to the desired rate verified by downstream measurements coming from a liquid mass flow meter. During the runs, outlet liquid samples are periodically withdrawn, measured for pH and then stored at 4°C prior to be analysed on phenol, COD or TOC content. At the end of each run, the carbon is dried overnight at 105°C under nitrogen flow, weighted and stored under inert atmosphere for further analysis. Results obtained in a similar way with the commercial carbon served as a reference to evaluate the CWAO performance of sludge based carbons in terms of activity and stability.

Materials and analytical methods

Analytical grade aromatics were all provided by Aldrich. The commercial carbons tested were from Chemviron Carbon Ltd. (AP4-X) and Merck. All sludge based carbons were provided from Imperial College of London. The synthetic air oxidant fed to the reactor has a purity of 99.995%.

Contaminant concentrations of liquid samples were determined by HPLC. To prepare the mobile phase, HPLC grade methanol and ultra-pure-water were mixed in proportion of 35:36 (v/v). The analysis was performed with a C18 reverse phase column (Hypersil ODS) and at a mobile phase flow rate of 1 ml/min and a wavelength in the diode array detector of 254 nm. COD was measured with the standard method 5220D, whereas TOC was quantified following the protocol 5310B.

At the end of the CWAO tests, the used carbons were collected, dried overnight and then weighted. Approx. 0.25 mg of the recovered sample was also subjected to thermogravimetric analysis carried out in a thermobalance heating the sample from 100 to 900°C under nitrogen flow at a heating rate of 10°C/min.

To measure the release of metals from the carbons during CWAO all batch reactor solutions and liquid samples from the continuous TBR were recovered and filtered to allow for detection of metals. Detection and quantification of metals are done by means of Atomic Absorption and standard solution of known metal concentration. Analysis includes the following metals: Fe, Cu, Mn, Mg, Zn, Co and Ca.

CFD simulation of CWAO in trickle bed reactor

The development of a TBR model was accomplished using a multiphase CFD approach incorporated in the FLUENT 6 software package by ensemble averaging of the local instantaneous balances for equations for mass, momentum and energy. Momentum transfer between the phases is modelled through an interphase drag term. The methodology consists in dividing the mesh domain in subdomains so that the decomposition of the catalytic bed was unstructured in order to manage with the geometric complexity. The momentum equations are solved with the coupling SIMPLE algorithm and the monotonic upwind discretization scheme. The pressure is computed by means of the PRESTO scheme and model equations were typically solved in a transient fashion with a time step of 10-3 s. The residuals convergence was accelerated by under-relaxation parameters: 0.4 for pressure and 0.8 for velocity vector field; inlet boundary conditions are assigned at the top distributor and outlet conditions at the free surface.

ACHIEVEMENTS RELATED TO THE STATE OF THE ART

Introduction

The main achievements of the work are: the collaboration in the development of original carbons from different sewage sludge feedstock, the assessment of their resulting catalytic activity and stability when subjected to batch and continuous oxidation of phenolic effluents and CFD simulations of CWAO conducted in a pilot scale trickle bed reactor.

Pioneering the use of sludge based carbons as a catalytic material in wet air oxidation of phenolic effluents

The production of active carbon from sewage sludge feedstock has been patented as early as 3 to 4 decades ago (Kemmer *et al.*, 1971; Nickerson *et al.*, 1975 and Lewis, 1978). However, its use as a direct solid catalyst for the treatment of organic waste water effluents has yet not been reported in the recent review literature (Stüber *et al.*, 2005). Most of the studies dedicated to this topic have primarily centred on the production of activated carbons with high surface areas and their applications to liquid phase adsorption of organic pollutants or heavy metals and to adsorption/oxidation of H_2S and NO_2 from contaminated gaseous effluents (Dias *et al.*, 2007 and Smith *et al.*, 2009). Therefore, we can state that our experimental study on CWAO of phenolic effluents using for the first time sludge based carbon as catalyst in a Trickle Bed Reactor is a pioneer work in the field of waste water treatment with sludge based active carbons.

WAO or CWAO are commercially established techniques for the treatment of diverse municipal and industrial waste water streams at large scale (Busca *et al.*, 2008; Levec and Pintar, 2007 and Bhargava *et al.*, 2006). Existing WAO processes (ZIMPRO, WETOX, VERTECH, KENOX, OXYJET) use more severe operating conditions of temperature and pressure than do CWAO processes (homogeneous: LOPROX, ATHOS, CIBA-GEIGER, WPO, ORCAM; heterogeneous: NIPPON SHOKUBAI, OSAKA GAS, CALIPHOX), leading to a retail in overall remediation costs per m^3 of effluent treated (see also Table 11.1 in sub-section 'Industrial and economic Impacts' of section 'Impact of the technology on its industry or research sector'). Nevertheless, relatively few works have been published to date on the catalytic liquid phase oxidation of organic compounds in continuous Trickle Bed Reactors and further experimental data, description of mass transfer processes and reactor simulation studies are required for the successful commercial exploitation of continuous CWAO technology (Levec and Pintar, 2007).

Table 11.1 Metal concentration measured after 4h batch reaction: P_{O2}: 4 bar, T: 160°C, $C_{Ph,0}$: 5 g/L, W: 2 g

SBAC	[Fe] (ppm)	[Cu] (ppm)	[Cr] (ppm)	[Mn] (ppm)	[Mg] (ppm)	[Ca] (ppm)
DM-SA	13.1	1.3	3.0	1.9	15.2	16.2
DM-CO2	21.9	1.2	0.4	1.1	61.0	4.0
DMK_A	0.5	0.0	0.0	0.3	3.1	0.7
DMK_W	0.2	0.0	0.0	0.0	56.9	3.2
DM1000 – C	4.4	0.2	0.0	2.8	59.7	21.2
DR-SA	1.9	1.2	0.0	0.1	16.7	n.a.
DR-CO2	3.5	0.6	0.0	0.1	60.7	6.0
DR250 – C	125.0	1.2	1.4	1.1	4.4	4.4
DR500 – C	9.2	0.2	0.2	0.6	67.1	12.8
DR1000 – C	0.1	0.0	0.0	0.0	4.9	75.0

Catalytic wet air oxidation over sludge based active carbons in stirred batch reactor

Effect of sludge origin on active carbon performance in phenol oxidation

Batch experiments revealed that the Dewatered RAW sludge is a better carbon feedstock than the Dewatered Mesophilic Anaerobically Digested sludge irrespectively of the preparation method applied (see Figures 11.4a and 11.4b). Dewatered undigested sludge has a higher organic content and less mineral ash content and is economically more attractive because it does not require any further pre-treatment. For example, after 4 h of phenol (5 g/L) oxidation at 160°C and 4 bar of oxygen partial pressure, DRAW sludge activated with CO_2 performed conversion of phenol and TOC of 69% and 61%, respectively, compared to 58% and 47% obtained in the case of DMAD sludge activated with CO_2 at the same conditions.

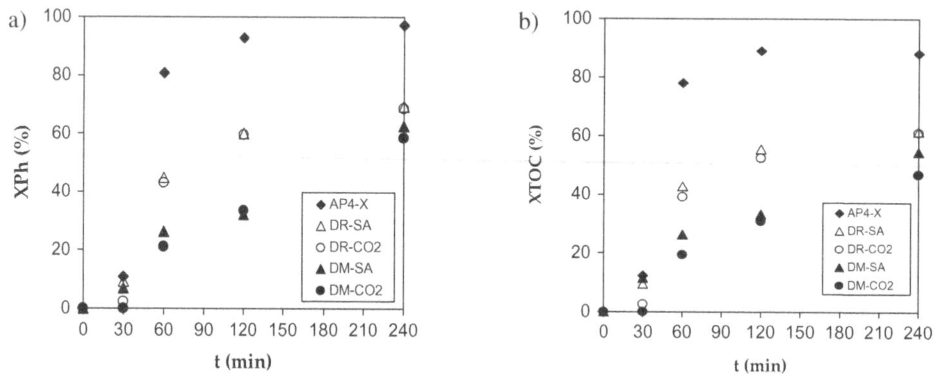

Figure 11.4 Influence of sludge precursor on a) Phenol and b) TOC removal; batch operating conditions: P_{O2}: 4 bar, T: 160°C, $C_{Ph,0}$: 5 g/L, W: 2 g

It is known that sewage sludge does exhibit some fluctuation in its composition (fixed carbon, volatiles, ash). Nevertheless, tests with carbons prepared from sludge loads that were recollected at different days showed small variation of only ±10% in their catalytic and adsorption performance. This finding is important because it confirms that seasonal variation in the sludge feedstock does not significantly alter the reproducibility of the carbon preparation and their adsorption-oxidation behaviour.

Effect of carbonisation temperature on active carbon performance in phenol oxidation

A mayor discovery has been that the production of stable DRAW carbons requires carbonisation temperatures in the range of 800 to 1,000°C. The so prepared carbons ensure acceptable phenol and TOC conversions (see phenol and TOC conversions reported in Figures 11.5a and 11.5b) and at the same time are resistant to metal leaching. Carbons carbonised at temperatures below 500°C were found to release into the reaction solution considerable amounts of both organic matter (see Figure 11.5b) and metals, in particular catalytic active Fe (125 ppm after 4 h of reaction for the carbon carbonised at 250°C).

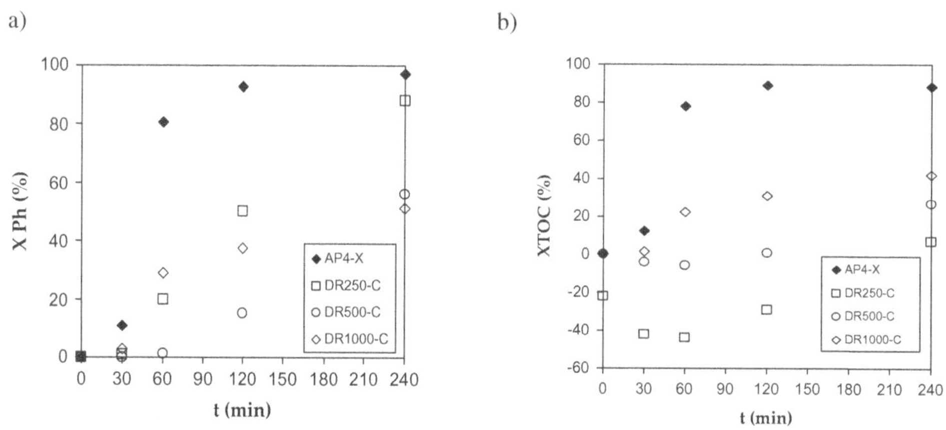

Figure 11.5 Influence of carbonisation temperature on a) Phenol and b) TOC removal; batch operating conditions: P_{O2}: 4 bar, T: 160°C, $C_{Ph,0}$: 5 g/L, W: 2 g

Effect of physical activation and posterior acid washing on active carbon performance in phenol oxidation

The oxidation of phenol over carbons prepared by steam or CO_2 activation led to very similar catalytic performance in terms of phenol and TOC conversions, although higher than those obtained for carbonised carbons (see differences in Figures 11.5 and 11.6). Enhancement of phenol demineralisation achieved by physically activated DRAW carbon is about 50% with a TOC conversion of 62% against 42% obtained with the DRAW carbon carbonised at 1,000°C. A further increase of the catalytic activity can be performed by posterior acid washing of the carbon with HCL solution. In case of the steam activated DRAW carbon, soaked for 24 h in 3% HCL solution, 80% of the TOC fraction corresponding to a 5 g/L solution of phenol were destroyed after 4 h of reaction. Acid washing is thus an efficient post treatment for improving the catalytic activity of SBACs. This result is significant as for a commercial Merck carbon, washed with HCl solution to eliminate the mineral matter of the carbon, a negative effect on conversion (50% decrease compared to the untreated carbon) was reported by Santiago *et al.* (2005). The major drawbacks of acid washing are related to the increasing carbon production costs and the environmental contamination due to the use of HCl solutions.

Effect of chemical activation and posterior acid washing on active carbon performance in phenol oxidation

The work conducted on K_2CO_3 activated carbons revealed that chemical activation is the most efficient but also the most expensive method to produce active carbons. The use of environmentally friendlier K_2CO_3 as the chemical activation reagent with subsequent water washing resulted in DMAD carbons (DMK_W) with both high surface areas and phenol and TOC removals of 87% and 76%, respectively (see Figures 11.6a and 11.6b). As for physically activated carbons, the chemically activated DMAD carbon subjected to acid washing with 3% of HCl solution (DMK_A) exhibited a superior oxidation potential reaching 93% and 82% of phenol and TOC destruction at the same reaction conditions (see Figures 11.6a and 11.6b). For comparison, the commercial Chemviron Carbon AP4-X performed similar phenol and TOC removals of 97% and 88%.

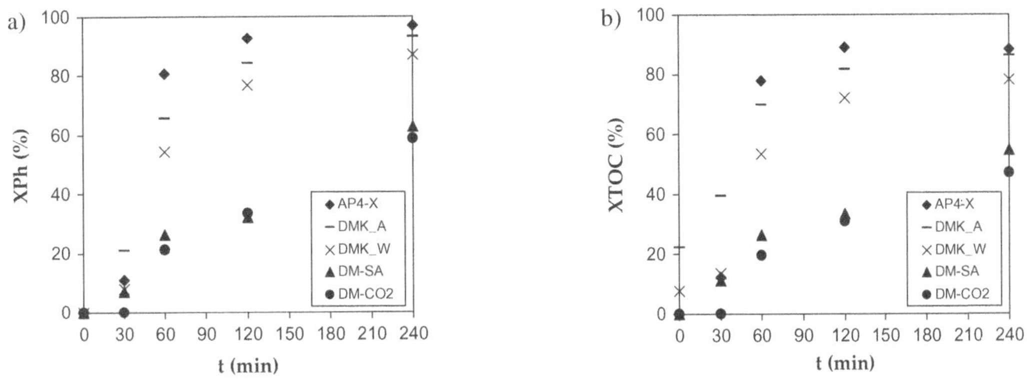

Figure 11.6 Influence of physical and chemical activation on a) Phenol and b) TOC removal; batch operating conditions: P_{O2}: 4 bar, T: 160°C, $C_{Ph,0}$: 5 g/L, W: 2 g

Quantification of metal leaching from sludge based active carbon during batch oxidation of phenol

SBACs are known to have a high residual mineral content that attained values of 70% for DRAW carbons and up to 80% for DMAD carbons. A key problem associated with the use of such carbons is their high propensity towards metal leaching at acidic pH values (Fitzmorris *et al.*, 2006).

At the reaction conditions studied in this work, final pH of the reaction solutions were ranging from 3 to 7, most probably due to a combined action of low chain carboxylic acids formation during CWAO and the more or less basic nature of the SBACs. All metals under investigation (Fe, Cu, Cr, Mn, Mg and Ca) leached into solution (see Table 11.1). Carbons exhibited a greater tendency towards leaching Ca (for DMAD: 1 to 20 ppm and for DRAW: 0 to 75 ppm) and Mg (for DMAD: 3 to 61 ppm and for DRAW: 5 to 67 ppm).

However, these metals were not considered to pose a significant problem. Catalytically active and more toxic metal ions such as Cu and Fe were leached in quantities below or close to the established threshold values for surface discharge (1 ppm for Cu and 10 ppm for Fe). Most stable, almost leaching free, carbons were the DRAW carbon carbonised at 1,000°C and the K_2CO_3 activated carbon leading to a final pH of solution of 7.1 and 6.9, respectively. This corroborates the findings of Fitzmorris *et al.* (2006) who reported that metal leaching is pH dependent and strongly reduced at neutral values. Control of pH of the reaction medium can therefore be a simple and efficient solution to prevent or reduce undesirable metal leaching of sludge based carbons, that otherwise would require a posterior treatment of the effluents coming from the CWAO process.

Batch oxidation of phenol, o-cresol, o-chlorophenol and p-nitrophenol over sludge based active carbons

Real waste water effluents are in general of complex composition and it is important to assess the catalytic potential of SBACs for other phenolic compounds. Oxidations tests of substituted phenols performed with a selected carbon prepared from each sewage sludge feedstock allowed establishing the following order of catalytic reactivity of compounds: o-cresol > phenol > o-chlorophenol >> p-nitrophenol.

The same order was found for all carbons tested (DRAW-SA, DMAD-K_2CO_3 and commercial AP4-X). Very close and high compound and TOC removals were observed for commercial AP4-X and sludge based carbons in the case of o-cresol, phenol and o-chlorophenol (see Figures 11.7, whereas p-nitrophenol proved to be very refractory to oxidation over active carbon at the given conditions of temperature and oxygen pressure. Recent results reported by Santos *et al.* (2006) and Suarez-Ojeda *et al.* (2005) on two other highly active commercial active carbons from Merck and Chemviron Company (Industrial React) are in line with our study and reaffirm the promising catalytic potential that have sludge based carbons in the oxidation of certain substituted phenolic compound families.

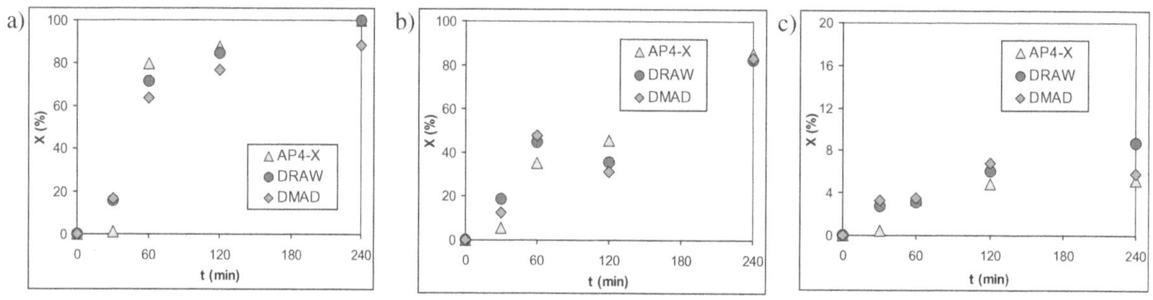

Figure 11.7 Oxidation of phenolic compounds over sludge based activated carbons: a) o-cresol, b) o-chlorophenol, c) p-nitrophenol. Batch operating conditions: P_{O2}: 4 bar, T: 160°C, $C_{Ph,0}$: 5 g/L, W: 2 g

Conclusions

A mayor conclusion of the oxidation experiments performed in a batch reactor is that the reactivity of sludge based carbons in CWAO of phenolic effluents is strongly dependent on the preparation method applied. The following order starting with the lowest catalytic performance can be established from the results: carbonised sludge based carbons < carbons prepared by steam or CO_2 activation < physically activated carbons with posterior acid washing < chemically activated carbons < chemically activated carbons with posterior acid washing ≈ commercial carbons.

However, an improvement in catalytic performance is associated with the use of more expensive activation reagents or 2 steps treatment (activation plus acid washing) and ergo adversely affects the overall production costs of the carbon. It has to be pointed out that the most efficient preparation method (chemical activation plus acid washing) is capable to provide sludge based carbons for the CWAO that are fully competitive with the most reactive carbons commercially available (among others Chemviron AP4-X, Chemviron Industrial React and Merck). These latter carbons perform on a par with many of the noble metal or (mixed) metal oxide supported catalysts tested in CWAO of phenol (see Table 11.2 in Eftaxias *et al.*, 2005).

Table 11.2 Weight loss of SBACs after 72 h of continuous CWAO (ΔW)

Experiment	DMSA-140-2	DMSA-160-2	DMSA-160-4	DRSA-H-140-4	DRSA-H-160-4
ΔW (%)	12.4	−18.3	−33.4	−12.1	−35.5

Catalytic wet air oxidation over sludge based active carbons in continuous trickle bed reactor

Industrial waste water treatment requires the use of active and stable supported catalysts in continuous reactors such as trickle bed reactors to treat the high volumetric quantities of waste water effluents generated. In order to assess the long term activity and stability of SBACs, oxidation tests of 3 days were conducted in a laboratory trickle bed reactor to determine adequate operating conditions for catalyst activity and stability, i.e. resistance to burn off, metal leaching and mechanical attrition. Selected carbons were the economically more attractive steam activated DMAD and steam activated hardened DRAW carbons.

Determination of operating conditions for CWAO of phenol in trickle bed reactor

A stable activity over 72 h can be only observed at 140°C and 4 bar of partial oxygen pressure. The two steam activated DMAD and DRAW carbons performed phenol conversions of 20–25% compared to 50% in case of the commercial

AP4-X carbon. Contrary to batch oxidation, both DMAD and DRAW carbons show now a similar activity in continuous operation (see Figure 11.8). To recover a phenol removal of 50% it is enough to raise the reaction temperature by 20°C to 160°C (at same oxygen partial pressure of 2 bar). However, at this temperature the carbons (commercial AP4-X included) exhibited a certain loss of activity over time certainly due to enhanced carbon burn off (see Figure 11.8b).

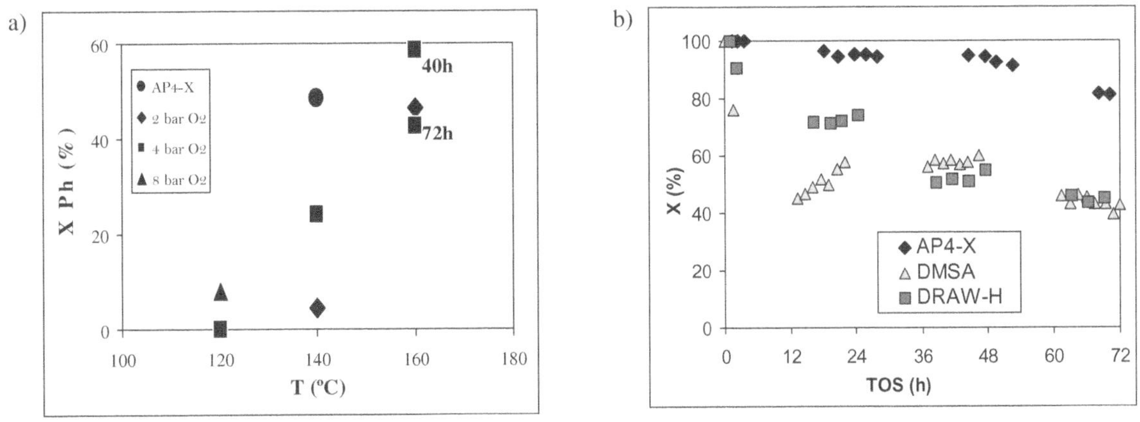

Figure 11.8 a) TBR conversion of phenol for DMAD obtained at different operating conditions after 72 h of time on stream and b) TBR conversion of phenol versus time on stream for commercial AP4-X, steam activated DMAD and hardened DRAW obtained at 160°C and 4 bar of oxygen partial pressure

Determination of carbon burn off during continuous CWAO of phenol

CWAO itself is a very soft surface treatment and depending on the operating conditions (temperature and oxygen partial pressure) some burn off of the carbon surface and carbon weight loss after oxidation has been observed (Fortuny *et al.*, 1999). In our TBR experiments carbon burn off becomes only noticeable at temperatures of 160°C or 4 bar of oxygen partial pressure, though attaining in the worst case approximately 35% after 72 h of operation (see Table 11.2).

Determination of carbon hardness and metal leaching during continuous CWAO of phenol

A significant finding is the low propensity of the SBACs towards toxic metal leaching during continuous operation despites their high mineral fraction. Lowest metal leaching underwent the second generation of the hardened DRAW carbon with metal concentration in the reactor exited samples of 1 to 5 ppm for Fe and below 1 ppm for Cu. The steam activated DMAD carbon leached somewhat more Fe and Cu with mean values of 6 ppm and 1 ppm, respectively. Surprisingly, the commercial AP4-X carbon exhibited the highest propensity towards Fe leaching with mean concentration of 8 ppm and 15 ppm at 140°C and 160°C, respectively.

Mechanical stability of carbons was only a problem in the case of the steam activated DRAW carbon. The hardness of pellets obtained from DRAW carbon powder was very low and attrition readily provoked undesired clogging of the Trickle Bed Reactor. Several hardened forms of the DRAW carbon were thus tested in continuous operation. It appears that lignosulphonate based binder material is the best option to pelletise DRAW carbons with an adequate mechanical strength for use in continuous reactor operation.

Conclusions

The two carbons prepared by steam activation from DRAW and DMAD sewage sludge feedstock performed a stable activity and high resistance to both metal leaching and mechanical attrition during 72 h of continuous CWAO of phenol at 140°C and 4 bar of oxygen partial pressure. However, at this temperature the catalytic activity of the carbon is too low for the efficient destruction of phenolic waste water contaminants (only 20–25% of phenol conversion). Higher conversions can be readily obtained by increasing the temperature to 160°C, but the improvement of activity is set off by a significant loss of stability due to enhancement of carbon burn off reactions. Nevertheless, there exists several possible solutions to mitigate this problem, i.e. the use of chemically activated or/and acid washed sludge based carbons that are much more active or the surface impregnation of carbonised or physically activated sludge based carbons with small amounts of iron (Fe). The latter alternative has been already demonstrated to strongly enhance the catalytic activity of commercial active carbons for the catalytic wet air oxidation of phenol in trickle bed reactors (Quintanilla *et al.*, 2005 and Baricot, 2008).

It must be stressed that the potential and economical perspective of CWAO lies in its smart utilisation as a pre-treatment (and not a total remediation) of organic wastewater to improve their biodegradability to safe levels for subsequent biological end treatment. The exact determination of the required oxidation deepness or biodegradability of the effluent stream is then the key factor that determines the process economics and ergo the successful application of CWAO. Biodegradability enhancement (sum of readily and slowly biodegradable COD) can be assessed by fast respirometric experiments (Orbon *et al.*, 1997). This technique only measures the readily biodegradable COD fraction of a specific effluent, but avoids the extremely time-consuming determination of the total biodegradable COD (Brouwer *et al.*, 1998). Further research work is required to obtain suitable indicators that establish the partial oxidation degree of a treated CWAO effluent necessary for safe discharge into the biological end treatment.

CFD investigation of pilot scale trickle bed reactor at cold and reacting flow conditions

CFD hydrodynamic predictions of moderate- to high-pressure trickle-bed reactors

Two Computational Fluid Dynamics multiphase frameworks were developed for the hydrodynamic analysis aiming to predict the liquid holdup and pressure drop in the trickling flow regime with a three-dimensional computational grid. Euler-Euler and Volume-of-Fluid multiphase models were optimized in terms of mesh density, time step, convergence criteria, discretization schemes and Reynolds Averaged Navier-Stokes turbulence models (Lopes and Quinta-Ferreira, 2008, 2009a, 2009b). Several computational runs were performed querying the effect of gas and liquid flow rates, operating pressure and temperature on liquid holdup and two-phase frictional pressure drop. Multiphase flow distribution was investigated with different gas-liquid distributors using the Eulerian framework (Lopes and Quinta-Ferreira, 2009c), whereas hysteresis phenomena and wetting efficiency in trickle beds were evaluated with the Volume-of-Fluid model (Lopes and Quinta-Ferreira, 2009d).

Quantification of axial and radial profiles bulk temperature and total organic carbon conversions in continuous trickle-bed reactors

The kinetic parameters of CWAO were further integrated within the Euler-Euler and Volume-of-Fluid models where the reaction aspects as well as the transport mechanisms were accounted for in a high-pressure trickle-bed reactor so that both multiphase CFD frameworks were evaluated for either cold flow or reacting flow conditions. Axial and radial profiles of hydrodynamic and reaction parameters were evaluated at different operating conditions (Lopes and Quinta-Ferreira, 2009e). The influence of gas and liquid flow rates as well as the effect of temperature and pressure were investigated in terms of total organic carbon (TOC) conversions. TOC removal efficiency depends heavily on the bed temperature while the operating pressure has minor influence in final conversion. Whereas the effect of liquid flow rate was mainly related with the reaction time, the influence of gas flow rate led to a maximum on TOC conversion. Moreover, higher values of inlet pollutant concentration led also to higher conversions (Lopes and Quinta-Ferreira, 2007a, 2007b). The multiphase Eulerian predictions handled agreeably better the effect of different numerical solution parameters either supported at non- and reacting flow conditions with respect to Volume-of-Fluid model.

IMPACT OF THE TECHNOLOGY ON ITS INDUSTRY OR RESEARCH SECTOR

Research impacts

Pioneering the utilisation of sewage sludge based carbons as catalytic materials for waste water treatment of phenolic effluents in trickle bed reactor

The work focus on the innovative utilisation of low costs carbons prepared from sewage sludge as novel catalysts for the waste water treatment of phenolic effluents. A key impact is that the catalytic activity of sludge based carbons can be readily tuned by adequate preparation methods to provide catalytic materials that are fully competitive with both commercial carbon catalysts and many metal or metal oxide supported catalysts. A mayor drawback that is the burn off of carbons occurring during continuous operation was evidenced, however strategies to mitigate this problem are proposed. It is therefore hoped that this work has made a convincing and stimulating case for the scientific and industrial community to continue investigating the perspective of sludge based carbons as catalysts for waste water treatment and ergo to develop low cost processes that can be integrated at industrial scale.

Social and environment impacts

In the European Union as a whole, the annual production of sewage sludge exceeds 10 millions tons a year (Laturnus *et al.*, 2007). Consequently, its safe handling and disposal has become the most pressing problem of the wastewater treatment sector. If this problem is not appropriately dealt with in the near future, it will exert a strong negative impact on the environment and human society.

Work have demonstrated in close collaboration that sewage sludge can be valorised into active carbons that among other applications as adsorbent may serve as an efficient catalyst for the remediation of phenolic wastewater. The conversion of sewage sludge into valuable products is therefore a highly viable and environmentally friendlier alternative to the existing disposal (land filling, agriculture) routes and thermal incineration. Some of the main advantages that if can offer are briefly described below.

- The production and subsequent use of SBACs is a sustainable and environmentally friendlier option than land filling, disposal and incineration of produced sewage sludge.
- SBAC production should receive better public acceptance than land filling, agricultural disposal and incineration of sewage sludge.
- A 60–80 wt% reduction in dried sludge mass by physical activation is possible.
- The presence of phenolic compounds in WWTP streams is harmful to the active biomass and generates extra sewage sludge during the biological treatment. The partial remediation of phenolic effluents prior to their discharge into a WWTP using sludge based active carbons would therefore benefit from an in-situ reduction of the production of sewage sludge. Moreover, the sewage sludge in its valorised form would return to the WWTP where it was originated and would help to close the life cycle of sewage sludge.
  The quick disposal of generated sewage sludge into production of active carbon could mitigate the problem of malodours in the surroundings of WTTP which stems from the release of hazardous gases (H2S) produced during anaerobic microbiological decomposition processes (Ansari *et al.*, 2005).
- Irrespectively of the benefits that will generate reduced sludge volumes, a mayor social and environment impact is the opportunity that low cost sludge based carbons combined with CWAO technology offer to decontaminate waste water that contain biorefractory organic compounds and ergo are not amendable to direct biological treatment.

Industrial and economic impacts

The industrial and economic impacts of this work go hand in hand and have been grouped together under a single heading. Probably the most viable option to utilise CWAO technology and sludge based carbons is their integration into the wastewater treatment facilities operated by the own plant to remediate the arising process effluents. The idea of this approach is to improve in-situ the biodegradability of phenolic effluents to levels that allow for posterior discharge into the existing biological end treatment of the industrial plant. Figure 11.9 gives a schematic illustration of the combined process of CWAO and biological WWTP.

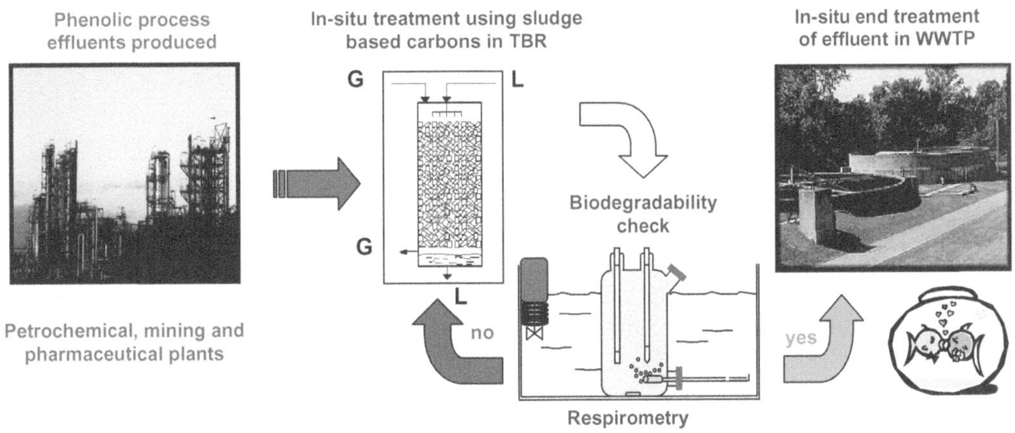

Figure 11.9 In-situ remediation of process water produced at the plant site: Integration of CWAO technology using TBR and sludge based carbon catalyst for the treatment of factory own phenolic effluents

Since waste water treatment is mandatory to preserve the environment, the treatment costs must be kept as low as possible in order not to adversely affect the net benefits of the industrial plant. It is in this sense that CWAO of organic effluents using sludge based active carbon in trickle bed reactor can positively interfere in the overall cost balance. However, to fully exploit the potential of this technology, some prerequisites are required at the plant site:

- Existence of a biological wastewater treatment in the industrial plant (Harf *et al.*, 1999), i.e. CWAO should be applied at the source of wastewater production (Levec and Pintar, 2007).
- Intelligent management of the different wastewater streams produced at the plant, i.e. process water polluted with biotoxic organic compounds should be directly discharged to the CWAO unit to avoid both undesired dilution of the TOC fraction and high liquid volumes to be treated.
- Availability of adequate system for the recollection and/or separation of wastewaters originated at the plant site.

In the aforementioned scenario, the following relevant items of cost savings/revenues would arise from the utilisation of CWAO-SBAC-TBR technology integrated in existing WWTP facilities.

It was assumed that if the phenolic wastewater effluents were pre-treated by CWAO then the cost of externally incinerating the effluents would be obviated. This will lead to substantial cost savings for the company given the low treatment costs of CWAO, in particular with respect to incineration. Among the commercial oxidation processes or thermal incineration applied at industrial scale, the estimated costs per m^3 of effluent treated are 10–15€/m^3 for CWAO (Nippon Sokubai process) compared to 15–30€/m^3 for WAO (ZIMPRO process) and 210–280€/m^3 for thermal incineration (see Table 11.3).

Table 11.3 Estimated treatment costs per m^3 effluent for WAO, CWAO and thermal incineration

Process	T (°C)	P (bar)	t_{res} (min)	Conversion (%)	Costs (€/m^3)
WAO (Zimpro)	200–325	20–210	10–90	80–99	15–30[a]
CWAO (Nippon)	130–250	20–50	10–60	98	10[a]
Incineration	900–1000°C	1	–	99	210–280[b]

[a]*As reported in Levec, 1997.*
[b]*Costs estimation obtained from incineration plant of Constanti (Tarragona) called Gestió Residual Especials de Catalunya, S.A.*

Moreover, CWAO conducted in TBR was evaluated to become economically attractive for treating organic effluents with an initial COD content superior than 10 g/L. Beyond this COD limit, adiabatic operation of the CWAO reactor is thermally self-sufficient due to the exothermic nature of the oxidation reactions (Katzer *et al.*, 1976 and Farha *et al.*, 1978). Then, CWAO is highly cost effective compared with other advanced oxidation processes (O_3) or biological processes (BIO) as shown in table 11.4 (Lenntech). This table also highlights the fact that according to economic evaluations of Lenntech Water Treatment & Air Purification Holding, B.V., the treatment costs per m^3 effluent are strongly dependent on the pollution level of the effluent stream.

Table 11.4 Treatment costs (€/m^3) versus initial COD

COD (g/L)	CWAO	BIO	O_3
10	5.1	8.4	45.4
50	15.9	41.3	227
100	29.6	82.5	454

Important savings in operation costs can be also expected from the replacement of more expensive supported noble metal/metal oxide catalysts or commercial carbons by low cost SBACs (see Table 11.5). CWAO requires only low grade application carbons, which therefore can be benefit from a significant retail of the market price of commercial carbons. An economic study of the production plant of a steam activated SBAC undertaken by ICL revealed thus an overall cost of only 51 € per tonne of carbon. In 2007 a commercial coal-based steam-activated granular activated carbon retailed in the region of 720 €/tonne (Roskill Information Services Ltd, 2008). From Table 11.5 it is also seen that the prices of supported metal oxide and noble metal catalysts compared to commercial carbons strongly increase by about one and two orders of magnitude, respectively. This therefore makes the use of low cost SBACs highly viable for CWAO of organic wastewater streams at industrial scale.

Table 11.5 SBAC production cost and market prices of CWAO catalysts

Catalyst	SBAC	AP4-X	1% CuO/Al_2O_3	1% Pt/C
Cost (€/tonne)	51	1500[a]	~9000[b]	~50000[b]

[a]*Bulk price as offered from Chemviron Company.*
[b]*Price offer from catalyst manufacturer.*

The treatment cost with integrated CWAO + WWTP technology will be further decreased if one takes into account that only partial oxidation of the TOC fraction of the organic wastewater effluent is required. This will result in significant savings of equipment and operation costs if lower temperatures and pressures can be applied to CWAO process.

Finally, integrated CWAO + WWTP technology should be oriented towards the remediation of process water for their potential reuse in or recycle to the industrial plant. The cost of tap water is constantly increasing and on average one cubic meter prices now about 1.5–2.5 €. Upon that, when extracting groundwater, a levy of 0.15–0.2 € must be added. The same trend holds for draining off waste water, which has risen by 20% in the past 5 years. The reuse of treated process water will thus further contribute to save a company additional expenses arising from the renewal of fresh tap water and draining off process water.

Conclusion

Overall, the utilisation of SBACs as a CWAO catalyst for the partial oxidation of phenolic effluents in continuous Trickle Bed Reactor combined with subsequent biological end treatment can extremely lower the company's costs to pay for external waste water treatment. Application of chemical activation or HCl washing to enhance the SBAC performance may be required and will augment the production costs of carbon; however the margin of cost reduction expected for CWAO + WWTP technology remains strong even under these circumstances.

REFERENCES

Ansari A., Bagreev A. and Bandosz T. J. (2005). Effect of adsorbent composition on H2S removal on sewage sludge-based materials enriched with carbonaceous phase. *Carbon*, **43**, 1039–1048.

Arena F., Trunfio G., Negro J. and Spadaro L. (2008). Optimization of the MnCeOx system for the catalytic wet oxidation of phenol with oxygen (CWAO). *Applied Catalysis B: Environmental*, **85**, 40–47.

Baricot Mendoza, M. (2007). Tailoring activated carbons as catalyst for catalytic wet air oxidation of phenol. PhD thesis, Universitat Rovira i Virgili, Tarragona (Spain).

Bhargava S. K., Tardio J., Prassad J., Föger J., Akollekar D. B. and Grocott S. C. (2006). Wet oxidation and catalytic wet oxidation. *Indsutrial Engineering and Chemistry Research*, **45**, 1221–1258.

Brouwer H., Klapwijk A. and Kessman K. J. (1998). Identification of activated sludge and wastewater characteristics using respirometric batch-experiments. *Water Research*, **32**, 1240–1254.

Busca G., Berardinelli S., Resini C. and Arrighi L. (2008). Technologies for the removal of phenol from fluid streams: A short review of recent developments. *Journal of haxardous Materials*, **160**, 265–288.

Dias J. M, Alvim-Ferraz M. C. M., Almeida M. F., Rivera-Utrilla J. and Sanchez-Polo M. (2007). Waste materials for activated carbon preparation and its use in aqueous-phase treatment: A review. *Journal of Environmental Management*, **85**, 833–843.

Eftaxias A., Font J., Fortuny A., Fabregat A. and Stüber F. (2005). Kinetics of phenol oxidation in a trickle bed reactor over active carbon catalyst. *Journal of Chemical Technology & Biotechnology*, **80**, 677–687.

Fitzmorris K. B., Lima I. M., Marshall W. E. and Reimers R. S. (2006). Anion and cation leaching or desorption from activated carbons from municipal sludge and poultry manure as affected by pH. *Water Environment Research*, **78**, 2324–2329.

Fortuny A., Miro C., Font J. and Fabregat A. (1999). Three-phase reactors for environmental remediation: catalytic wet oxidation of phenol using active carbon. *Catalysis Today*, **45**, 323–328.

Harf J., Hug A., Vigel F. and von Rhor R. (1999). Scale-up of catalytic wet air oxidation under moderate conditions. *Environmental Progress*, **18**, 14–20.

Kemmer F. N., Robertson S. R. and Mattix R. D. (1971). Sewage Treatment Process, Nalco Chemical Company. US Patent Office, Patent No. 3,619,420.

Laturnus F., von Arnold K. and Gron C. (2007). Organic contaminants from sewage sludge applied to agricultural soils – false alarm regarding possible problems for food safety? *Environmental Science and Pollution Research*, **14**, 43–60.

Lenntech: http://www.lenntech.com/cwao.htm.

Levec J. (1997). Wet oxidation processes for treating industrial wastewaters. *Chemical and Biochemical Engineering Quartely*, **11**, 47–58.

Levec J. and Pintar A. (2007). Catalytic wet air oxidation processes: A review. *Catalysis Today*, **124**, 172–184.

Lewis F. M. (1978). Method of pyrolysing sewage sludge to produce activated carbon. US Patent Office. Patent No. 4,122,036.

Lopes R. J. G., Silva A. M. T. and Quinta-Ferreira R. M. (2007a). Kinetic modelling and trickle-bed CFD studies in the catalytic wet oxidation of vanillic acid. *Industrial and Engineering Chemistry Research*, **46**(25), 8380–838.

Lopes R. J. G. and Quinta-Ferreira R. M. (2007b). Trickle-bed cfd studies in the catalytic wet oxidation of phenolic acids. *Chemical Engineering Science*, **62**(24), 7045–7052.

Lopes R. J. G. and Quinta-Ferreira R. M. (2008). Three-dimensional numerical simulation of pressure drop and liquid holdup for high-pressure trickle-bed reactor. *Chemical Engineering Journal*, **145**(1), 112–120.

Lopes R. J. G. and Quinta-Ferreira R. M. (2009a). Turbulence modelling of high-pressure trickle-bed reactor. *Chemical Engineering Science*, **64**(8), 1806–1819.

Lopes R. J. G. and Quinta-Ferreira R. M. (2009b). Numerical simulation of trickle-bed reactor hydrodynamics with RANS-based models using a volume of fluid technique. *Industrial and Engineering Chemistry Research*, **48**(4), 1740–1748.

Lopes R. J. G. and Quinta-Ferreira R. M. (2009c). CFD modelling of multiphase flow distribution in trickle beds. *Chemical Engineering Journal*, **147**(2–3), 342–355.

Lopes R. J .G. and Quinta-Ferreira R. M. (2009d). VOF based model for multiphase flow in high-pressure trickle-bed reactor: optimization of numerical parameters. *A.I.Ch.E. Journal*, DOI: 10.1002/aic.11862.

Lopes R. J. G. and Quinta-Ferreira R. M. (2009e). Evaluation of multiphase CFD models in gas-liquid packed-bed reactors for pollution abatement: euler-euler and volume-of-fluid. *Chemical Engineering Science*. DOI: 10.1016/j.ces.2009.06.039.

Nickerson R. D. and Messman H. C. (1975). Making active carbons from sewage sludge. Combustion Engineering Inc. US Patent Office. Patent No. 3,887,461.

Nousir S., Keav S., Barbier Jr J., Bensitel M., Brahmi R. and Duprez D. (2008). Deactivation phenomena during catalytic wet air oxidation (CWAO) of phenol over platinum catalysts supported on ceria and ceria-zirconia mixed oxides. *Applied Catalysis B: Environmental*, **84**, 723–731.

Pintar A., Batista J. and Tisler T. (2008). Catalytic wet-air oxidation of aqueous solution of formic acid, acetic acid and phenol in a continuous-flow trickle-bed reactor over Ru/TiO2 catalysts. *Applied Catalysis B: Environmental*, **84**, 30–41.

Orbon D., Ates E., Sözen S. and Ubuy Çokgör E. (1997). Characterisation and COD fractionation of domestic wastewaters. *Environmental Pollution*, **95**, 191–204.

Quintanilla A., Casas J. A., Zazo J. A., Mohedano A. F. and Rodriguez J. J. (2005). Wet air oxidation of phenol at mild conditions with a Fe/activated carbon catalyst. *Applied Catalysis B: Environmental*, **62**, 115–120.

Resini C., Catania F., Berardelli S., Paladino O. and Busca G. (2008). Catalytic wet air oxidation of phenol over lanthanum strontium manganite. *Applied Catalysis B: Environmental*, **84**, 678–683.

Roskill Information Services Ltd. (2008). The Economics of Activated Carbon, London, p. 322.

Santiago M., Stüber F., Fortuny A., Fabregat A. and Font J. (2005). Modified activated carbons for catalytic wet air oxidation of phenol. *Carbon*, **43**, 2134–2145.

Santos A., Yustos P., Rodriguez S. and Garcia-Ochoa F. (2006). Wet oxidation of phenol, cresols and nitrophenols catalyzed by activated carbon in acid and basic media. *Applied Catalysis B: Environmental*, **65**, 269–281.

Smith K. M., Fowler G. D., Pullket S. and Geaham N. J. D. (2009). Sewage sludge-based adsorbents: A review of their production, properties and use in water treatment applications. *Water Research*, **43**, 2569–2594.

Stüber F., Font J., Fortuny A., Bengoa C., Eftaxias A. and Fabregat A. (2005). Carbon Materials and catalytic wet air oxidation of organic pollutants in wastewater. *Topics in catalysis*, **33**(1–4), 3–50.

Suarez-Ojeda M. E., Stüber F., Fortuny A., Fabregat A., Carrera J. and Font J. (2005). Catalytic wet air oxidation of substituted phenol using activated carbon as catalyst. *Applied Catalysis B: Environmental*, **58**, 105–114.

Wang J., Zhu W., Yang S., Wang W. and Zhou W. W. (2008). Catalytic wet air oxidation of phenol with pelletized ruthenium catalysts. *Applied Catalysis B: Environmental*, **78**, 30–37.

Chapter 12

Utilisation of activated carbon as catalyst in AD-OX process

Henri Delmas, Carine Julcour, Caroline Andriantsiferana, Hélène Chaumat and Anne-Marie Wilhelm

SUMMARY

Project objectives

This study aims to investigate the feasibility of a new potential water treatment process: AD-OX, with Sludge Based Carbonaceous Materials (SBCM). AD-OX is a sequential process based on: first room temperature adsorption and second regeneration by catalytic wet air oxidation at higher pressure and temperature. The main advantage of this process would be to provide in situ regeneration of the spent activated carbon which is a waste in industrial waste water treatment.

Work performed

First both commercial Activated Carbons (AC) and Sludge Based Carbonaceous Materials (SBCM) were thoroughly analysed and compared before and after being used as adsorbents and oxidative catalysts. The SBCM were provided by Imperial College (London) and GEPEA (Ecole des Mines de Nantes).

The sequential AD-OX process being based on two major steps, adsorption and oxidation, these two steps have been extensively investigated, first separately, with various commercial AC and sludge based materials and with various organic pollutants: phenol, phenol derivatives, mixtures of these phenols, and real effluents issued from industrial wastewater treatment plants.

In parallel two reactors have been built up: a small automated reactor for multiple adsorption-oxidation cycle tests and a semi industrial pilot plant reactor for preliminary scale up. They have been used to carry out AD-OX process with selected commercial AC and sludge based carbons.

As oxidative regeneration by pressurised air was not as efficient as expected, several other alternative advanced oxidation techniques have been tested. Especially Fenton oxidation was more intensively investigated.

End results

SBCM investigated have very high ash content including metal oxides and low surface area compared to commercial AC. Nevertheless they offer rather acceptable adsorption capacity and catalytic activity for pollutant oxidation (about

 Reduction, Modification and Valorisation of Sludge (REMOVALS). Edited by Christophe Bengoa, Azael Fabregat, Josep Font and Frank Stüber. ISBN: 9781843393450. Published by IWA Publishing, London, UK.

half that of commercial AC). Most of pollutants investigated in this study were conveniently adsorbed then oxidized on AC and SBCM in the AD-OX process. The main problems are due to:

- Relatively low regeneration of AC and SBCM by Catalytic Wet Air Oxidation (CWAO): as soon as the first oxidation has been carried out, the adsorption capacity drops to about 25%, but is quasi stabilized for other AD-OX cycles.
- Poor mechanical properties of the best SBCM (produced from steam activation of dewatered raw sludge). This SBCM is not strong enough to support high pressure and shear stresses of a fixed bed and was crushed leading to bed clogging. Trials for strengthening it were not successful due to foaming and metal leaching. More work is still necessary to design and test convenient SBCM in oxidation.

Other regenerative oxidations (photocatalytic, UV, ultrasonic, Fenton...) are only slightly better than air oxidation (standard AD-OX). Preliminary tests on electrochemical regeneration are very promising. Concerning economic aspects standard AD-OX would be highly profitable for actual industrial effluent treated with commercial AC (1,350 €/ton), if corrosion is prevented at reasonable cost. Concerning SBCM, due to its low cost (about 250 €/ton), it does not deserve uneasy partial regeneration and should be very profitably used as adsorbent for single use and, to less extend, as oxidative catalyst in CWAO.

METHODOLOGIES AND APPROACHES EMPLOYED

AD-OX is a sequential water treatment process, based on two main steps: 1 – adsorption on activated carbon at room temperature to remove pollutants from aqueous effluents, 2 – batch oxidative regeneration of AC by pressurised air (or other stronger oxidant) (Polaert *et al.*, 2002). The advantage of wet air oxidation is that AC is the catalyst of its own regeneration.

The methodology selected to get a deep knowledge of such process was to investigate separately the two steps in simplified conditions and equipments with respect to the pressurized AD-OX reactors to be built in parallel.

Adsorption was first characterized by isotherms at room temperature for a variety of commercial AC. Adsorption capacity of different non biodegradable pollutants was linked to AC physicochemical properties (surface area, micro-meso porosity, metal content, surface function acidity...).

Similarly wet air oxidations were carried out in aerated autoclave reactor containing AC in a basket. Oxidation rates were compared for same AC and pollutants, on several successive runs, to observe the decay of adsorption capacity. The pollutant concentration was followed but also the Total Organic Carbon or the Chemical Oxygen Demand. The evolution of AC properties after several uses was also linked to adsorption and oxidation deactivation.

The results of these two investigations gave orientations to select convenient operation conditions (temperature, pressure, reaction time), and best SBCM to be used in the AD-OX fixed bed process.

An important aspect concerning the significance of this work with respect to actual industrial waste water has been addressed in this work: in addition to phenol as the usual pollutant model, other non biodegradable pollutants have been investigated, separately or in mixtures – actual effluents being always complex mixtures. Finally true effluents, withdrawn at the exit of industrial waste water treatment plants, have been treated.

As the economy of AD-OX process is strongly connected with the number of reuses of AC it is important to be able to carry out long series of AD-OX cycles in pre-selected convenient conditions. For this a mini automated AD-OX reactor has been designed and built up. It can move from (1) adsorption step to (2) liquid batch heating and pressurization, then (3) oxidation and (4) cooling and depressurization, by following UV absorbance of the liquid effluent at adsorption exit (→2), temperature and pressure (→3) in the reactor, CO_2 concentration at gas exit (→4), temperature in the reactor (→1). The small size of the fixed bed of AC will limit the adsorption time and give faster results.

In addition a much larger pilot plant, previously used as continuous fixed bed for CWAO, has been modified to be used for AD-OX cycles with batch oxidation and liquid recycle in a loop. This equipment is of less easy use but performs a first scale up for predictive design of a future industrial plant.

As regeneration by wet air oxidation was not as efficient as expected several other advanced oxidation processes have been tried to be involved in a modified AD-OX process.

ACHIEVEMENTS RELATED TO THE STATE OF THE ART

Introduction: state of the art on activated carbon use and regeneration

One of the major techniques for the treatment of low concentrations of toxic organics in water is based on adsorption by Activated Carbon (AC), either in powdered or, most often, in granular form. AC adsorption can fully replace biological treatment but better it can be effective for eliminating biorefractory organics from water leaving conventional biological waste water treatment plants. AC adsorption as end step appears as the best technology according to Sarp Industries-Veolia (private communication, 2009).

However, the saturated AC must be regenerated before reuse. The first commercial-scale adsorption plants used to burn the spent AC. Over the years a variety of regeneration techniques have been suggested, evaluated and applied. These methods are based either on desorption, induced by increasing temperature or by displacement with a solvent, or on chemical decomposition of the adsorbed species induced by thermal, chemical, electrochemical or microbial processes, as presented on Figure 12.1 according to Sheintuch and Matatov-Meytal (1999).

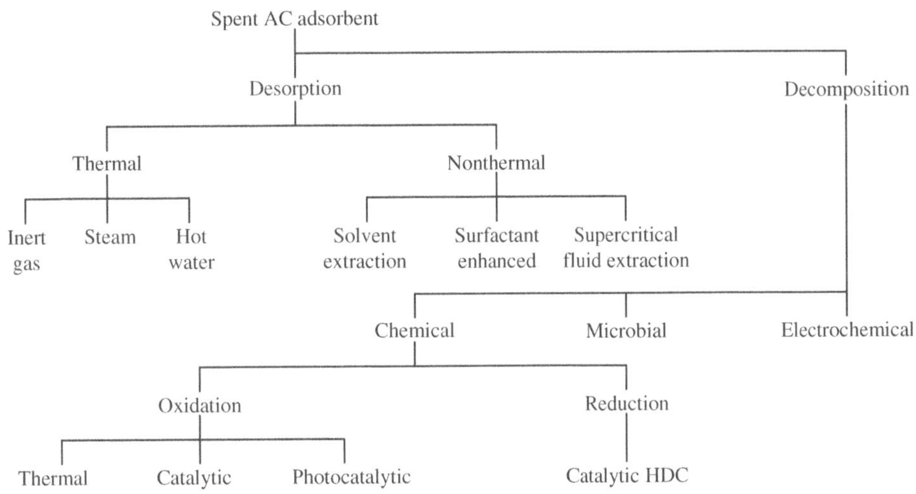

Figure 12.1 Overview of available techniques for regeneration of spent AC adsorbents. (Reprinted from Catalysis Today, vol. 13, no. 1, M. Sheintuch, Y. I. Matatov-Meytal, Comparison of catalytic processes with other regeneration methods of activated carbon, 73–80, 1999, with permission from Elsevier)

The most commonly used regeneration, thermal regeneration, based on either desorption with inert gas or reactive decomposition by water vapour or flue gas, needs very high temperature (700–1,000°C) and then should only be achieved in few very large specialized units.

Extractive regeneration methods that use volatile solvents (solvent regeneration) have been investigated, but current applications of this technology with AC are not known. Ethanol regeneration has been proposed but in any case a subsequent separation is still needed to recover the solvent. Supercritical fluid extraction exhibits enhanced mass-transfer properties over liquid solvent extraction. However, complete regeneration of AC loaded with organics like phenolic compounds, with supercritical fluid extraction using CO_2, could not be achieved and a large amount of CO_2 would be required. As main drawback, these solvent regeneration serves only to transfer the contaminant from one location (AC) to another (solvent or surfactant) and do not destroy the pollutants which have to be treated subsequently.

Reactive regeneration methods appear much more attractive as they destroy the adsorbed organics by chemical processes (including acid–base or oxidative transformations, microbial or electrochemical processes). The chemical methods may involve complete mineralization of the adsorbed species (to carbon dioxide and water), transformation of complex molecules into simpler compounds and the conversion of hazardous materials to more desorbable, water-soluble or more biodegradable compounds. Chemical oxidation using various oxidants – chlorine, chlorine dioxide, hydrogen peroxide, Fenton, ozone and potassium permanganate (Horng and Tseng, 2008; Liang *et al.*, 2009; Bento and Rein, 2007; Huling *et al.*, 2005; Arnold *et al.*, 2006; Toledo *et al.*, 2003 and Alvarez *et al.*, 2005) – has been tested and at least partial restoration of AC adsorption capacity was demonstrated. Similarly unconventional techniques including ultrasound (Weng and Hsu, 2008), electrochemistry (Zhang, 2002; Zhou and Lei, 2006; Weng and Hsu, 2008 and Narbaitz and Karimi-Jashni, 2009) microwaves (Ania *et al.*, 2005) electrodialysis (Drouiche *et al.*, 2006) and plasma discharges (Qu *et al.*, 2009) have been tried.

These advanced oxidation processes, however, have not fully proven technically feasible for continuous operation and nor to be economically viable.

An established technology for regeneration of spent AC with organics or oxidizable inorganic compounds is wet air regeneration (WAO) but this method is rather expensive because of the required large investments in high pressure equipment (~200 bar) and the high cost of running the regeneration at high pressure and temperature (Gonzalez *et al.*, 2002 and Shende and Mahajani, 2002).

Catalytic regeneration methods may enhance the decomposition rates and enable to carry out the regeneration at lower temperatures and lower residence times. They could use either desorption followed by catalytic liquid-phase oxidation, or direct oxidation of the adsorbed species on an AC-adsorbent modified with a catalyst. In the former approach, the water is not contacted with the catalyst and metal dissolution is not a problem. In the latter approach, metal dissolution may occur and should be controlled.

The great improvement to be expected would be to achieve direct oxidation only with activated carbon which in this case catalyses its own regeneration. This is the purpose of AD-OX process proposed by Institut National Polytechnique de Toulouse.

Characterization of the sludge based carbonaceous materials and comparison with commercial activated carbons

First a complete comparison of usual characteristics of 10 carbons (4 commercial ones and 6 sludge based ones) tested in this process have been performed using the same equipment and methods. Table 12.1 presents the textural properties of commercial AC and SBCM (sludge based carbonaceous materials), showing much less surface area of sludge based carbons (from 90 to 265 m^2/g) when compared to commercial ones (from 980 to 1,860 m^2/g).

Table 12.1 Textural properties of commercial AC and SBCM. (Reprinted from Journal of Environmental Management, vol. 91, no. 12, C. Julcour Lebigue, C. Andriantsiferana, N'G. Krou, C. Ayral, E. Mohamed, A. M. Wilhelm, H. Delmas, L. Le Coq, C. Gerente, K. M. Smith, S. Pullket, G. D. Fowler, N. J. D. Graham, Application of sludge-based carbonaceous materials in a hybrid water treatment process based on adsorption and catalytic wet air oxidation, 2432–2439, 2010, with permission from Elsevier)

Carbon	BET surface area[a] (m^2/g)	Mesoporous volume[b] (cm^3/g)	Microporous volume[c] (cm^3/g)	Mean pore diameter[d] (Å)	Apparent density[e] (kg/m^3)
PICA F22	985	0.11	0.41	20	993
PICA L27	1860	0.48	0.77	25	552
PICA S23	1230	0.04	0.49	17	1013
MERCK 2514	980	0.20	0.37	22	1032
C_DRAW	180	0.08	0.07	27	1710
C_DMAD	125	0.11	0.05	44	1837
SA_DRAW	265	0.17	0.11	35	1471
SA_DMAD	155	0.15	0.06	45	1730
Hardened SA_DRAW	200	0.16	0.08	37	1519
CA_DSBS	90	0.03	0.03	25	1932

[a]*BET surface area and porous volume measurements on Micromeritics ASAP 2010 Analyser.*
[b]*Mesoporous volume calculated from Barrett-Joyner-Halenda model.*
[c]*Microporous volume calculated from Horvath-Kawazoe model.*
[d]*Deduced from the total porous volume at* $p/p_0 = 0.98$ *and the BET surface area.*
[e]*Calculated from porous volume (micro- and meso-) and true solid density (helium pycnometer).*

Three different sewage sludges have been used by Imperial College (London) and GEPEA (Nantes) as feedstock for production of the carbons: DMAD (Dewatered, Mesophilic Anaerobically Digested), DRAW (Dewatered, RAW) and DSBS (Dewatered, Secondary Biological Sludge). DSBS was collected in a municipal WWTP in Nantes, whereas DMAD and DRAW originated from the UK (Ashford and Little Marlow respectively).

From the UK sludges, both charred and steam activated samples were prepared as detailed in Chapter 10. Prior to carbonisation, the sludge samples were sterilised, dried and ground. Subsequently, they were heated at 5 or 10°C/min under a flowing (500 mL/min) nitrogen atmosphere. After the desired temperature (from 250 to 1,000°C) was reached, the furnace was automatically cooled down to produce carbonised samples. When steam activation was applied, the only difference was that once the target temperature was reached, the equivalent of 0.7 g/min of steam was mixed with the nitrogen flow (acting as carrier) and the temperature was maintained for the whole duration of the prescribed activation period. The flow of the activation reagent was arrested and the furnace was left to cool at the end of the activation period.

A carbonisation temperature of 900°C achieved an optimal BET and microporous surface area of the carbon series. For steam activation procedure the maximum BET surface area was attained when the activation temperature was 838°C and dwell time was 80 and 73 min for DRAW and DMAD based carbons, respectively (Pullket *et al.*, 2009).

The DSBS sludge underwent a different activation procedure, conducted with CO_2: the sample was first carbonised at 600°C for 1 hr under 2 L/min of N_2; then the temperature was increased to an optimum activation temperature of 875°C and the gas flow was switched to CO_2 (1.5 L/min) for one more hour.

In this chapter, the Carbonised series of carbons are referred to as 'C_DMAD' or 'C_DRAW' according to the sludge origin, while Steam Activated samples are labelled with the prefix 'SA_' and the CO_2 Activated DSBS is addressed as 'CA_ DSBS'.

An additional batch of SA_DRAW particles was prepared by pelletisation of the milled steam activated sludge together with a binder (PVA, 5 wt%), to increase the hardness of the material.

The commercial AC, used as reference, were provided by MERCK or PICA (VEOLIA group). PICA S23 was made from coconut shell, MERCK 2514 and PICA L27 from wood and PICA F22 from coal.

An additional chemical characterization of these SBCM has been performed by elemental analysis, as metal content (especially iron) is supposed to play a very important role on catalyst activity. Results are reported on Table 12.2 showing much higher ash and metal content in SBCM than in AC, which may be explained by the way of production and the material origin respectively.

Table 12.2 Chemical analysis[a] of commercial AC and SBCM. (Reprinted from Journal of Environmental Management, vol. 91, no. 12, C. Julcour Lebigue, C. Andriantsiferana, N'G. Krou, C. Ayral, E. Mohamed, A. M. Wilhelm, H. Delmas, L. Le Coq, C. Gerente, K. M. Smith, S. Pullket, G. D. Fowler, N. J. D. Graham, Application of sludge-based carbonaceous materials in a hybrid water treatment process based on adsorption and catalytic wet air oxidation, 2432–2439, 2010, with permission from Elsevier)

Carbon	%Fe	%Cu	%Zn	%Mn	%ash[b]
PICA F22	0.058	0.003	0.001	0.001	12.0
PICA L27	0.008	0.011	0.003	0.006	8.3
PICA S23	0.012	0.007	0.001	0.001	3.2
MERCK 2514	0.014	0.007	<0.001	<0.001	<6
C_DRAW	4.4	0.058	0.026	0.012	68.5
C_DMAD	5.4	0.072	0.11	0.058	79.0
SA_DRAW	4.8	0.067	0.061	0.015	65.0
SA_DMAD	6.2	0.069	0.12	0.063	78.2
Hardened SA_DRAW	5.0	0.073	0.066	0.016	–
CA_DSBS	0.66	0.055	0.053	0.042	59.0

[a]*From analysis of acid lixiviates (HCl:HNO_3 = 3:1 v/v) by inductively coupled plasma atomic emission spectroscopy (Jobin Yvon – Ultima 2R).*
[b]*From PICA, MERCK or IC (London)/GEPEA (Nantes).*

Adsorption isotherms on SBCM, comparison with two commercial AC

Phenol has still been selected as model molecule for adsorption characterisation. Adsorption isotherms at 25°C were measured batchwise using powdered particles sieved to a size between 200 and 400 μm to reduce equilibration time. A mass of 0.5 g of adsorbent was added to 100 mL of unbuffered phenol solution (0.005–0.05 mol/L concentration range), and the suspension was left under stirring in a thermostated bath for three days to reach equilibrium as determined by preliminary kinetic experiments. Then the solutions were filtered using 0.25 μm nylon filter membranes before analysis. The amount of adsorbed phenol was deduced from initial and final concentrations in the liquid phase, measured by high performance liquid phase chromatography with UV detection (HPLC-UV).

Adsorption isotherms are reported on Figure 12.2 where also the data corresponding to two commercial AC (PICA S23 and F22) are recalled. MERCK 2514 and PICA L27 show similar phenol adsorption capacity as PICA F22 in the investigated range.

As expected, the SBCM appear to be less attractive adsorbents than commercial AC. First they have less adsorption capacity than commercial AC, from 30% to 70% depending on what SBCM and what commercial AC, but this loss in adsorption capacity is moderate compared to the surface area ratio. The most dramatic loss concerns the initial slopes which are much lower for SBCM, with the exception of SA_DRAW which appear as a potential convenient material for adsorption. These data have been analysed by fitting the well known Langmuir and Freundlich models (Langmuir fitting shown in Figure 12.2).

As a conclusion of this preliminary work on phenol adsorption on the six investigated SBCM:

- These SBCM have low surface area. They have very high ash content which is not desired for adsorption but might have a positive effect on oxidation
- Adsorption performances on these SBCM are low compared to commercial AC but not as low as expected from surface area ratios.
- Only SA_DRAW, the Steam Activated carbon from Dewatered RAW sludge, shows conveniently shaped isotherm (Langmuir type) and adsorption capacity for potential use in AD-OX process.
- This SA_DRAW when hardened with PVA is still a convenient adsorbent though about 20% less efficient.

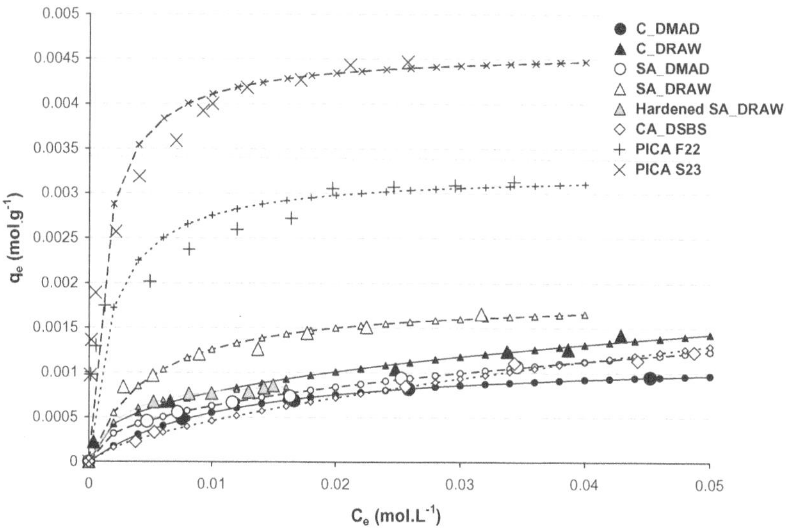

Figure 12.2 Adsorption isotherms of phenol on various commercial activated carbons and sludge based carbonaceous materials at 25°C. Experimental (large symbols) and model fitting (curves with small symbols): Freundlich (C_DRAW, SA_DMAD and CA_DSBS) or Langmuir (other carbonaceous materials). (Reprinted from Journal of Environmental Management, vol. 91, no. 12, C. Julcour Lebigue, C. Andriantsiferana, N'G. Krou, C. Ayral, E. Mohamed, A. M. Wilhelm, H. Delmas, L. Le Coq, C. Gerente, K. M. Smith, S. Pullket, G. D. Fowler, N. J. D. Graham, Application of sludge-based carbonaceous materials in a hybrid water treatment process based on adsorption and catalytic wet air oxidation, 2432–2439, 2010, with permission from Elsevier)

Autoclave oxidation on SBCM, comparison with two commercial AC

Phenol as the reference model pollutant has been selected. All tests have been performed with same temperature (150°C) and pressure (3.2 bar of oxygen partial pressure), using 200 mL of 5 g/L phenol solution and 2 g of carbon (0.8–1 mm sieved fraction maintained in a fixed basket).

In order to separate adsorption and oxidation effects the aqueous solution of phenol was heated to the reaction temperature and pressure under nitrogen during 12 hours to reach adsorption equilibrium before starting the oxidation step by starting the pressurized air flow (60 L/h NTP to ensure constant oxygen pressure). A liquid sample was taken for initial oxidation concentration before air was provided in the reactor. The stirrer speed was set to 800 rpm, enough to avoid external mass transfer resistance. Liquid samples were periodically withdrawn for analysis by HPLC-UV and remaining Chemical Oxygen Demand (COD) according to the closed-type reflux colorimetric method.

Due to different adsorption capacities of the various SBCM the 'initial' concentrations – after preliminary adsorption and before oxidation – were no longer identical and results are reported as variations of normalized concentrations by dividing by the 'initial' oxidation concentration.

The poor mechanical quality of most of these SBCM did not give the opportunity to use many times these materials as adsorbents then catalysts in order to follow their ageing and ability to be used in successive AD-OX cycles.

Figure 12.3 shows first oxidation tests of the six SBCM and, for comparison, of the two aforementioned commercial AC (PICA S23 and PICA F22, the other commercial AC showing intermediate activity). PICA S23 gave better oxidation kinetics but, as for adsorption, the SA-DRAW – Steam Activated carbon from Dewatered RAW sludge performed rather well, and even better than PICA F22.

The four other SBCM were clearly less active for catalytic oxidation and in addition the less active was also the less efficient for adsorption and the less porous (carbonated and CO_2 activated). Hardening SA_DRAW with PVA 5% results in lower catalyst activity but still acceptable.

It can be concluded that the catalytic behaviour of the SBCMs is a complex interplay between their adsorption capacity and metal content.

Regarding the COD conversion calculated at the end of the oxidation run, the results were in agreement with the phenol conversion showing the lowest value for CA_DSBS (3.3%) and a much higher one for SA_DRAW (47.9%) and F22 (52.5%).

As it was shown on commercial AC that catalyst activity was significantly reduced when reusing the same AC sample in successive oxidations (Creanga Manole *et al.*, 2007), deactivation has also been tested on the best hard SBCM, hardened SA_DRAW. Very peculiar features were observed: nearly no oxidation at the first recycling run and very fast reaction rate increase in the second recycling run due to delayed iron leaching favouring oxidation. However it is not clear if this leaching came from the lixiviation of the starting carbonaceous material itself or from the release of micron particles due to PVA dissolution, as explained in the next section.

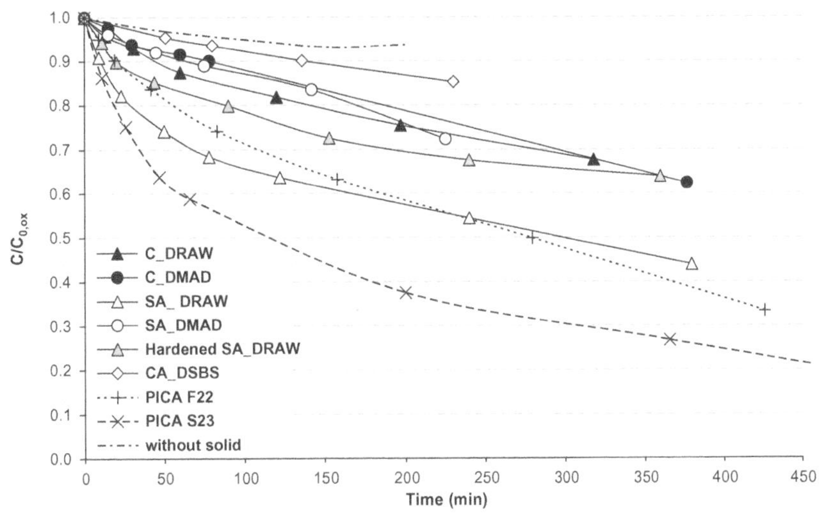

Figure 12.3 Batch oxidation of phenol on activated carbons and sludge based carbonaceous materials (T=150°C, p_{O2} = 3.2 bar): normalised concentration – time profiles *($C_{0,ox}$ refers to the concentration after adsorption)*. (Reprinted from Journal of Environmental Management, vol. 91, no. 12, C. Julcour Lebigue, C. Andriantsiferana, N'G. Krou, C. Ayral, E. Mohamed, A. M. Wilhelm, H. Delmas, L. Le Coq, C. Gerente, K. M. Smith, S. Pullket, G. D. Fowler, N. J. D. Graham, Application of sludge-based carbonaceous materials in a hybrid water treatment process based on adsorption and catalytic wet air oxidation, 2432–2439, 2010, with permission from Elsevier)

AD-OX fixed bed experiments in a mini automated set-up

The mini automated plant for fast AD-OX tests has first been used with commercial AC for verification then for a preliminary selection of carbons and operation conditions. The small fixed bed reactor was 18 cm high and 1 cm internal diameter. The schematic diagram of the set-up is shown in Figure 12.4.

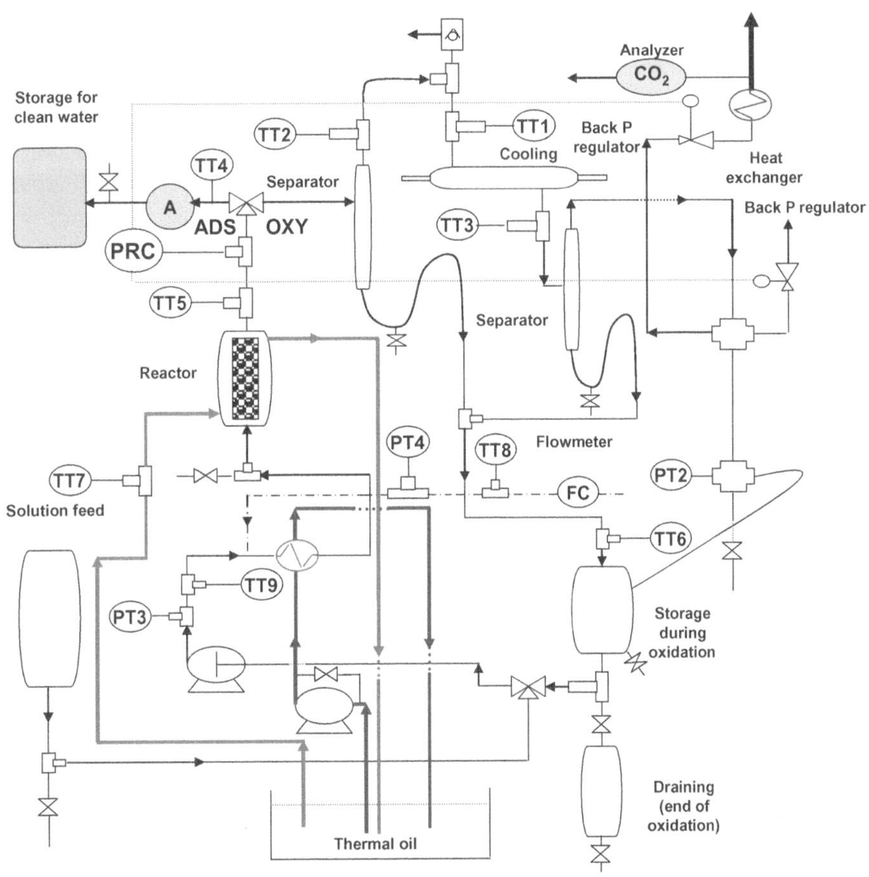

Figure 12.4 Scheme of the mini automated AD-OX set-up

After the adsorption step, the storage tank connected to the reactor outlet was filled with 400 mL of the solution circulating through the fixed bed.

To start the oxidative regeneration of the adsorbent, the reactor and the recycling loop including the storage tank were pressurised to 50 bar with air flowing concurrently upward at 30 L/h NTP. Simultaneously the thermofluid circulating in the reactor jacket and in the gas-liquid preheater was heated to the required temperature. A first sample was taken in the pressurised tank when the reaction temperature of 150°C was reached (after about 45 min). Then, as for the adsorption step, samples were periodically withdrawn for analysis (every 30 min). CO_2 concentration was also analysed at the outlet gas stream.

The first AC been loaded in this mini automated plant was PICA L27. It was selected intuitively due to its mesoporous type which was supposed to be less modified by phenol oligomer deposition in the pores. In fact, this AC was not strong enough to undergo large pressure variations in the 50 bar runs and it was dramatically crushed leading to bed clogging, abnormal pressure drop and finally no more flow through the bed.

New experiments were carried out with microporous PICA S23 mechanically much stronger. In accordance with the real AD-OX process, the adsorption step, performed using an aqueous solution of 4-Hydroxyl-Benzoic Acid (4HBA), was stopped before complete breakthrough. The oxidation step was then investigated during three days. A complete recording of a three day oxidation run is presented on Figure 12.5.

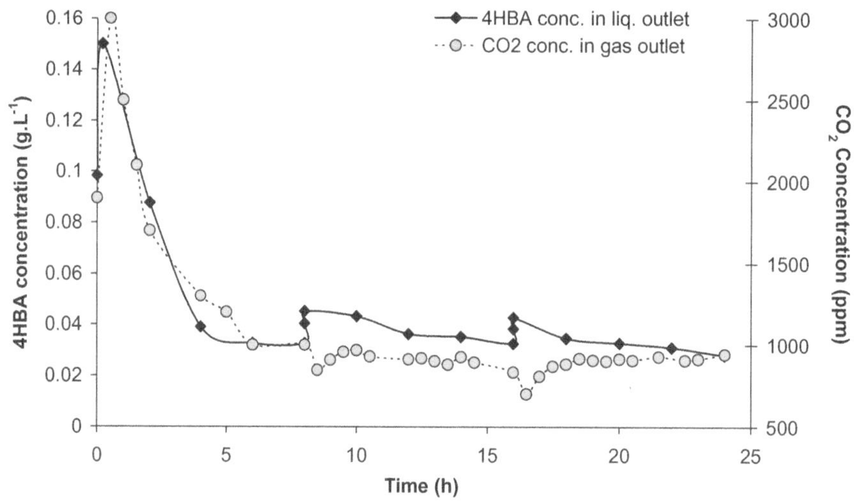

Figure 12.5 Oxidation step of 4HBA in AD-OX process in the mini automated pilot plant on PICA S23: T = 150°C, p_{total} = 50 bar (p_{O2} = 9.5 bar), V_L = 400 mL, Q_L = 0.6 L/h, Q_{air} = 30 NL/h, w_{AC} = 8.53 g. Solution treated by adsorption = 4HBA 0.5 g/L

A clear agreement was found by the two independent analyses. The main feature is the strong decrease of CO_2 and 4HBA concentrations after a short peak corresponding to 4HBA desorption due to heating. Then the oxidation rate slows down after about 5 hrs which may be considered a convenient oxidation time.

New experiments have then been carried out with the hardened SA_DRAW, supposed to be the best SBCM after preliminary adsorption and first oxidation runs and able to undergo high pressure conditions. Phenol has been selected as a model pollutant. Surprisingly a lot of foam was formed during the oxidation step which was leaving the reactor by the gas exit, damaging the CO_2 analyzer and leading to the stop of the run and the rejection of this SBCM for AD-OX process.

The AD-OX tests have been finally performed with one of the less efficient, but sufficiently stable carbons, C_DMAD (0.8–1.6 mm sieved fraction). The results in terms of successive adsorption breakthrough curves and concentration profiles during oxidation are presented on Figure 12.6 and 12.7 respectively. Here the adsorption step was operated until complete bed saturation occurred, so that the adsorption capacity of fresh/regenerated carbon could be calculated from numerical integration of the breakthrough curves.

Figure 12.6 clearly shows a dramatic decrease of the adsorption capacity – or breakthrough time, after the first oxidation (only 15% of original capacity is recovered). This was yet observed in successive batch oxidation experiments but not clearly quantified.

Phenol conversion in the liquid phase (Figure 12.7) calculated between 30 and 420 min varied from 89% at oxidation 1 to 70% at oxidation 3, while within the same time interval COD conversion varied from 71% (oxidation 1) to 53% (oxidation 3) suggesting that a large part of the pollutant was oxidised to carbon dioxide and water.

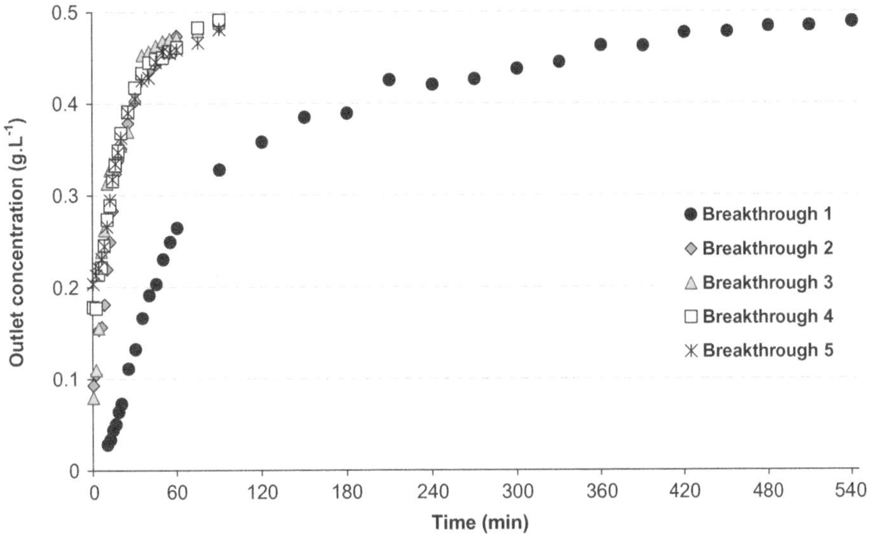

Figure 12.6 Successive breakthrough curves during AD-OX process using C_DRAW: T = 25°C, C_{phenol} = 0.5 g/L, Q_L = 0.23 L/h, w_{AC} = 10.05 g. (Reprinted from Journal of Environmental Management, vol. 91, no. 12, C. Julcour Lebigue, C. Andriantsiferana, N'G. Krou, C. Ayral, E. Mohamed, A. M. Wilhelm, H. Delmas, L. Le Coq, C. Gerente, K. M. Smith, S. Pullket, G. D. Fowler, N. J. D. Graham, Application of sludge-based carbonaceous materials in a hybrid water treatment process based on adsorption and catalytic wet air oxidation, 2432–2439, 2010, with permission from Elsevier)

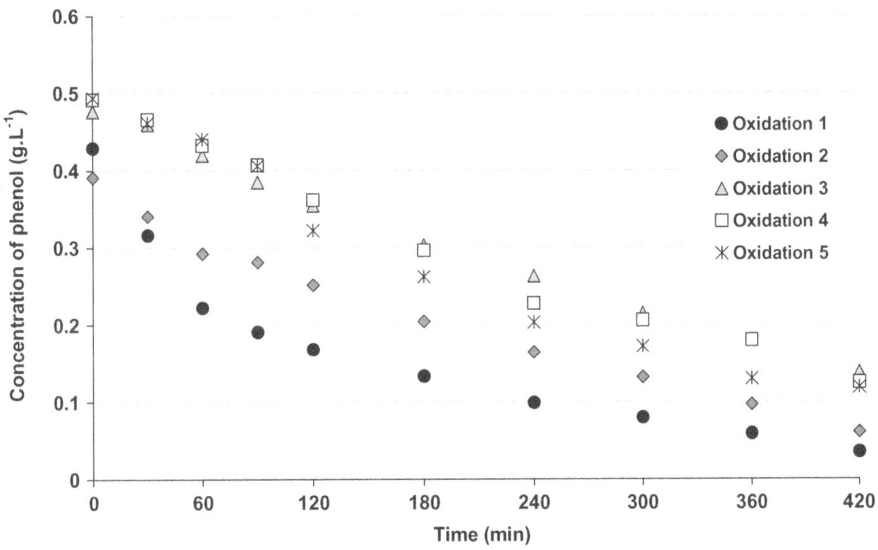

Figure 12.7 Successive oxidations in AD-OX process using C_DRAW: T = 150°C, p_{total} = 50 bar (p_{O2} = 9.5 bar), $C_{phenol,0}$ = 0.5 g/L (V_L = 400 mL), Q_L = 2.44 L/h, Q_{air} = 30 NL/h, w_{AC} = 10.05 g. (Reprinted from Journal of Environmental Management, vol. 91, no. 12, C. Julcour Lebigue, C. Andriantsiferana, N'G. Krou, C. Ayral, E. Mohamed, A. M. Wilhelm, H. Delmas, L. Le Coq, C. Gerente, K. M. Smith, S. Pullket, G. D. Fowler, N. J. D. Graham, Application of sludge-based carbonaceous materials in a hybrid water treatment process based on adsorption and catalytic wet air oxidation, 2432–2439, 2010, with permission from Elsevier)

AD-OX fixed bed experiments in a semi industrial pilot reactor

The experimental set-up has been derived from a continuous CWAO equipment, yet detailed in Suwanprasop *et al.* (2005). The main change has concerned the liquid flow which may successively enter and leave the reactor, during the adsorption step, or be recycled through a batch loop including a pressurized storage tank, during the heating, reaction and cooling steps.

The fixed bed reactor was packed with 325 g of MERCK 2514 (1.25–1.60 mm sieved fraction). This AC has yet undergone long CWAO experiments (>300 hrs) in the same reactor with successively 4HBA, phenol and an equimolar mixture of these pollutants. Its catalytic activity was checked to have reached a quasi-steady state by repeating experiments in reference conditions.

An equimolar phenol-4HBA mixture was treated in this AD-OX set-up. As usual, the adsorption step was operated at ambient conditions (25°C, atmospheric pressure).

In order to verify the adsorption capacity the complete breakthrough curve was first derived, even the proposed AD-OX process does not follow the same procedure, stopping the adsorption step before complete breakthrough to maintain low pollutant concentration at the AC bed exit. The complete breakthrough curve gives much more information on the bed capacity and, when comparing several successive runs, on the loss of adsorption capacity due to AC deactivation during oxidation.

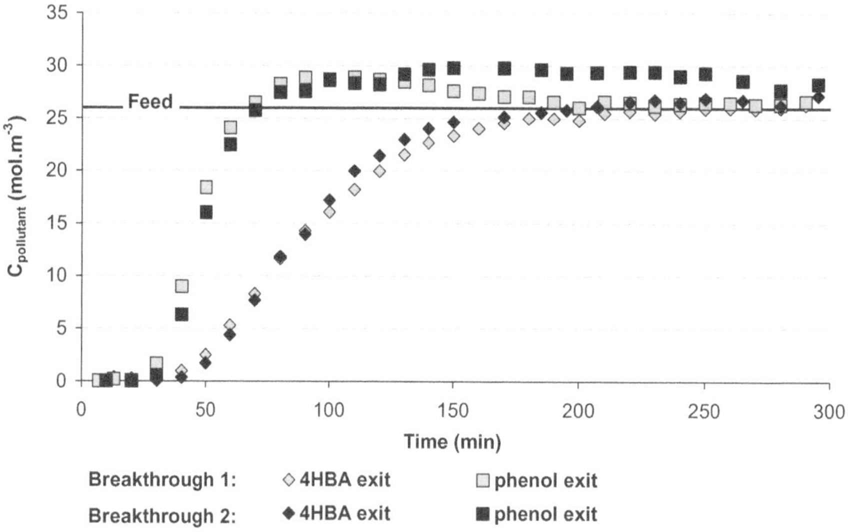

Figure 12.8 Breakthrough curves for 4HBA and phenol (equimolar mixture) in AD-OX process using aged MERCK AC: T = 25°C, $C_{4HBA} = C_{phenol} = 25.8\,mol/m^3$, $Q_L = 3\,L/h$, $w_{AC} = 325\,g$. (Reprinted from Chemical Engineering Journal, vol. 152, no. 1, H. Delmas, C. Creanga, C. Julcour-Lebigue, A. M. Wilhelm, AD–OX: A sequential oxidative process for water treatment—Adsorption and batch CWAO regeneration of activated carbon, 189–194, 2009, with permission from Elsevier)

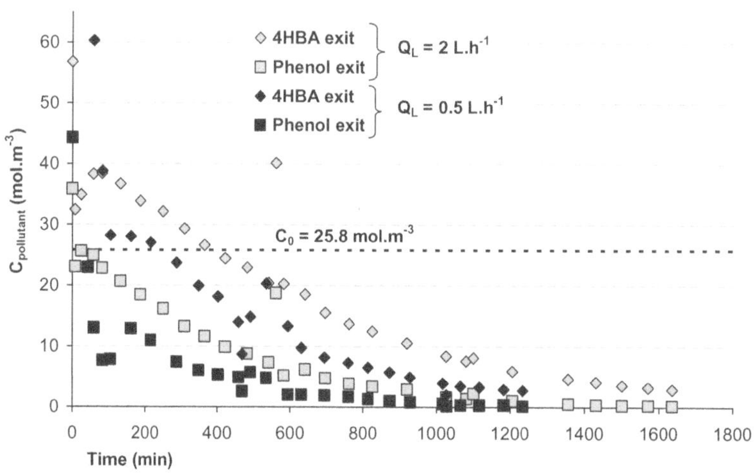

Figure 12.9 1st and 2nd oxidation steps of 4HBA-phenol mixture in AD-OX process using aged MERCK AC: T = 140°C, p_{O2} = 2 bar, $C_{4HBA,0} = C_{phenol,0} = 25.8\,mol/m^3$ (V_L = 2 L), Q_G = 100 NL/h, w_{AC} = 325 g. (Reprinted from Chemical Engineering Journal, vol. 152, no. 1, H. Delmas, C. Creanga, C. Julcour-Lebigue, A. M. Wilhelm, AD–OX: A sequential oxidative process for water treatment—Adsorption and batch CWAO regeneration of activated carbon, 189–194, 2009, with permission from Elsevier)

The results of two successive adsorption runs are presented on Figure 12.8. The two consecutive experiments gave reproducible breakthrough curves, confirming the stabilization of the activated carbon both as adsorbent and as oxidation catalyst. This stable behaviour was only observed on used catalyst while as shown previously fresh activated carbon underwent a severe drop of adsorption capacity after the first oxidation.

Concerning adsorption selectivity, phenol adsorbs faster but is progressively removed by 4HBA in agreement with competitive adsorption data showing preferential 4HBA adsorption (Creanga Manole, 2007). This selectivity can be explained by the much lower solubility of 4HBA in water (about 20 times less than phenol).

Two successive batch oxidation steps have been carried out after complete adsorption steps (presented just above). The wall temperature was here set to 140ºC, and the oxygen partial pressure (p_{O2}) was adjusted to 2 bar. The gas flow rate (85:15 O_2:N_2 mixture) was 100 L/h (NTP) and the liquid flow rate was varied from 2 to 0.5 L/h between the two oxidations. Gas and liquid circulated cocurrently upward through the fixed bed.

Due to significant initial desorption at higher temperature the concentration profiles start at higher concentrations than the adsorption feed concentration of pollutants (25.8 mol/m^3). This batch CWAO being carried out during three days with stops by night, some peaks were found when starting heating again (e.g. at 550 min) due to subsequent desorption.

Phenol is apparently degraded faster than 4HBA, which is thought to be due to lower adsorbed amount and not to faster intrinsic kinetics, as previously observed in autoclave reactor (Creanga Manole *et al.*, 2007).

Figure 12.9 also shows faster oxidation at low liquid recycle flow rate. This might be explained by steeper temperature profile in this condition, resulting in a larger part of the bed being catalytically active.

AD-OX on real effluents

Finally a long range of successive AD-OX cycles (20) has been carried out on real water effluent at the exit of conventional biological treatment for industrial wastewaters. Despite the chemical complexity of this effluent, only characterized by its COD and its significant salt concentration, very similar features were observed: after the first severe cut of its adsorption capacity at about 25%, the activated carbon appeared approximately stabilized without further significant loss of adsorption capacity (about 20% of fresh AC after 20 cycles).

This similar behaviour has also been observed when investigating separately adsorption and oxidation: salt has no noticeable effect on the two steps of AD-OX process. This set of data has no equivalent in the open literature as industrial effluents are very rarely investigated and due to tedious repetitions of many cycles. It provides very important information when thinking to the economy of this process which is now to be investigated in details. At the actual cost of the corresponding AC (1350 €/kg) AD-OX process is highly profitable with regards to consumable costs (pressurized air and water vapour instead of AC). The main economical issue is the equipment cost, as true wastewaters are highly chlorinated and then highly corrosive at temperature conditions required by CWAO regeneration. In addition a synergetic corrosion effect occurs with AC: three usual stainless steel reactors (316 L) were perforated during the 20 AD-OX runs with this chlorinated effluent.

Fenton and other advanced oxidation processes for AC regeneration

As shown in the previous sections, AC regeneration by CWAO is not perfectly convenient as about 25% only of adsorption capacity of the carbon may be recovered whatever the selected AC or SBCM and the model pollutant (phenolic compound or actual industrial effluent).

In order to improve this adsorption capacity regeneration, several other advanced oxidation techniques have been tested: ultrasound, UV, photocatalysis, (photo-)Fenton oxidation, electrochemical oxidation. Among them the best results have been obtained with preliminary tests of electrochemistry, which seems also to perform according to Weng and Hsu (2008).

Fenton oxidation (H_2O_2 + Fe^{2+}) gives interesting results as compared to US, UV and photocatalysis. First it has been verified that Fenton reagent does not interact with activated carbon and does not alter its adsorption capacity. Then successive adsorption runs followed by Fenton oxidation at room temperature, on the same AC sample, have been carried out in different concentration solutions. For simplicity the regeneration efficiency has been defined as the ratio of phenol amount re-adsorbed on AC after Fenton treatment to the initial amount adsorbed.

A typical result is shown on Figure 12.10 (for batch adsorption system) proving the Fenton oxidation to be very fast, the resulting Total Organic Carbon (TOC) variation being probably controlled by pore diffusion phenomena. Best conditions (iron and hydrogen peroxide concentrations) were obtained with fast stabilization (1st recycle) at about 50% adsorption capacity.

These results are not as promising as reported by Huling *et al.* (2005). The main drawback of Fenton regeneration is the progressive consumption of Fe^{2+} which could be avoided by photo regeneration (photo-Fenton). Preliminary work on photo-Fenton proved it to be the most efficient AC regeneration technique, with 65% adsorption efficiency at the 2nd use, better than about 50% with usual Fenton and 25% with air. Nevertheless the technical and economical feasibility of this photo Fenton process (not mentioned in literature) at industrial scale appears questionable, due to both uneasy scale up of UV irradiation and much higher cost of H_2O_2 than compressed air and vapour.

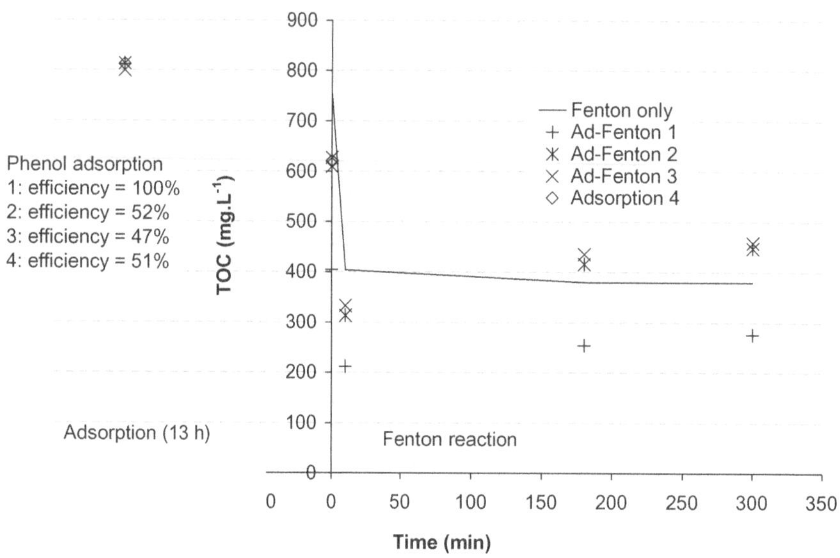

Figure 12.10 L27 AC adsorption capacity regeneration by Fenton oxidation. T = 30°C, $C_{phenol,0}$ = 1 g/L (500 mL), w_{AC} = 1.7 g, Fe^{2+} = 20 mmol/L, H_2O_2 = 300 mmol/L, pH = 3. (Reprinted from Industrial & Engineering Chemistry Research, vol. 49, no. 3, C. T. Muranaka, C. Julcour, A. M. Wilhelm, H. Delmas, C. A. O. Nascimento, Regeneration of Activated Carbon by (Photo)-Fenton Oxidation, 989–995, 2010, with permission from American Chemical Society)

Some experiments have been conducted on Fenton regeneration of SBCM, using SA_DMAD carbon (sieved at 1–1.6 mm). Very similar trends were obtained with this SBCM and with commercial AC. Fenton regeneration is again slightly better than usual air oxidative regeneration (Figure 12.11).

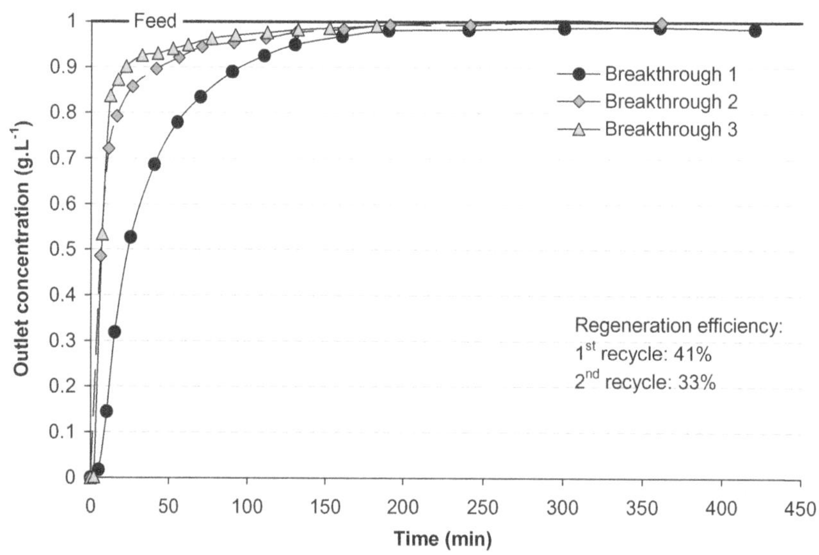

Figure 12.11 Successive breakthrough curves for phenol on SA_DMAD (fresh or regenerated by Fenton oxidation): T = 20°C, C_{phenol} = 1 g/L, Q_L = 0.5 L/h, w_{AC} = 8.52 g

Conclusion

Six different SBCM have been analyzed and used as adsorbents, oxidation catalysts and finally in AD-OX process. All the specifications for convenient use of SBCM in oxidation step have not been fulfilled and improvements of mechanical strength, prevention of leaching and of foaming are still to be achieved. Adsorption capacities were not as low as it could be expected from surface area and especially the steam activated carbon from dewatered raw sludge showed promising adsorption performances: capacity of about half of commercial AC and isotherm Langmuir shape, with stiff initial slope.

Due to their current inadequacy to undergo regenerative oxidation and to the relatively high cost of this regeneration a single adsorption process followed by incineration seems the most appropriate use of these SBCM.

The other contributions of this work have shown that AD-OX might be used in profitable conditions as an end process after biological or physicochemical treatment of industrial wastewater when corrosion issues will be managed. It was validated for model phenolic compounds and industrial effluents.

Among the variety of new AC regeneration techniques tested Fenton oxidation process gives interesting results especially when coupled with UV as a photo-Fenton process. An electrochemical regeneration has been also successfully tested and should be investigated in much more details.

IMPACT OF THE TECHNOLOGY ON ITS INDUSTRY OR RESEARCH SECTOR

Adsorption on granular Activated Carbon (AC) is the best process for reducing hazardous compound concentration and global Chemical Oxygen Demand (COD) in aqueous effluents leaving water treatment plants (either biological or physicochemical). The main drawback of adsorption is the uneasy, ex situ and non sustainable, thermal regeneration of AC. In industrial waste water treatment this ex situ regeneration is even forbidden. This work has succeeded in performing economic in situ partial regeneration of AC in a sequential water treatment process: AD-OX (room temperature adsorption – catalytic wet air batch oxidative regeneration at 150ºC). Technical feasibility was proved with actual aqueous effluent from industrial wastewater treatment using commercial AC and with model bio refractory pollutants using sludge based carbonaceous materials. Economical analysis shows AD-OX is profitable with commercial AC as soon as it may be used more than ten times at 20%. The main issue concerns the severe corrosion due to chlorinated compounds in actual effluents: a convenient material selection has to be performed to estimate the equipment cost. Concerning SBCM, if they are actually produced at low cost (about 250 €) they would be more profitably used in single adsorption without AD-OX regeneration.

This study has two major impacts: first concerning the technical and economic feasibility of AD-OX process aiming to save Activated Carbon (AC) by improving its in situ regeneration, second concerning AD-OX application with Sludge Based Carbonaceous Materials (SBCM).

Clean water is vital for public health and ecosystems. Water treatment is thus a major activity for mankind welfare, as well for producing drinkable water as for reducing pollution of sewage effluents to be returned in natural environment. The European Community considers that about 20% of its superficial waters are threatened by pollution. Improving water treatment is then more and more necessary due to both increasing agricultural and industrial activities which produce large amounts of chemical pollution in aqueous effluents in addition to urban wastewater and to more drastic regulations on water quality. The Water Framework Directive aims to ensure the good chemical status of both surface water and groundwater bodies across Europe. For surface waters this goal is defined by limits on the concentration of specific pollutants of EU relevance, known as priority substances (or micropollutants). To date, 33 priority substances have been identified. A new Directive, Directive 2008/105/EC of 16 December 2008, establishes limits, known as Environmental Quality Standards (EQS), for these 33 substances and for an additional 8 substances regulated under previous legislation.

For eliminating most of these priority substances generally present at low or very low concentration the best economical technology is undoubtedly the widely spread process of adsorption on Activated Carbon (AC). Though very commonly used for more than a century in water treatment, either in powdered or, most often, in granular form, adsorption on AC indeed remains the most adequate technology for eliminating biorefractory compounds not destroyed by usual biological water treatment plants. According to Sarp Industries-Veolia (private communication, 2009) AC adsorption will be more and more used as end treatment after biological or physicochemical treatment, due to more severe regulations on TOC (Total Organic Carbon) of aqueous effluents. The main advantages of AC adsorption are its flexibility, being able to face large concentration and flow rate variations, its simplicity and its large spectrum of compounds to be adsorbed.

According to 'The Economics of Activated Carbon' (8th edition, published 01/10/2008, Roskill metals and minerals reports) *world activated carbon consumption is estimated at 650,000t in 2007. Growth in consumption in current markets is forecast to be 5%py through 2015. Growth will be led by water treatment applications both in the USA, to control disinfection by-products in drinking water, and in the industrialising countries, to upgrade the quality of drinking and wastewater. In China, 10,000 wastewater treatment plants are scheduled for construction by 2010, raising the proportion of wastewater treated from 29% to 50%. The next five years could see the emergence of the largest-ever market for activated carbon with the use of powdered activated carbon to control mercury emissions from coal-fired power plant flue gas in North America. Driven by state environmental legislation in the USA and limitations on new power plant construction, this market is estimated to increase from 5,000t powdered activated carbon in 2007 to 125,000t in 2010. Growth in demand for activated carbon could accelerate to close to 400,000t in 2015 if US federal legislation, requiring installation of 600–700 activated carbon injection systems, is introduced. Total world demand for activated carbon therefore has the potential to rise by nearly 10%py to 1.36Mt in 2015, with mercury emission control accounting for 30% of projected total consumption.*

Nevertheless these very positive perspectives are seriously counterbalanced by the most important drawback of adsorption processes i.e. the adsorbent saturation: the saturated AC must be regenerated before reuse. The first commercial-scale granular carbon adsorption units used to burn the spent AC. Applied research has proposed several regeneration techniques but only few applications at industrial scale are reported. These methods are based either on desorption, mainly thermal, or on chemical decomposition of the adsorbed species.

The most commonly used regeneration, thermal regeneration, is based on either desorption with inert gas or reactive decomposition by water vapour or flue gas and needs very high temperature (about 850°C). It should only be achieved in few very large specialized units (50,000 tons/year), requiring withdrawal of spent AC from the adsorption tower, long transportations, up to 1,000 km, to produce 85–90% of 'new' (slightly degraded) AC.

Furthermore in the case of industrial wastewater treatment by specialised companies, ex situ regeneration is not allowed due to hazardous compounds and the spent AC itself should be considered a hazardous waste needing a special treatment facility, based on uneasy incineration needing special care for environment. The last situation is the worst case not only from economy point of view but also environmentally as it generates large quantities of CO_2 (about ten times what would be produced by the complete mineralisation of adsorbed species).

Thus developing water treatment processes based on AC adsorption with 'in situ' regeneration at moderate cost will have a strong economical impact in the field of waste water business. As explained earlier saving activated carbon results also in reduction of carbon dioxide emission produced when burning the spent AC. So from environmental point of view such processes would be twice profitable by improving water quality and also reducing CO_2 emission.

Concerning the possible use of Sludge Based Carbonaceous Materials in AD-OX process, the environmental balance would be even more positive as a waste (raw sludge) is transformed in useful sustainable material for water treatment. Nevertheless as AD-OX regeneration by WAO autocatalyzed by AC is rather expensive compared to SBCM itself (estimated at 250 €/ton) it appears more profitable to use SBCM in a single adsorption process before incineration. It should also be noticed that, due to very large amounts of available raw sludge, transform it in sustainable AC (able to be regenerated) would rapidly exceed the AC demand. As sludges are continuously produced by bio treatments their valorised by products should also be consumed.

As a conclusion this study results suggest a very attractive utilisation of SBCM as usual adsorbents in end waste water treatment. Despite moderate adsorption capacity, SBCM economical efficiency is much higher than commercial AC. Due to their moderate cost SBCM do not deserve regeneration.

Another potential use of SBCM is as catalyst of continuous Catalytic Wet Air Oxidation (CWAO), but quantities needed are much smaller. In addition prior improvement of their mechanical strength is required and long range oxidation runs are to be checked.

Concerning standard AD-OX with commercial AC and WAO regeneration, its use is profitable on specific effluents (e.g. phenols) but corrosion problems are to be solved at convenient cost to treat usual waste water issued from water treatment plants. On another hand the preliminary promising tests with cheap electrochemical regeneration should be confirmed with highly salted effluents which are very difficult to treat. If successful this electrochemical regeneration at room temperature and pressure for concentrated chlorine effluents could have the main impact on wastewater industry.

REFERENCES

Alvarez P., Garcia-Araya J., Beltran F., Masa F. and Medina F. (2005). Ozonation of activated carbons: Effect on the adsorption of selected phenolic compounds from aqueous solutions. *Journal of Colloid and Interface Science*, **283**, 503–512.

Ania C. O., Parra J. B., Menéndez J. A. and Pis J. J. (2005). Effect of microwave and conventional regeneration on the microporous and mesoporous network and on the adsorptive capacity of activated carbons. *Microporous and Mesoporous Materials*, **85**(1–2), 7–15.

Arnold R. G., *et al.* (2006). In-situ Regeneration of Granular Activated Carbon (GAC) Using Fenton's Reagents. http://superfund.pharmacy.arizona.edu/research%20translation/Regeneration_GAC_using_FentonsFinal%20Report%2706.pdf.

Bento L. R. S. M. and Rein P. W. (2007). US Patent Application #: 20080286193 WO/2007/032998 International Application No.: PCT/US2006/034925. Chemical regeneration of activated carbon.

Creanga Manole C. (2007). AD-OX process for remediation of non biodegradable organic pollutants (by adsorption then catalytic oxydation). Ph-D thesis, INP Toulouse, France, http://ethesis.inp-toulouse.fr/archive/00000561/.

Creanga Manole C., Ayral C., Julcour-Lebigue C., Wilhelm A. M. and Delmas H. (2007). Catalytic wet air oxidation of aqueous organic mixtures. *International Journal of Chemical Reactor Engineering*, **5**, A65, http://www.bepress.com/ijcre/vol5/A65.

Drouiche N., Mahmoudi M., Mameri N., Lounici H. and Ghaffour N. (2006). Utilization of electrodialysis for the regeneration of granular activated carbon packed in beds saturated with H_2S. *Desalination*, **200**(1–3), 629–631.

Gonzalez J. F., Encinar J. M., Ramiro A. and Sabio E. (2002). Regeneration by wet oxidation of an activated carbon saturated with p-nitrophenol. *Industrial and Engineering Chemistry Research*, **41**(5), 1344–1351.

Horng R. S. and Tseng I. C. (2008). Regeneration of granular activated carbon saturated with acetone and isopropyl alcohol via a recirculation process under H_2O_2/UV oxidation. *Journal of Hazardous Materials*, **154**(1–3), 366–372.

Huling S. G., Jones P. K., Ela W. P. and Arnold R. G. (2005). Fenton-driven chemical regeneration of MTBE-spent GAC. *Water Research*, **39**, 2145–2153.

Liang G., Lin Y. T. and Shin W. H. (2009). Persulfate regeneration of trichloroethylene spent activated carbon. *Journal of Hazardous Materials*, **168**, 187–192.

Narbaitz R. M. and Karimi-Jashni A. (2009). Electrochemical regeneration of granular activated carbons loaded with phenol and natural organic matter. *Environmental Technology*, **30**(1), 27–36.

Polaert I., Wilhelm A. M. and Delmas H. (2002). Phenol wastewater treatment by a two-step adsorption-oxidation process on activated carbon. *Chemical Engineering Science*, **57**, 1585–1590.

Pullket S., Smith K. M., Fowler G. D. and Graham N. J. D. (2009). Influence of source and treatment method on the properties of activated carbons produced from sewage sludge. *Journal of Residual Science and Technology*, **6**(1), 43–49.

Qu G. Z., Li J., Wu Y., Li G. F. and Li D. (2009). Regeneration of acid orange 7-exhausted granular activated carbon with dielectric barrier discharge plasma. *Chemical Engineering Journal*, **146**, 168–173.

Sheintuch M. and Matatov-Meytal Y. I. (1999). Comparison of catalytic processes with other regeneration methods of activated carbon. *Catalysis Today*, **53**, 73–80.

Shende R. V. and Mahajani V. V. (2002). Wet oxidative regeneration of activated carbon loaded with reactive dye. *Waste Management*, **22**, 73–83.

Suwanprasop S., Eftaxias A., Stuber F., Polaert I., Julcour-Lebigue C. and Delmas H. (2005). Scale up and modeling of fixed bed reactors for the catalytic phenol oxidation over adsorptive active carbon. *Industrial and Engineering Chemistry Research*, **44**(25), 9513–9523.

Toledo L. C., Silva A. C. B., Augusti R. and Lago R. M. (2003). Application of Fenton's reagent to regenerate activated carbon saturated with organochloro compounds. *Chemosphere*, **50**, 1049–1054.

Weng C. H. and Hsu M. C. (2008). Regeneration of granular activated carbon by an electrochemical process. *Separation and Purification Technology*, **64**(2), 227–236.

Zhang H. (2002). Regeneration of exhausted activated carbon by electrochemical method. *Chemical Engineering Journal*, **85**(1), 81–85.

Zhou M. H. and Lei L. C. (2006). Electrochemical regeneration of activated carbon loaded with p-nitrophenol in a fluidized electrochemical reactor. *Electrochimica Acta*, **51**(21), 4489–4496.

Chapter 13

Utilisation of activated carbon in adsorption processes

Laurence Le Coq, Claire Gerente, Laurence Guihéneuf, Reena Shrestha and Rengaraj Selvaraj

SUMMARY

The main objectives of this chapter were to characterize adsorption performances of Sludge Based Adsorbants (SBA) regarding synthetic wastewaters treatment containing molecules representative of industrial pollution. These adsorbants have been produced by physical activation (activating gas: steam and CO_2) of two activated sludge coming from British (DRAW, Dewatered Raw Sludge) and French (DSBS, Dewatered Secondary Biological Sludge) WWTP. This experimental work included batch and dynamic processes. Three different pollutant classes have been tested: dyes (AR18, BV4 and BB9) and aromatic compounds (phenol and phthalate), metal and metalloid ions (nickel, copper and arsenic V).

In batch reactors, optimal contact times and corresponding adsorption capacities have been determined. Some of the best adsorption performances (capacity and kinetic) have been obtained with phthalate pollutant (HPK) and steam activated SBA (DSBS-H_2O and DRAW-H_2O): maximum fixation capacities reached after 21 days of contact time and maximum fixation capacities (deduced from Langmuir model) close to 187 mg/g. For comparison, the commercial activated carbon Filtrasorb 400 provided by Chemviron Carbon exhibited a capacity of 344 mg/g.

Considering the best batch performance (phthalate adsorption on steam activated SBA), dynamic breakthrough curves have been determined from fixed beds, including the two SBA and Filtrasorb 400, and HPK as the pollutant. The design of the reactor was conducted taking into account the following parameters: the amount of SBA in relation with the available mass of adsorbent (one batch of production corresponded roughly to 90 g of SBA), the maximum fixation capacities and the initial concentration of pollutant, and the flowrate of wastewater according to the optimal contact times. Two main results could be extracted from these dynamic breakthrough curves: the dynamic capacities directly deduced from the curves were similar to batch ones indicating that the reactor design was relevant, and a very fast pollutant breakthrough was observed with SBA and delayed with Filtrasorb 400.

METHODOLOGIES AND APPROACHES EMPLOYED

The methodology used to reach the objectives of this work is a classical and integrated approach dedicated to adsorption processes. It involved first a detailed characterisation of the adsorbent, in order to find relationships between the properties and the adsorption performances.

Then, experiments in batch reactors have been conducted including on one hand, the evaluation of the optimal contact times for each couple adsorbant/adsorbate from kinetic decay curves, and on the other hand, the determination of adsorption capacities from adsorption isotherms. The use of some equilibrium models (for example Langmuir model) allows to estimate the maximum fixation capacity and then, to compare the performances of the different sorbents. Based on these results, the design of a dynamic process like a fixed bed was attainable. Three beds have been performed and breakthrough curves have been obtained until the saturation of the sorbent. Dynamic fixation capacities could be extrapolated and compare to the previous ones: if the design is correctly realised, these values should be similar. Secondly, they also give some information about the behaviour of the bed in terms of breakthrough time (potential leak). With all these information, a conclusion can be made about the feasability to scale-up towards an industrial trial. All these experiments conducted with SBAs, have been compared with those realised with a commercial activated provided by Chemviron Carbon.

To comply with this methodology, different approaches have been used. In this chapter, the production of adsorbents is not presented and their selection has focused primarily on DSBS-H_2O and DRAW-H_2O, and as a second choice on DSBS-CO_2 and DRAW-CO_2. Regarding the physico-chemical characterisation, following experiments were realised:

- determination of BET surface area from nitrogen adsorption, micro and meso pore volumes,
- determination of ash content,
- determination of C, H, N, S and O content by elemental analysis,
- determination of the pH after washing step: 400 mg of adsorbent were placed in 40 mL of deionised water for 5 days. Each day, the supernatant of the adsorbent's suspension was removed (for pH determination) and replaced by 40 mL of deionised water, until the pH remains constant. It has been shown that 24 hours are sufficient for proton exchange in solution,
- determination of pHpzc: 50 mL of 0.01 M NaCl solution was placed in closed bottle. The initial pH was adjusted to a value between 3 and 12 by adding HCl or NaOH solutions. Then 150 mg of adsorbent was added and the final pH measured at 48 h under agitation at constant temperature. The pHpzc is the point were the curve pHfinal versus $pH_{initial}$ crosses the line $pH_{initial} = pH_{final}$,
- lixiviation tests: lixiviation tests have been realised in order to determine the quantity of calcium, copper, iron and zinc ions in adsorbent sample. S_o, 500 mg of sorbent were placed in 50 mL of HCl 0.5 M and were agitated during 5 days. In the liquid phase, calcium, copper, iron and zinc ions have been measured by atomic absorption. Phosphate has been measured in triplicate, according to a protocol adapted to the norm NF T 90-023/EN ISO 6878. 0.7 g of SBA has been dried at 105°C during the night, and it has been burned off at 550°C during 4 hours. Then, samples were placed in 5 mL of HCl 0.5 M in a water bath (100°C) for 15 min, in order to solubilize all phosphates. A mixture was added to sample: 85.7% $(NH_4)Mo_7O_{24}{\cdot}4H_2O$ + 14.3% ascorbic acid (10%). With the presence of phosphate, a bluish coloration appeared after 1h at 37°C. The concentration was determined by UV-visible spectrometry at 820 nm.

As far as batch adsorption studies, different pollutants have been tested in synthetic wastewaters. Two metal ions (copper(II) and nickel(II)), three dyes (one acid called Acid Red 18 – AR18, and two basic named Basic Violet 4 – BV4 and Basic Blue 9 – BB9), one metalloid element (arsenic(V)) and two organic compounds (phenol and potassium hydrogenophtalate) have been selected. The procedures were:

- Sorption of metal ions: experiments were carried out with 500 mg of adsorbent and 1 L of copper(II) solution (C_0 = 30 mg/L) or nickel(II) (C_0 = 50 mg/L). Initial pH was fixed at 4 and 5 with HCl 0.05 M for the kinetic study in order to avoid Cu/Ni precipitation. At regular interval times, some milliliters of solution were removed for analysis. Copper and nickel concentrations were determined by spectrometric atomic absorption.
- Sorption of dyes: adsorption kinetics were carried out with 500mg of adsorbent and 1 L of AR18 (C_0,SBA = 30 mg/L, C_0,Filtrasorb 400 = 50 mg/L), BV4 (C_0,Filtrasorb 400 / DRAW-H_2O / DRAW-CO_2 = 80 mg/L, C_0,DSBS-CO_2 = 15 mg/L) or BB9 solutions (C_0,SBA = 65 mg/L, C_0,Filtrasorb 400 = 150 mg/L). Dyes concentrations were measured by UV-Visible spectrometry. The wavelengths are 507 nm for AR18, 596 nm for BV4 and 663 nm for BB9. Adsorption isotherms experiments were undertaken in 125 mL flasks and with 50 mg of adsorbent and 100 mL of AR18 solution. The equilibrium time was fixed at 1200 h.
- Sorption of arsenic V: kinetics experiments were carried out with 50 mg of adsorbent and 100 mL of arsenic(V) solution (C_0 = 1 mg/L). pH was initially fixed at 5.5 with HCl 0.01 M for the kinetic study. Arsenic(V) concentrations were determined by spectrometric atomic absorption (Spectrometer Perkin Elmer AAnalyst 600). Batch adsorption isotherms experiments were undertaken in 125 mL flasks and with 50 mg of adsorbent and 100 mL of As(V) solution. The equilibrium time was fixed at 144 h.
- Sorption of phenol and potassium hydrogenophtalate (HPK): studies were carried out with 500mg of adsorbent and 1 L of phenol and HPK (C_0,SBA = 50 mg/L, C_0,Filtrasorb 400 = 200 mg/L). Organic compounds concentrations are measured by UV-Visible spectrometry. The wavelengths are 270 nm for phenol and 281 nm for HPK.

Finally, fixed beds have been designed following this procedure: a glass column of 24 cm length and 4.1 cm internal diameter was used. The columns were packed with adsorbents (DRAW-H_2O, DSBS-H_2O and Filtrasorb 400) for different heights (2, 6, 8 cm). The bed was rinsed with deionized water until the pH of effluent was equal to the contact pH determined previously. The polluted water was prepared at 100 mg/L of HPK. The flowrate in the column was performed from bottom to head of the column (up flow mode) with volumetric flow rate close to 3 mL/min. A peristaltic pump was used to control the flow rate. Samples of effluents were collected periodically and the remaining concentration (C) was analyzed using UV spectrophotometer. The breakthrough curves of C/C_0 versus time were drawn. The flow through the column was continued until the concentration of column effluent approached the initial concentration.

ACHIEVEMENTS RELATED TO THE STATE OF THE ART

Activated carbon (AC) is a highly effective adsorbent because of its extended specific surface area (exceeding 600 m^2/g), high pore volume, and for some of them the presence of surface functional groups. AC is commonly obtained from various organic precursors such as peat, wood, bituminous coal, coconut shells, polymers (Bansal *et al.*, 1988). In recent years, a growing research interest consists in the production of cost-effective carbon-based adsorbents from a range of residues, mainly industrial or agricultural by-products. In this context, sewage sludge generated during wastewater treatment processes has been identified as potential low cost precursor for activated carbon production (Martin *et al.*, 2003; Otero *et al.*, 2003; Rio *et al.*, 2005 and Rozada *et al.*, 2003). Recent studies have shown the feasibility of this conversion, but they rarely determine the effects of activation conditions on surface chemistry and adsorption properties (isotherm and kinetics) of standard industrial pollutants. Considering industrial wastewaters, the sludge based adsorbents show good adsorption capacities of dyes and metal ions compared with commercial ones (Rio *et al.*, 2005 and Rozada *et al.*, 2003), but in specific operating conditions (pH), some leachings could occur. The main achievements of this WP are presented below in terms of physico-chemical characterisation, pollutant adsorption performances in batch reactors and in fixed beds.

Physico-chemical characterisation

From Table 13.1, it is shown that steam activated SBA (DSBS-H_2O and DRAW-H_2O) exhibit the highest BET surface and carbon content (excepted for DRAW-H_2O) and the lowest ash content. These values are roughly 4 times lower than Filtrasorb 400. None of the other obtained properties has clearly led to exploitable results, due to many reasons.

For example, the leaching of these particular SBA in acid and base solution (shown by the lixiviation tests) prevented the determination of surface chemical function and altered the values of pH and pHpzc. An other example was the presence of high mineral content conducting to technical problems with elemental analyser (not adapter for this kind of materials) and not favourable for adsorption performance. The last restriction is the low mechanical resistance of these materials, to overcome this drawback other hardener SBA have been produced but not presented in this chapter. Finally, four SBA (DSBS-H_2O, DRAW-H_2O, DSBS-CO_2 and DRAW-CO_2) have been selected for adsorption tests, but only the steam activated ones seem to be favourable for use in aqueous pollutant adsorption.

Table 13.1 Main properties of SBAs compared with Filtrasorb 400

	Adsorbent				
	DSBS-CO_2	**DSBS-H_2O**	**DRAW-CO_2**	**DRAW-H_2O**	**Filtrasorb 400**
S_{BET} (m^2/g)	181	218	215	282	1090
Mesopore vol. (cm^3/g)	0.065	0.084	0.154	0.135	0.097
Micropore vol. (cm^3/g)	0.080	0.067	0.085	0.049	0.312
Ash content (% wt)	59	41	70	65	6
pH after washing step	8.1	7.7	8.9	9.3	8.0
pHzpc		7.6	8.2	8.6	9.3
C (% wt)	34.2	37.2	32.4	27.9	88.3

Sorption of dyes

The sorption of different dyes, namely AR18, BB9 and BV4, is presented on Table 13.2. Considering 2 days of contact time, adsorption capacities on BV4 onto DRAW are better than Filtrasorb 400 (in the same operating conditions) and are interesting regarding AR18. DSBS exhibit clearly poor capacities and BB9 is the less efficiently removed dye.

The nature of the precursor (DRAW compared with DSBS) seems to be a predominant parameter. However after the equilibrium is reached (>50 days for all dyes and all SBA), the performances of DSBS-CO_2 become better than DRAW for BV4 fixation. These performances remain low considering AR18 and BB9. Isotherm curves shown a multilayer adsorption (DRAW and DSBS with AR18) which prevents to use the Langmuir model to a wide range of concentration. The maximum fixation capacities can not be deduced which prevents to compare the performance of each sorbent at equilibrium.

Table 13.2 Partial results of dyes adsorption onto SBAs and Filtrasorb 400

Adsorbent	AR18 (t = 2 days)		BB9 (t = 2 days)		BV4 (t = 2 days)	
	C_0 (mg/L)	Adsorption capacities (%)	C_0 (mg/L)	Adsorption capacities (%)	C_0 (mg/L)	Adsorption capacities (%)
Filtrasorb 400	50	45	150	67	80	20
DSBS-CO_2	30	7	60	5	15	21
DSBS-H_2O	–	–	–	–	15	32
DRAW-CO_2	30	38	60	31	80	33
DRAW-H_2O	30	48	60	34	80	32

Sorption of metal ions and metalloid

No adsorption occurs with copper and nickel (cations) in the operating conditions described previously, so it was decided to replace cations by anions: Arsenic V.

Regarding AsV kinetic decay curves, the equilibrium time was found to 6 days for SBA. The comparison of the adsorption capacities after 2 days showed that only 10% of arsenic was removed with the two tested DSBS, 30 to 40% with the two DRAW and 64% with Filtrasorb 400.

Adsorption isotherms have been realised and the use of Langmuir model has well described the experimental data (good correlation coefficient as shown in Table 13.3). The maximum fixation capacities were the following (Table 13.3): near 4 mg/g for Filtrasorb 400, 2 mg/g for DRAW-H_2O and 1.4 mg/g for DRAW-CO_2. The nature of the precursor (DRAW compared with DSBS) seems once again to be the predominant parameter, as well as the activating gas. From the comparison of q_m values, SBA exhibits performances 25 to 50% less than Filtrasorb 400.

Table 13.3 Langmuir parameters for adsorption of As(V) onto DRAW-H_2O, DRAW-CO_2 and Filtrasorb 400

Adsorbent	Langmuir model (linear)		
	q_m (mg/g)	b (L/mg)	R^2
Filtrasorb 400	3.93	12.84	0.968
DRAW-CO_2	1.39	0.81	0.969
DRAW-H_2O	1.98	0.73	0.997

Sorption of aromatic compounds

This section presents the adsorption tests on 2 aromatic molecules: phenol and HPK. It was clear from Table 13.4, that the activating gas was the predominant parameter regarding the fixation capacities. Indeed, better performances were obtained with steam activation, in comparison with CO_2. Moreover, DSBS sorbents presented similar or better efficiencies than DRAW ones in the same conditions. The sorption of HPK seemed to be higher than phenol. The equilibrium times have been found to 14 days for HPK and 28 days for phenol. Adsorption isotherms were then performed with each pollutant regarding the selected SBA : DSBS-H_2O and DRAW-H_2O.

Table 13.4 Partial results of phenolic compounds adsorption onto SBAs and Filtrasorb 400

Adsorbent	C_0 (mg/L)	HPK (t =12 days)	Phenol (t =14 days)
		Adsorption capacities (%)	Adsorption capacities (%)
Filtrasorb 400	200	56	55
DSBS-CO_2	50	56	30
DSBS-H_2O	50	67	52
DRAW-CO_2	50	53	36
DRAW-H_2O	50	62	47

The use of Langmuir model has well described the experimental data (good correlation coefficient as shown in Table 13.5). It was obvious that the sorption of HPK was highly favourable compared to phenol. The deduced maximum fixation capacities were 171 mg/g for DRAW-H_2O, 187 mg/g for DSBS-H_2O and 344 mg/g for Filtrasorb 400. DSBS-H_2O was slightly superior to DRAW-H_2O. From the comparison of q_m values, SBA exhibits again performances 25 to 50% less than Filtrasorb 400 as far as phenol and HPK are concerned.

Table 13.5 Langmuir parameters for adsorption of HPK and phenol onto DRAW-H_2O, DSBS-H_2O and Filtrasorb 400

Adsorbent	HPK			Phenol		
	q_m (mg/g)	b (L/mg)	R^2	q_m (mg/g)	b (L/mg)	R^2
Filtrasorb 400	344	0.0270	0.996	278	0.42	0.959
DSBS-H_2O	187	0.0346	0.987	74	0.93	0.999
DRAW-H_2O	171	0.0260	0.951	65	1.22	0.933

For the last part of this study, a selection of the better results has been made. It has been decided to use steam activated SBA because in most experiments, it was the better activating gas regarding the adsorption performances (As(V), aromatic compounds and dyes), and because this gas allow easier and cheaper industrial production of SBA than CO_2. Depending of the pollutants, DRAW-H_2O was more or less efficient than DSBS-H_2O. So, DRAW-H_2O and DSBS-H_2O were the best SBA.

Fixed bed experiments

The selected pollutant was HPK: equilibrium time of 14 days (in comparison with 6 for As(V), 28 for phenol, and more than 50 for dyes) and best q_m values (half of Filtrasorb 400) for the 2 SBA. The results were expressed in terms of breakthrough curves (Figures 13.1, 13.2 and 13.3).

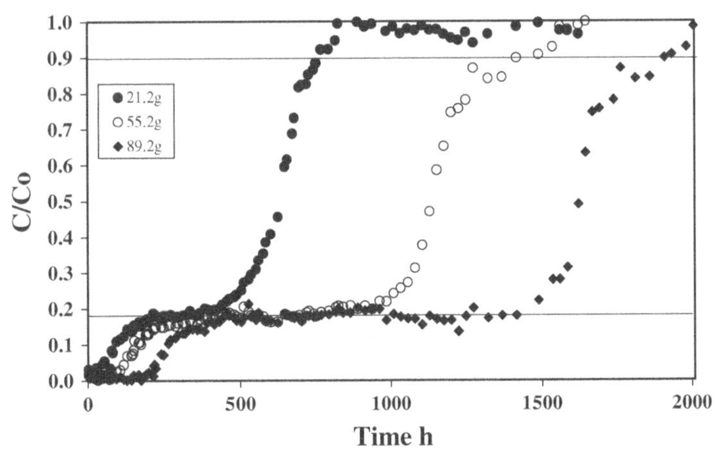

Figure 13.1 Adsorption breakthrough curves of HPK onto Filtrasorb 400 in column

The dynamic capacities at saturation deduced from the curves, were approximatively close to the maximum capacity at equilibrium obtained in batch reactors (300 mg/g for Filtrasorb 400, 170 mg/g for DSBS-H_2O and 130 mg/g for DRAW-H_2O) indicating that the design of the beds was correct.

Moreover, it can be seen in Figures 13.2 and 13.3 and considering a breakthrough for $C/C_0 = 0.1$, that the breakthrough time was immediate in comparison with Filtrasorb 400 (Figure 13.1). It means that the leak of pollutant is not delayed and occurs at the beginning of the test. A way to improve these results is either to optimise the design of the fixed bed, either to change the SBA using other produced ones.

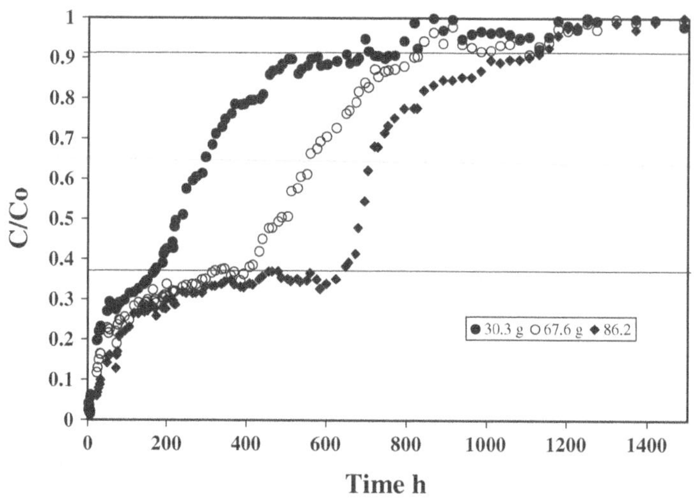

Figure 13.2 Adsorption breakthrough curves of HPK onto DRAW-H_2O in column

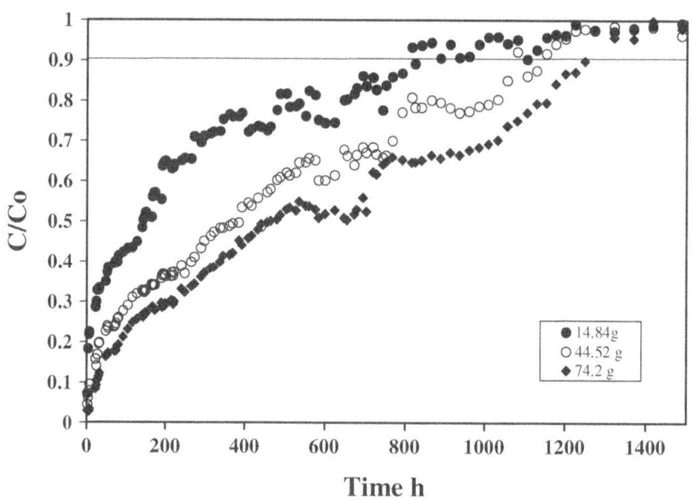

Figure 13.3 Adsorption breakthrough curves of HPK onto DSBS-H_2O in column

CONCLUSIONS AND IMPACT OF THE TECHNOLOGY ON ITS INDUSTRY OR RESEARCH SECTOR

From these adsorption experiments (kinetic decay curves, adsorption isotherms and breakthrough curves), some conclusions can be presented:

- adsorption tests have been performed, but the duration of experiments has been underestimated due to very long kinetic reactions.
- as far as metal ions are concerned, no adsorption has been noticed. Some different results have been published in the literature, but it seems greatly depend on the nature of sludge (calcium or lime content), the conditions of production, the technology of the oven... To confirm this, SBA coming from UK (DRAW) show in the most cases, better performances in comparison with French sludge (DSBS). Moreover, activation of chars with H_2O leads to greater adsorption capacities than CO_2 activation.
- in terms of results, the optimum contact times are high in all cases excepted for arsenic(V). Considering the adsorption capacities, performance of SBA are generally ranged between 25 to 50% of those obtained with the tested commercial activated carbon.

For all these reasons, the application of the studied SBA in industrial adsorption processes for wastewater treatments does not appear as a promising development. Others SBA has been produced, with different operating conditions (chemical activation, de-ashing procedure, addition of hardening additives...) and for which, some specific applications could be found. But the SBA selection made in this study has focused of parameters like: low production costs in term of operating conditions for their production, available amounts of SBA... In terms of structure (leaching of minerals within different pH conditions, low mechanical resistance and low carbon content), other SBA are required (de-ashing, hardening...) for their applicability to industrial scale. In terms of performance, results show capacities ranged between 25 to 50% of those obtained with Filtrasorb 400.

REFERENCES

Bansal R. C., Donnet J. B. and Stockli F. (1988). Active Carbon, Marcel Decker, New York.

Martin M. J., Artola A., Balaguer M. D. and Rigola M. (2003). Activated carbons developed from surplus sewage sludge for the removal of dyes from dilute aqueous solutions. *Chem. Eng. J.*, **94**, 231–239.

Otero M., Rozada F., Calvo L. F., Garcia A. I. and Moran A. (2003). Elimination of organic matter pollutants using adsorbents obtained from sewage sludge. *Dyes and Pigments*, **57**, 55–65.

Rio S., Faur-Brasquet C., Le Coq L. and Le Cloirec P. (2005). Production and characterization of adsorbent materials from industrial waste. *Adsorption*, **11**, 793–798.

Rozada F., Calvo L. F., Garcia A. I., Martin-Villacorta J. and Otero M. (2003). Dye adsorption by sewage sludge-based activated carbons in batch and fixed bed systems. *Biores. Technol.*, **87**, 221–230.

Chapter 14

Utilisation of activated carbon in biofilters for elimination of industrial waste gases

David Gabriel, Óscar Jesús Prado, Javier Lafuente, Xavier Gamisans, Juan Pedro Maestre, Jerónimo Hernández and Antonio David Dorado

SUMMARY

Project objectives

The main target of this research has been to evaluate the feasibility of employing activated carbon produced from sludge obtained from a wastewater treatment plant as packing material in biological reactors for the abatement of odours generated in industrial facilities, as, for instance, in solid waste or wastewater treatment plants. Mainly, an evaluation of biofilters' performance parameters and modelling of the biofiltration units was carried out.

Work performed

During the development of the study an in-depth characterization of several common packing materials as well as carbon-based packing materials was performed. Also, a series of bench-scale biofilters packed with four selected organic packing materials and with sludge-based (non-activated) carbon (SBC) were designed and tested. It is important to mention that sludge-based activated carbon (SBAC) and commercial activated carbon (CAC) were not tested during biofilters operation. SBC was used instead. However, after analysis of the characteristics of the packing materials, it is expected that very similar results in terms of biofilter performance would have been obtained if SBAC or CAC had been employed.

Also, a general dynamic biofiltration model applied to toluene removal was developed, calibrated and validated. Mathematical equations employed are based on discretized mass balances which take into account the main chemical and physical phenomena involved in the system. Previous experimental results, in which a toluene degrading biofilter inoculated with bacterial populations evolved to a fungal biofilter (Maestre *et al.*, 2007), were used herein as input data to calibrate and validate the biofiltration model. In addition, a statistical procedure was applied to check the confidence intervals of the parameters obtained during the model calibration procedure. Finally, a detailed economical evaluation on the use of sludge-based carbons was performed.

End results

A series of organic and inert packing materials, including different carbons, have been characterized. SBC, when mixed with coconut fibre, has proven to be an appropriate packing material in a conventional biofilter. Toluene removal

 Reduction, Modification and Valorisation of Sludge (REMOVALS). Edited by Christophe Bengoa, Azael Fabregat, Josep Font and Frank Stüber. ISBN: 9781843393450. Published by IWA Publishing, London, UK.

efficiency was commonly above 90% when an inlet concentration around 6 ppmv (similar to what can be found in actual industrial odorous waste gases) was fed to the bioreactor.

Also, it was proved that the use of SBC may represent an economically viable alternative to other commonly employed inert packing materials, as CAC. Total costs may be reduced from 40 to 50% if SBAC or SBC is selected as packing material instead of CAC. Also, a dynamic model to simulate toluene abatement by bacterial and fungal biofilters has been developed, calibrated and validated through a set of different experimental conditions for biofilters with switching populations from bacteria to fungi. The model was able to satisfactorily predict different operation periods, including bacteria-, transition from bacteria to fungi-, and fungi-based operation by means of a relatively small number of parameters. The model developed is a valuable tool for the design of odour treatment biofilters.

METHODOLOGIES AND APPROACHES EMPLOYED

Bioreactor setup

A biofiltration setup that had been previously employed for the depuration of toluene-containing vapours and consisting of two biofilters operated in parallel (Maestre *et al.*, 2007) was redesigned and restructured to fit four PVC columns (8.8 cm internal diameter) with a packing height of 50 cm, giving a total bed volume of 3 L per reactor. Each biofilter was fed with pre-humidified air regulated with calibrated flowmeters as primary gas flow. A secondary air stream was pumped by a peristaltic pump into a glass bubbler unit containing pure liquid toluene. Both gaseous streams were mixed, and the resulting gas mixture was fed from the bottom of the reactor. Each reactor had four gas sampling ports located, respectively, at 0, 15, 30, and 45 cm from the gas inlet. Tap water or a nutrient solution was automatically sprinkled on a daily basis over the biofilter beds to provide additional moisture, the necessary nutrients for the microorganisms and to wash dead cells and the end-products of toluene degradation.

Methodologies

For the analysis of the bioreactors system operation, gas samples were collected from sampling ports and stored using Tedlar® bags prior to their analysis. Toluene concentration was measured by gas chromatography in a GC 6890N (Agilent Tech., Spain) equipped with a HP-5 capillary column and a flame ionization detector. Leachate samples were periodically collected from the bottom of each reactor. Conductivity and pH of these samples were measured by means of microCM 2100 and MicropH 2001 lab probes (Crison, Spain), respectively. Nitrite, nitrate and phosphate contents in the liquid samples were determined by capillary electrophoresis in a Quanta 4000E unit (Waters, Spain). Ammonia concentration was measured by means of a continuous flow analyzer (Baeza *et al.*, 2002). The porosity and the specific surface area of the packing materials were determined in an external laboratory (Serveis Científico-Tècnics, UB), by BET absorption isotherms in an ASAP2000 porosimeter (Micromeritics, U.S.A.). Materials densities were measured in a helium picnometer. Elementary analysis was performed by combustion under standard conditions using sulphanilamide as standard (EA-1108 ThermoFisher Scientific). Extractable phosphorus was determined by ICP in a Thermo Jarell-Ash model 61E multichannel analyser in standard conditions (Polyscan, U.S.A.). Filter media moisture content, organic matter content and pH were determined according to Standard Methods (APHA, 1995). Water holding capacity was measured by keeping the material wet (sparkling constantly tap water for 100 min) and determining the weight changes. Water retentivity was measured by keeping wet material in constant contact with dry air flow circulating through the bed and measuring the loss of weight of the bed. Conductivity, pH and buffer capacity for the materials leachate were determined after submerging them in water for 1 h.

Approaches employed

The mathematical model was built considering the most relevant phenomena occurring during the biofiltration process, namely convection, absorption, diffusion and biodegradation. The theoretical model describing the elimination of toluene in a biofilter bed is based on the mass balance in the gas phase and within the biofilm (Dorado *et al.*, 2008). Important assumptions underlying the model were:

1) Gas phase circulation regime is modelled as plug flow pattern. Thus, axial dispersion is not considered.
2) Gas-biofilm interface equilibrium is described by Henry's law.
3) Planar geometry and perpendicular diffusion in biofilm are used to derive model equations considering that the solid support size is significantly higher than the biofilm thickness. Diffusion in the biofilm is described by Fick's law.
4) Biofilm is formed on the external surface of the packing material. Thus, biomass does not grow in the pores of the packing material and reactions only take place in the biofilm phase.

5) Physical properties of the species in the biofilm are assumed to be the same as in water since this is the main component.
6) There is no accumulation of biomass in the filter bed in each period and biomass properties (thickness, specific surface area and kinetic coefficients) are uniform along the bed. This assumption was experimentally verified by an almost constant pressure drop and reactor weight in the whole studied operation period.
7) Adsorption of pollutant onto the support is neglected due to the low pollutant concentration and the low adsorption capacity of the packing material. Moreover, under steady-state conditions, the adsorption-desorption process is considered to be in equilibrium.

A procedure developed by Devinny *et al.* (1999) in terms of the economical analysis was revised and updated (Prado *et al.*, 2009) for the comparison of investment and operation and maintenance costs of biofilters packed with different packing materials. Such approach was based on the analysis of open-bed biofilters considering all cost related to their construction and operation including packing material replacement.

ACHIEVEMENTS RELATED TO THE STATE OF THE ART

Evaluation of biofilters' performance parameters

Even though it has been widely reported that the packing material is one of the main operational parameters, as well as the main one influencing the investment and operational costs in biofilters (Prado *et al.*, 2009), there is still controversy on which are the most important requirements that a packing material must fulfil if long-term, stable performance are desired.

In order to clarify this and to value the potential use of SBC, SBAC or other carbons, a complete characterization of several organic and inorganic packing materials as well as SBC and CAC carbons was performed (Dorado *et al.*, 2010a). Afterwards, the performance of four biofilters packed with organic packing materials (coconut fibre, peat, pine leaves and compost) was assessed and was later compared to that of SBC. In addition, the four biofilters were characterized for typical parameters such as bed porosity or water content before inoculation. It must be stressed that the pore size, the specific surface area, the material density and the C, H, N, S and P content are inherent to the material and, therefore, comparable to other materials characterized in the literature (Bohn, 1996). It was shown that the pore size showed a high correlation with the toluene adsorption capacity determined by breakthrough curves performed in the biofilters. Adsorption tests performed at an inlet toluene load of 100 g/m^3h showed that the bed was saturated after 6, 15, 25 and 34 minutes for the coconut fibre, peat, pine leaves and compost biofilters, respectively. In any case, it is important to mention that high surface areas, with deep pores as those exhibited by compost, may be counteracted due to the biomass growth over the surface of the packing material, thus reducing the sorption capabilities to buffer transient inlet loads. Also, highly mineralized materials such as compost and peat showed low organic matter contents compared to other usual biofilter packing materials, which may lead to a higher resistance to long-term deterioration. Compost seemed *a priori* the most suitable material out of those tested in terms of physical and chemical characteristics. However, packing materials suitability is highly dependant on the biofiltration scenario, which is mostly related to the characteristics of the off-gases to be treated (Dorado *et al.*, 2010a).

A previous study performed under similar conditions as the ones expected to be employed in the current study showed that unacceptably long start-up periods could happen when the biofilters were inoculated with sludge from an urban wastewater treatment plant (Maestre *et al.*, 2007). This same effect has been found in a number of previous studies (Veiga and Kennes, 2001 and Prado *et al.*, 2005), as it is a typical operational problem both in biofilters and in biotrickling filters. In order to shorten this start-up time, an alternative start-up strategy was applied. In this study, instead of inoculating the four biofilters with a specific microbial culture able to perform the biodegradation process, they were packed with filter bed materials that had been previously employed in another study. The latter were, accordingly, already colonized by toluene-degrading strains. The materials selected were pine leaves, a mixture of compost and pine leaves in a proportion 3:1 (v:v), coconut fibre and peat. This would allow to characterize the performance of these packing materials, serving as a first step before the operation with SBC. Subsequently, the performance of all systems was compared.

Effect of the EBRT

The empty bed residence time (EBRT) is one of the main operational parameters in both biological and physico-chemical gas-treating systems. Previous works have highlighted the influence of the EBRT on the performance of VOC-degrading bioreactors (Deshusses and Hamer, 1993 and Zhou *et al.*, 1998). According to the literature, it can be hypothesised that bioreactors packed with different materials can show different responses to EBRT variations. However, there is still controversy on the degree of influence of variations in the EBRT on the performance of bioreactors packed with organic materials. In order to clarify this, four conventional biofilters, each packed with one of the four abovementioned organic materials were operated at EBRT values ranging from 90 to 5 s, corresponding to inlet air flow rates from 120 to 2,000 L/h. These results will be of special value for the industrial implementation of bioreactors, allowing for a more precise scale-up of the systems and, hence, reducing costs. Although there are a few studies regarding the biodegradation of VOCs at low EBRT (Popov *et al.*, 2005), no previous works have been found in the literature on the removal of toluene in bioreactors at EBRT values as low as 5 s.

Each EBRT change was performed after steady-state conditions had been reached and kept for at least three hours to ensure consistent results. A toluene inlet concentration of $1.1 \pm 0.3\,g/m^3$ was set during all the experiment. The reactors were periodically fed with 100 mL of a standard nutrient solution twice per day. It was observed that a fungal population remained dominant inside the reactor during all the study. Figure 14.1 shows that a decrease in the removal efficiency was found in all four reactors as the EBRT decreased, even though slight differences in their performance could be observed. However, removal efficiencies above 75% were found in all systems when operated at EBRT values above 60 s. In the reactor packed with pine leaves (Figure 14.1a), only a small effect of the EBRT on the removal efficiency was found when operating at EBRT values between 20 and 90 s. Under such conditions, the removal efficiency remained between 75 and 85%. However, at EBRT values as low as 5 s the removal efficiency dropped to 35%. The biofilter packed with a mixture of compost and pine leaves (Figure 14.1b) decreased its performance from 77% at an EBRT of 60 s to below 20% at an EBRT of 5 s.

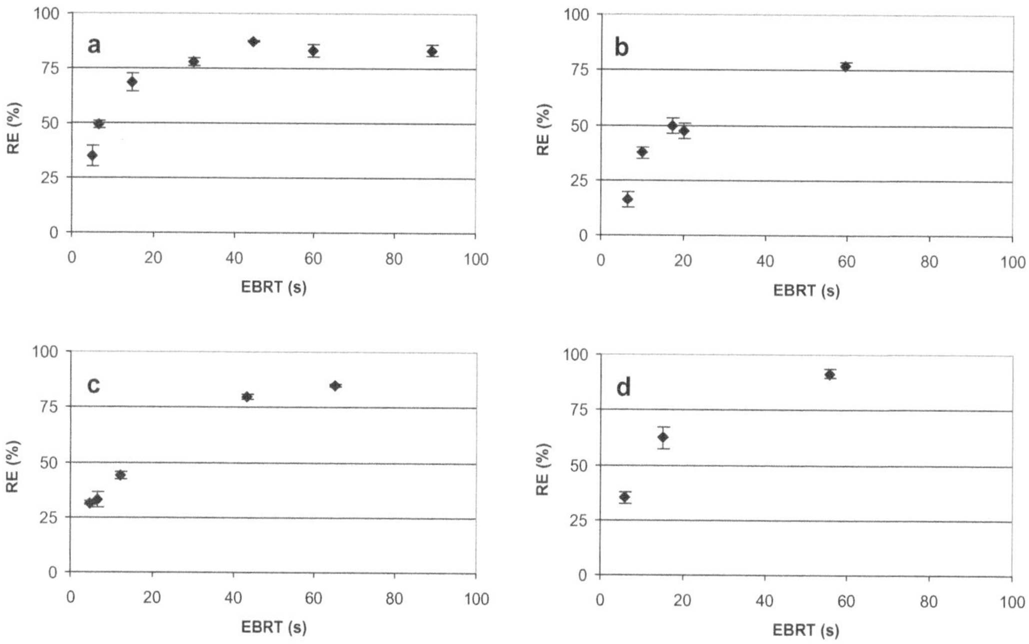

Figure 14.1 Influence of the EBRT on the toluene removal efficiency of four biofilters packed with different organic materials. (a) pine leaves biofilter; (b) compost + pine leaves biofilter; (c) coconut fibre biofilter; (d) peat biofilter

Both the biofilters packed with coconut fibre and with peat (Figures 14.1c and 14.1d, respectively) showed a similar behaviour to that found in the biofilter packed with pine leaves. Except for the reactor packed with compost plus pine leaves, all other reactors showed removal efficiencies between 30 and 35% at EBRTs as low as 5 s, showing that the elimination of toluene at EBRT values around 5 s can be performed with moderate efficiencies.

These results will allow for an optimal scale-up when the systems are to be applied at industrial scale, reducing costs and maximizing efficiency. From these results it was hypothesized that the design of biofilters with SBC or SBAC for toluene degradation would require EBRTs above 40 s to achieve removal efficiencies above 80%. This was subsequently checked in a biofilter packed with SBC (see next sections).

Influence of nutrients supply

The supply of a nutrient solution plays an important role in the biofilter performance, providing the adequate moisture level and the nutrient content needed by the microorganisms. Although it has been stated that organic packing materials usually offer the necessary nutrients to maintain a proper microbial population, an extra-nutrient addition has been demonstrated to be needed when high pollutant loads are being fed (Devinny *et al.*, 1999). Also, this addition is a key factor for pH control in bioreactors, providing the necessary wash out of by-products. However, considering that a significant fraction of the nutrients of the packing material may be in a non-biologically available form, specific experiments need to be undertaken in order to assess the need of an external nutrient supply to the bioreactors. Even though a number of studies have been performed with this purpose (Zhu *et al.*, 1996; Deshusses and Cox 1999; Jorio *et al.*, 2000; Moe and Irvine 2001 and Prado *et al.*, 2002), this is the first to date to evaluate the need of nutrient supply in reactors packed with coconut fibre, which may be a reliable and cheap option. In the present study, the reactor packed with coconut fibre was operated during 100 days under different watering conditions in order to determine the effect of the addition of a nutrient solution on its performance. The EBRT was kept at a constant value of 60 s during all the study. Initially, the reactor was fed a nutrient solution under the conditions previously described. The toluene load was set to 107 ± 14 g/m^3 h during the first 25 days operation. An elimination capacity of 37 ± 11 g/m^3 h was found during this stage.

As Figure 14.2 shows, the removal efficiency did not overcome 50%. On day 26 of operation, the toluene load was decreased to a value of 28 ± 6 g/m^3 h with the purpose of increasing the removal efficiency prior to the nutrient limitation stage. Immediately, an increase in the removal efficiency was observed, up to values around 90%. pH remained slightly below 2 during this stage. Once a steady-state had been reached on day 39, the nutrient solution was replaced with distilled water. Thus, the biofilter did not receive any external nutrients between days 39 and 88. The toluene load during this stage was 28 ± 7 g/m^3 h. Figure 14.2 shows that the toluene removal efficiency remained above 90% during more than one month, dropping sharply after day 80. A progressive increase of the leachate pH was observed, which was related to a decrease in the production of acidic metabolites. Even though an increase in pH could lead to the appearance of bacteria overgrowing the fungal population, which, in turn, would decrease the performance of the biofilter, in the present case the biofilter was predominantly fungal-based during all the study.

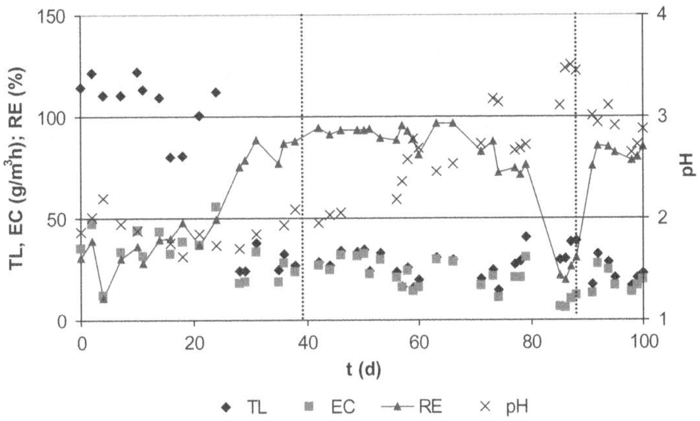

Figure 14.2 Performance of the biofilter under different toluene inlet loads and nutrient supply conditions. The dotted lines point the days when the liquid solution supplied was switched. TL: Toluene load. EC: Elimination capacity. RE: Removal efficiency

A previous work performed under similar conditions (Maestre *et al.*, 2007) proved that the performance of a series of fungal bioreactors packed with organic media were very stable at pH values up to 4–5. Hence, it can be concluded that the decrease in the removal efficiency was not due to the pH increase. Subsequently, in order to assure that it was due to a lack of nutrients, on day 88 the nutrient solution feeding was restored. Figure 14.2 shows that this led to an immediate increase in the removal efficiency, which reached values comparable to the original ones just after a few days. A concomitant pH drop was then observed. The results presented herein point out that the operation of toluene-degrading biofilters colonized with fungi may be carried out temporarily without any nutrient addition, even though a N-containing nutrient solution must be added when no other nitrogen source is available to keep reactor performance.

Similarly to the results shown herein, the biofilters filled with SBC and CAC showed a similar performance in terms of nutrients requirements both indicating that the nutrient content of SBC could contribute to the growth and the maintenance of the biomass immobilized, even if external nutrients addition may improve biofilter performance.

Effect of the watering rate

As the biodegradation processes require of high moisture content in the filter bed material, it is a common practice in biofilters to supply water in an intermittent manner (Devinny *et al.*, 1999). This commonly leads to an increase in the pressure drop of the bioreactor. In the present section, an assessment of the pressure drop variations as a function of different operational parameters is given. In order to determine this, a specific lab-scale setup was constructed and operated (Dorado *et al.*, 2010a, 2010b). Several packing materials, namely compost, coconut fibre, pine leaves, peat and heather, polyurethane foam, immature coal, an innovative material composed of an inner core of clay externally covered by compost, CAC and SBC were tested. The study was performed at seven different air flow rates, five moisture content values and, when possible, different material porosities (Dorado *et al.*, 2010b). The air flow rates selected were so as to achieve EBRT values between 5 and 40 s, considered typical in actual bioreactors. Figure 14.3 shows the results obtained, respectively, for coconut fibre, compost, polyurethane foam and SBC. It can be seen that, in all four cases, the pressure drop increases when the air flow rate or the water content increase as well or when the porosity decreases. In organic materials, the results revealed small differences for all packing materials at the various possible porosities and, in general, the pressure drop was well below 250 mm wc m-1/m in all conditions. As an exception, peat and heather showed a higher head loss value (up to 350 mm wc m-1/m). In general, organic materials led to lower maximum pressure drop values compared to inorganic packing materials. The lowest pressure drop among the non-organic and organic packing materials was found in the CAC (below 100 wc m-1/m), while SBC showed the highest pressure drop detected among the organic and non organic support media because of its small particle size.

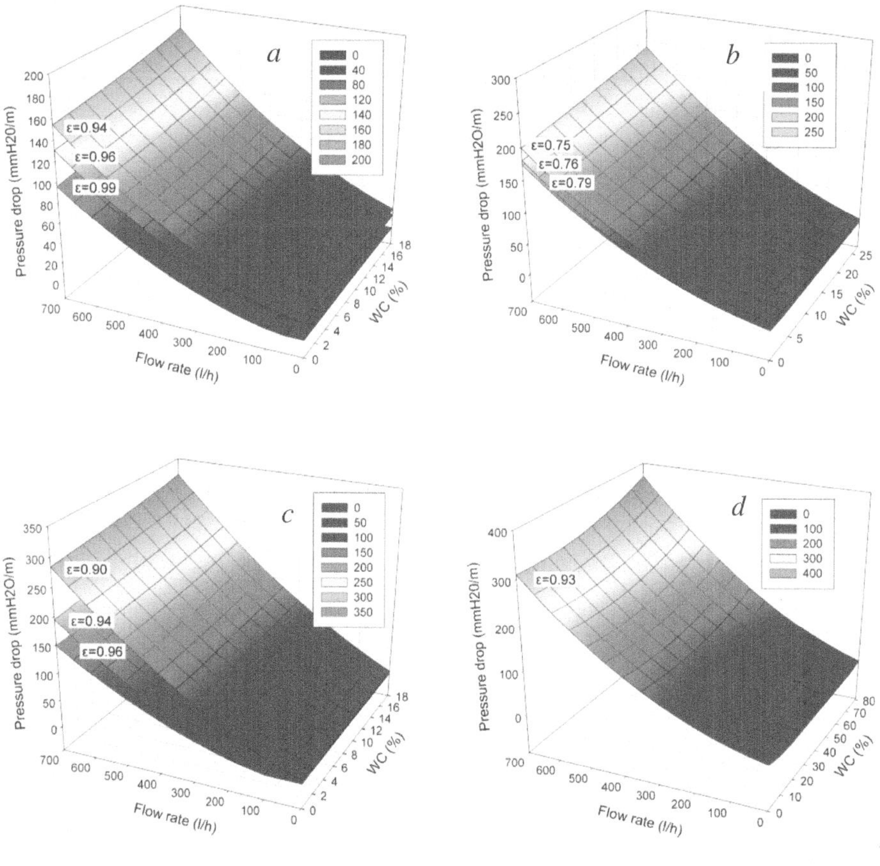

Figure 14.3 Influence of the operating parameters on the pressure drop in a bioreactor packed with (a) coconut fibre; (b) compost; (c) polyurethane foam; (d) SBC

It is important to mention that pressure drop in a fixed bed has been already described by different semi-empirical mathematic expressions. Most studies use Ergun's expression (Ergun and Orning, 1952), shown in Equation 1, to describe the pressure drop.

$$\frac{\Delta P}{H} = a \cdot \frac{\mu \cdot v_0}{d_\varphi^{\,2}} \cdot \frac{(1-\varepsilon)^2}{\varepsilon^3} + b \cdot \frac{\rho \cdot v_0^{\,2}}{d_\varphi} \cdot \frac{(1-\varepsilon)}{\varepsilon^3} \qquad \text{Eq. 1}$$

In Equation 1, ΔP is the pressure drop in Pa; H is the filter bed height in m; μ is the air viscosity in Pa·s; ν_0 is the superficial air velocity in m/s; ε is the filter bed porosity; d_φ is the spherical equivalent diameter of the particle in m and a and b are, respectively, the first and the second parameters of Ergun's equation.

These parameters are related to the friction factor. Some authors have satisfactorily adjusted experimental data to Ergun's equation by adapting these parameters by means of a correction factor (Delhoménie *et al.*, 2003). Some other authors have developed specific relationships, due to the heterogeneity of the material and to the inherent difficulty to model the pressure drop using Ergun's equation (Comiti and Renaud, 1989).

In the present work, parameters a and b from Ergun's equation were adjusted as a function of the material, the porosity and the water content (Dorado *et al.*, 2010b). The pressure drop ($\Delta P/H$) was adjusted by linear regression as a function of the gas superficial velocity. In all cases, the correlation coefficient was above 0.99, showing a clear linear trend between the operational parameters and the pressure drop. The experimental study incorporates the effect of the water content in the pressure drop, even though this parameter is not present in Ergun's equation. Accordingly, this allows establishing a modified version of Ergun's equation which incorporates the effect of the water content on the pressure drop for different packing bed materials. This is a key factor to properly estimate operating costs during biofilters operation (Dorado *et al.*, 2010b).

Evaluation of SBC as filter bed material in a biofilter

After proving that SBC (and, accordingly, SBAC) showed acceptable characteristics in terms of several properties and provided an acceptable pressure drop in a wide range of operating conditions, the behaviour of this material as packing bed was assessed in a biofilter treating a mixture of six organic volatile compounds (VOCs) such as hexanal, butyric acid, limonene, methylisobutylketone (MIBK), dimethylsulfide (DMS) and toluene. The experimental time was divided in 2 phases. In the first phase, the inlet concentration was 6 ± 2 ppmv per VOC, corresponding an inlet loading rate of 3.2 ± 1.0 g/m^3 h, while in the second phase the VOCs inlet concentration was doubled to 10 ± 1 ppmv, corresponding to an inlet loading rate of 5.6 ± 0.6 g/m^3 h. The biofilter was packed with a mixture of coconut fibre plus SBC from a waste water treatment plant (1:1, v:v), provided by the Department of Civil and Environmental Engineering of the Imperial College (London, U.K.). According to the pressure drop analysis and packing materials characterization, the coconut fibre was added as bulking agent in order to decrease the pressure drop in the biofilter (McNevin and Barford, 2001 and Burgess *et al.*, 2001), and aided to maintain a suitable water content. Growth and immobilization of biomass on SBC were evaluated inoculating the lab-scale plant with activated sludge and studying the abatement of VOCs in the reactor, during 95 days of operation (Figure 14.4). The gas flow rate was kept at 425 L/h (EBRT of 25 s). The bioreactor was sprinkled once a day with tap water.

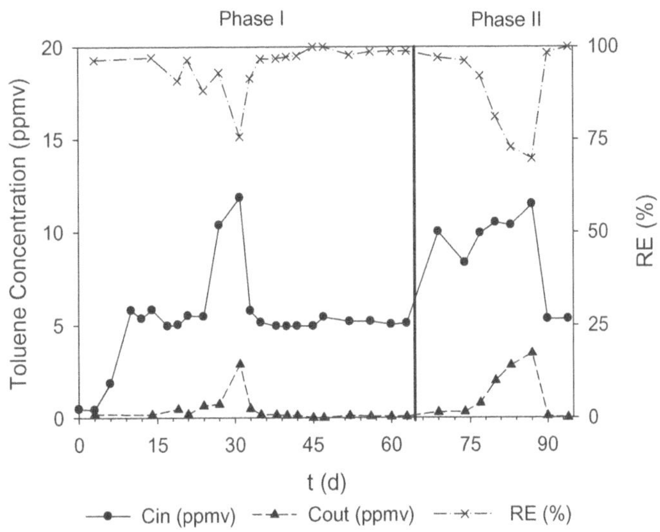

Figure 14.4 Evolution of toluene inlet and outlet concentration and toluene removal efficiency (RE) in the SBC-coconut fiber biofilter

In Figure 14.4, the operation and performance of the biofilter in terms of toluene inlet and outlet concentration, and toluene removal efficiency (RE) is represented. The first 10 days correspond to the biofilter start-up phase, in which SBC was basically adsorbing toluene and the immobilized biomass was being acclimatized. Toluene removal efficiencies above 90% were achieved from the first day, corresponding to an EC between 2.5 and 3 g/m^3 h.

Results demonstrate that SBC was able to develop and maintain an active biomass on the surface of the material after a short time of acclimatization, due especially to the high surface area and to the high water holding capacity. In the first phase, while the biomass was adapting to the operating conditions of the reactor, toluene was being adsorbed on the material. Moreover, the presence of nutrients in the SBC could contribute to the growth and the maintenance of the biomass immobilized even if this was limited during the treatment of higher loads of toluene. On day 69, the inlet toluene concentration was increased from 5 to 10 ppmv (corresponding to an inlet load increase from 3 to 5 g/m^3h. and toluene RE was practically kept at 90% during the following days. As soon as the inlet toluene concentration was decreased again to 5 ppmv on day 88 of operation, high RE were reached again. According to the toluene sorption capacity (Dorado *et al.*, 2010a), the material is able to adsorb such load increase during an interval of time between 1 and 4 days depending on the degree of surface covered by water and biomass. Thus, results demonstrate some capacity of the biofilter packed with SBC-coconut fibre to smooth inlet load changes. The RE reported in the analysed operation time shows that SBC is an excellent support media to fix, grow and keep active the biomass responsible to degrade toluene as an example of abatement of a commonly VOC present in many gas waste flows. Removal efficiencies for all six VOCs were maintained above 70% during the whole experimental time (results not shown).

Model development, calibration and validation

A dynamic model to simulate toluene abatement by bacterial and fungal biofilters was developed, calibrated and validated through a set of different experimental conditions for biofilters with switching populations from bacteria to fungi (Dorado *et al.*, 2008). Simple and complex models have been employed to simulate toluene biofiltration under both steady-state and dynamic operating conditions (Abumaizar *et al.*, 1997; Metris *et al.*, 2001; Ottengraf and Van der Oever, 1983 and Zarook *et al.*, 1997). In all modeling works reported in the literature toluene removal in biofilters is based on bacterial degradation activity without taking into account fungal operation. The approach developed herein allows to model the biofilter considering the variation of the biological activity due to the development of different populations of microorganisms, approach that has not been previously described in the literature.

The model was able to satisfactorily predict different operation periods, including bacteria, transition from bacteria to fungi, and fungi based operation by means of a small number of parameters. Results show that a highly complex model is not necessary to properly describe the performance of a biofilter. However, it is required that the model parameters are determined for each situation, instead of extracted from literature, as they might vary significantly. Moreover, the physico-chemical model was checked by model predictions in the operation where biological degradation was not the predominant process. Estimation of kinetic parameters corroborated that biomass evolved from bacteria to fungi in both biofilters in concordance to pH changes reported along the entire experiment and microscopic observation. Kinetic parameters confirmed that fungi provide a better capacity to degrade toluene from gas phase. Moreover, a higher affinity was observed between fungi and the pollutant in comparison to pollutant and bacteria consortium. Thus, determination of volumetric maximum growth rate and semi saturation constant for both biofilters with their corresponding confidence interval pointed out that biofilters colonized by fungi showed a better performance than those colonized by bacteria. The model calibration was checked by a novel procedure in gas biofilters modeling based on the Fisher Information Matrix (FIM). Model validation was verified using a rigorous statistical test. Deviations on model predictions are explained by biomass modification in the bioreactors obtaining better results at the end of each period due to the pseudo biomass steady-state achieved.

In order to determine the influence of model parameters on model predictions and to have a practical guide to what parameters are more important to determine from a practical point of view, a sensitivity analysis was performed for each parameter, including all kinetic, stoichiometric and physicochemical parameters. The analysis was performed by examining the effects of parameter variation over the elimination capacity (EC). Sensitivity was assessed by increasing and decreasing by 10% the values of the parameters and comparing the relative change of the state variables to a relative change of the value of the parameter according to Equation 2. Results obtained from sensitivity analyses are shown in Table 14.1.

$$\text{sensitivity} = \frac{\dfrac{\Delta V}{V_d}}{\left|\dfrac{\Delta P}{P_d}\right|} \qquad \text{Eq. 2}$$

where ΔV is the difference between the simulated variable under the new conditions and the value of the variable in the default conditions (V_d). Similarly, ΔP represents the difference between the value of the parameter at the $\pm 10\%$ change and the value of the default parameter (P_d).

The major impact was obtained for physic-chemical parameters such as the specific surface area, the gas/biofilm partition coefficient and the diffusion coefficient in the biofilm. On the other hand, dispersion coefficient and bed porosity present a nil influence on model predictions. These results are consistent with others found in the literature (Baquerizo *et al.*, 2005 and Deshusses *et al.*, 1995). Thus, specific surface area, the gas/biofilm partition coefficient and the diffusion coefficient in the biofilm are the most influencing parameters in the model and should be specifically determined in all studies performed with the developed model.

Table 14.1 Sensitivity analysis results for outlet concentration and elimination capacity of the biofilter for the all model parameters in the general model

Parameter	Definition	Δ	Sensitivity C_g^{out}	Sensitivity EC
C_g^{in}	Pollutant inlet concentration	10%	1.57	0.79
		−10%	−1.49	−0.83
Q_g	Inlet air flow rate	10%	1.51	0.40
		−10%	−1.56	−0.48
EBRT	Empty bed residence time	10%	−1.49	−0.44
		−10%	1.67	0.45
a	Specific surface area	10%	−1.47	0.52
		−10%	1.66	−0.59
H_P	Pollutant gas/biofilm partition coefficient	10%	1.00	−0.36
		−10%	−1.08	0.38
$D_{b,P}$	Pollutant diffusion coefficient in the biofilm	10%	−0.78	0.28
		−10%	0.88	−0.31
μ^*	Maximum volumetric rate	10%	−0.70	0.25
		−10%	0.80	−0.29
$K_{S,P}$	Pollutant half saturation constant	10%	0.21	−0.07
		−10%	−0.22	0.08
δ	Biofilm thickness	10%	0.10	−0.04
		−10%	−0.05	0.01
k_g	Mass transfer coefficient	10%	0.02	−0.01
		−10%	−0.02	0.00
$K_{S,O}$	Oxygen half saturation constant	10%	0.00	0.00
		−10%	0.00	0.00
$D_{g,P}$	Pollutant dispersion coefficient in the air	10%	0.00	0.00
		−10%	−0.01	0.00
ε	Clean bed porosity	10%	0.00	0.00
		−10%	−0.01	0.00

IMPACT OF THE TECHNOLOGY ON ITS INDUSTRY OR RESEARCH SECTOR

Social, environmental and industrial benefits of the use of sludge-based carbon as filter bed in a biofilter

As already exposed, the production of activated sludge is one of the main problems generated by the activity of solid waste and waste water treatment plants. Only in the United Kingdom, the yearly generation of this sludge is above 2 billion kg, which increases year after year. This leads to issues associated to the management of this material. Currently, processes such as sludge digestion by ozonation and thermophilic microbial digestion have proved their efficiency in reducing sludge generation. However, recalcitrant compounds are accumulated in the remnant sludge, reason why these materials can no longer be biologically treated. Their conversion in activated carbon and posterior utilization as packing bed in gas-treating biofilters is a promising alternative to their disposal. This alternative comprises two benefits which make it especially interesting both from an industrial and an environmental point of view:

1) It means an efficient and environmental-friendly way of managing the sludge generated in solid waste or waste water treatment facilities, valorising a material sludge with no economic interest and converting it into a product that can be employed in a wide series of applications.
2) A material that can be efficiently employed as packing bed in bioreactors is produced at a relatively low cost. The present work has proved that sludge-based activated carbon can serve as an efficient and stable packing material in conventional biofilters, leading to relatively low pressure drop values and serving at the same time as support for the microorganisms and as buffer, eventually helping to control shock loads.

From a social point of view, the use of bioreactors for waste gases treatment means an effective way of controlling gaseous emissions from a wide variety of industries, reducing and even eliminating their pollutant load and the odour associated to them. During past decades, the social concern regarding gaseous toxic and odorous emissions has significantly increased. Only in Spain, more than 18,000 industries were denounced for this reason between 2007 and 2008. More than 51% of the news published in press during that period regarding environmental issues were referred to conflicts due to the emission of odours from industrial facilities. Also, around 78% of the wastewater treatment plants operating in Spain were denounced at least once in 2007 for reasons associated to the emission of odours (data supplied by a specialised odour-treating company). Currently, it can be affirmed that around 12% of the operating costs of a wastewater treatment plant are due to the treatment of odours generated in the same facility. If costs due to fines, specific treatments, activity stops, etc. due to problems generated by odours are added, it results that more that 25% of the operating cost of wastewater treatment plants are due to odour management.

The increase in the social concern regarding environmental issues has lead to a concomitant rise in the legislative pressure. In the last three decades, the number of laws controlling gaseous emissions in Spain has risen six-fold. Currently, the European Union is in the process of developing a law aimed at controlling the emission of odours in industrial sources. By the time that law is approved, a technology that can handle the treatment of such emissions with a high efficiency, robustness and low costs will be needed. The results contained in the present work show that biofiltration may very well be that technology, at least in the cases where the physico-chemical conditions of the gas current allow for the development of a stable microbial community. Among the different packing materials that can be employed in these systems, SBC has proven to be an especially adequate one, due to its mechanical and chemical properties, its buffer capacity and its low cost. Many industries already have, or can have at a relatively low cost, the tools to produce this carbon from the sludge produced in their own facilities. In this case, two problems can be solved at the same time: the disposal of the sludge and the purchase of the packing material for the biological reactors. However, a series of technical and scientific questions have to be answered in advance in order to optimize their use, economically and technically.

In a first stage of the present study, the performance of a series of conventional biofilters packed with typically used organic bed materials and employed for the removal of a model volatile organic compound (toluene) was assessed during pseudo-stationary and transient conditions. Breakthrough values for the removal efficiency as a function of the empty bed residence time were established for all four packing materials. Moreover, the influence of the watering rate and of the nutrient supply was assessed. These results will allow for a most precise scale-up of industrial bioreactors, with a consequent decrease in the investment and operating costs. In this sense, one of the main operating parameters is the pressure drop, which may lead to increased maintenance costs after just a few months of operation. The present work has allowed for the characterization of the pressure drop as a function of the packing material, bed porosity, gas flow rate and water content. Results suggest that a standardized method for pressure drop control can be developed just by regularly controlling these operating parameters.

Subsequently, the performance of a biofilter packed with a mixture of coconut fiber and SBC from a waste water treatment plant (1:1, v:v) and treating a mixture of six volatile organic compounds was studied. The results show that, at an empty bed residence time of 25 s, a removal efficiency above 70% could be warranted for all the pollutants. This proves that SBC can serve as an adequate packing material for gas treatment in bioreactors. The environmental and industrial implications of this have been highlighted in the previous sections.

Additionally to experimental results, a theoretical model based on mass balances for pollutants in a biofilter was successfully developed, calibrated and validated. The biofiltration model revealed as a promising tool for further design of biofilters and to intensively test different biofiltration scenarios at low cost compared to experimental testing of biofiltration units. Modelling results additionally showed that the specific surface area of the packing material was the parameter that largely influenced model outputs. Thus, experimental determination of sludge-based carbons specific surface area appeared as a key factor for successful model usage. Apart from the improvement of the design of industrial biofilters, the programming environment and interface of the model developed may allow for training of plant operators from industrial facilities at low cost in short periods of time. This is particularly important since the expertise of plant operators in this type of emerging technologies such as biofiltration is significantly limited.

Summarizing, the use of bioreactors packed with SBC means a reliable and environmental-friendly alternative to traditional, physico-chemical technologies for waste gas remediation. The results obtained will allow for the design and implementation of cheaper, more reliable and more robust bioreactors.

Economical benefits of the use of sludge-based carbon as filter bed in a biofilter

Together with the already described social and environmental benefits, the economical advantages of employing sludge-based carbons as packing material in bioreactors are expected to significantly contribute to generate an impact in the sector. Two main economical benefits can be distinguished here: the ones related to the disposal of the waste water treatment facilities sludge, for which no optimal alternative exists, and those related to the savings in the purchase of the bioreactor packing material.

With the purpose of assessing the economical viability of using activated carbon as filter bed in bioreactors, a specific cost-assessment model which includes the main investment, operating and maintenance costs of a conventional biofilter was developed. Specifications on this model can be found in Prado *et al.* (2009). The total investment costs have been calculated as the sum of the initial site preparation costs, the packing and support materials costs, the equipment costs, the piping costs, the electrical costs, the equipment installation costs, the engineering design costs, the liner costs and the miscellaneous costs. The annual operating costs were calculated as the sum of the electricity and water annual consumption costs, the labor costs and the overhead costs. Among the medium replacement costs, two actions were considered: the disposal of the old packing material and the addition of the new one. All other eventual costs have been considered negligible compared to these. Although a very simple bioreactor was selected as model reactor, the protocol can be applied to different scenarios (as, e.g., biotrickling filters) after few modifications. No nutrient supply was considered for the economical evaluation. The protocol presented herein was applied to a series of biofilters packed with different filter bed materials and operating under the same conditions. A gas flow rate of 20,000 m^3/h and an EBRT of 60 s were chosen for the biofilters, which were considered to be operating for 15 years. The EBRT value of 60 s was selected as previous studies showed that removal efficiencies above 75% may be assured at that EBRT, irrespective of the packing material. Seven packing materials were selected for the study: compost, agricultural residues, a mixture of pig manure and sawdust, Biosorbens™ and three different types of activated carbon (CAC, SBAC and SBC). The cost and the estimated durability of each material are shown in Table 14.2. All costs presented herein are estimations for Spain, year 2008. Due to the inherent difficulty to quantify many of the parameters included in the protocol, an error around ±20 % was estimated for costs calculated in the present section.

Table 14.2 Average cost and estimated durability of all seven packing materials considered

Material	Cost (€/m^3)	Estimated durability (years)
Compost	25	2
Agricultural residues	75	1
Pig manure + sawdust	50	1
Biosorbens™	550	15
Commercial activated carbon	475	10
Sludge-based activated carbon	120*	10
Sludge-based non-activated carbon	80*	5

**Values calculated considering a market cost of 600 €/Tm for SBAC and 400 €/Tm for SBC and a carbon density of 0.2 Kg/L in both cases.*

Table 14.3 summarizes the main costs implied in all seven cases. It can be seen that the use of SBC represents an economically viable alternative to other inert packing materials, as Biosorbens™ or commercial activated carbon. Total annualized costs of 43,360 and 42,910 €/year have been estimated for SBAC and SBC, respectively, while, in the case of CAC, a total annualized cost of 79,710 €/year results from the application of the model. This reduction in the costs is a consequence of the much lower purchase price of sludge-based carbons, which, in turn, results of their reduced production costs. The investment cost for this average-sized system ranges from around 56,000 € when selecting compost as packing material to around 380,000 €, if Biosorbens™ is the chosen bed. Among these, biofilters packed with sludge based carbons fit in the medium cost range (around 115,000 and 90,000 € for activated and non-activated carbon-based biofilters, respectively). However, the large durability of these materials causes the annualized medium replacement costs to be low in comparison to the others (4,200 and 6,600 €/year). Only Biosorbens™ has an estimated durability of the same range as the biofilter total operating time, reason why no medium replacement costs need to be considered, which implies lower annualized medium replacement costs. Regarding operating costs, as all the scenarios contemplate the same air flow rate and empty bed residence time, no differences are found. Overall, only compost is a better option from an economical point of view. However, as previously stated, the use of organic packing materials as compost as packing bed in biofilters may lead to operational problems which may decrease performance. Accordingly, there is evidence that SBC may be a promising alternative to traditional packing materials both from an operational and an economical point of view. This is expected to lead to a clear impact in the biofiltration market.

Table 14.3 Investment, operating and medium replacement costs estimated for a series of average-size conventional biofilters packed with different packing materials, with Q = 20 000 m^3/h, Vb = 400 m^3, EBRT = 60 s

COST	BF1	BF2	BF3	BF4	BF5	BF6	BF7
Total investment costs (€)	56,500	87,260	71,890	379,570	333,420	114,950	90,340
Annualized investment costs (€/year)	6,600	10,190	8,400	44,350	38,950	13,430	10,550
Annual operating costs (€/year)	19,340	19,340	19,340	19,340	19,340	19,340	19,340
Annualized operating costs (€/year)	25,750	25,750	25,750	25,750	25,750	25,750	25,750
Packing material replacement costs (€/action)	16,640	36,640	26,640	0	196,640	54,640	38,640
Annualized packing material replacement costs (€/year)	9,820	46,460	33,780	0	15,010	4,170	6,600
Total annualized costs (€/year)	42,170	82,400	67,930	70,100	79,710	43,360	42,910

BF1: Compost biofilter. BF2: Agricultural residues biofilter. BF3: Pig manure + sawdust biofilter. BF4: BiosorbensTM biofilter. BF5: Commercial activated carbon biofilter. BF6: Sludge-based activated carbon biofilter. BF7: Sludge-based non-activated carbon biofilter.

REFERENCES

Abumaizar R., Smith E. H. and Kocher W. (1997). Analytical model of dual-media biofilter for removal of organic air pollutants. *J. Environ. Eng.*, **123**, 606–614.

APHA. (1995). Standard methods for the examination of water and wastewater. 19th Ed. Washington DC, USA: American Publishers Health Association.

Baeza J. A., Gabriel D., Béjar J. and Lafuente J. (2002). A distributed control system based on agent architecture for wastewater treatment. *Computer-Aided Civil and Infrastructure Engineering*, **17**, 93–103.

Baquerizo G., Maestre J. P., Sakuma T., Deshusses M. A., Gamisans X., Gabriel D. and Lafuente J. (2005). A detailed model of a biofilter for ammonia removal: Model parameters analysis and model validation. *Chem. Eng. J.*, **113**, 205–214.

Bohn H. L. (1996). Biofilter media. In: 'Proceedings of the Air & Waste Management Association Annual Meeting & Exhibition'. Nashville (USA).

Burgess J. E., Parsons S. A. and Stuetz R. M. (2001). Developments in odour control and waste gas treatment biotechnology: A review. *Biotechnol. Adv.*, **19**, 35–65.

Comiti J. and Renaud M. (1989). A new model for determining mean structure parameters of fixed beds from pressure drop measurements: application to beds packed with parallelelepipedal particles. *Chem. Eng. Sci.*, **44**, 1539–1545.

Delhomenie M. C., Bibeau L., Gendron J., Brzezinski R. and Heitz M. (2003). A study of clogging in a biofilter treating toluene vapors. *Chem. Eng. J.*, **94**, 211–222.

Deshusses M. A. and Cox H. H. J. (1999). A cost benefit approach to reactor sizing and nutrient supply for biotrickling filters for air pollution control. *Environ. Progr.*, **18**, 188–196.

Deshusses M. A. and, Hamer G. (1993). The removal of volatile ketone mixtures from air in biofilters. *Bioproc. Eng.*, **9**, 141–146.

Deshusses M. A., Hamer G. and Dunn I. J. (1995). Behaviour of biofilters for waste air biotreatment. 1. Dynamic model development. *Environ. Sci. Tecnol.*, **29**, 1048–1058.

Devinny J. S., Deshusses M. A. and Webster T. S. (1999). Biofiltration for air pollution control. CRC Lewis Publishers.

Dorado A. D., Baquerizo G., Maestre J. P., Gamisans X., Gabriel D. and Lafuente J. (2008). Modeling of a bacterial and fungal biofilter applied to toluene abatement: Kinetic parameters estimation and model validation. *Chem. Eng. J.*, **140**, 52–61.

Dorado A. D., Lafuente J., Gabriel D. and Gamisans X. (2010a). A comparative study based on physical characteristics of suitable packing materials in biofiltration. *Environ. Technol.*, **31**, 193–204.

Dorado A. D., Lafuente J., Gabriel D. and Gamisans X. (2010b). The role of water in the performance of biofilters: Parameterization of pressure drop and sorption capacities for common packing materials. *J. Hazard. Mat.*, **180**, 693–702.

Durham D. R., Marshall L. C., Miller J. G. and Chmurny A. B. (1994). Characterization of inorganic biocarriers that moderate system upsets during fixed-film biotreatment processes. *Appl. Env. Microbiol.*, **60**, 3329–3335.

Ergun S. and Orning A. A. (1952). Fluid flow through packed columns. *Chem. Eng. Progr.*, **48**, 89–94.

Jorio H., Bibeau L. and Heitz M. (2000). Biofiltration of air contaminated by styrene: effect of nitrogen supply, gas flow rate and inlet concentration. *Environ. Sci. Technol.*, **34**, 1764–1771.

Maestre J. P., Gamisans X., Gabriel D. and Lafuente J. (2007). Fungal biofilters for toluene biofiltration: Evaluation of the performance with four packing materials under different operating conditions. *Chemosphere*, **67**, 684–692.

McNevin D. and Barford J. (2001). Biofiltration as an odour abatement strategy. *Biochem. Eng. J.*, **5**, 231–242.

Metris A., Gerrard A. M., Cumming R. H., Weigner P. and Paca J. (2001). Modelling shock loadings and starvation in the biofiltration of toluene and xylene. *J. Chem. Technol. Biotechnol.*, **76**, 565–572.

Moe W. M. and Irvine R. L. (2001). Effect of nitrogen limitation on performance of toluene degrading biofilters. *Wat Res*, **35**, 1407–1414.

Ottengraf S. and Van der Oever H. C. (1983). Kinetics of organic compounds removal from waste gases with a biological filter. *Biotechnol. Bioeng.*, **25**, 3089–3102.

Popov V., Khomenkov V. and Zhukov V. (2005). Design, construction and long-term performance of novel type of industrial biotrickling filters for VOC and odour control. In 'Proceedings of the International Congress Biotechniques for Air Pollution Control'. Eds. Kennes C. and Veiga M.C. pp. 257–262. La Coruña, Spain.

Prado Ó. J., Gabriel D. and Lafuente J. (2009). Economical assessment of the design, construction and operation of open-bed biofilters for waste gas treatment. *J. Environ. Manage. ASCE*, **135**, 359–366.

Prado Ó. J., Mendoza J. A. and Veiga M. C., Kennes C. (2002). Optimization of nutrient supply in a downflow gas-phase biofilter packed with an inert carrier. *Appl. Microbiol. Biotechnol.*, **59**, 567–573.

Prado Ó. J., Veiga M. C. and Kennes C. (2005). Treatment of gas-phase methanol in conventional biofilters packed with lava rock. *Wat. Res.*, **39**, 2385–2393.

Veiga M. C. and Kennes C. (2001). Parameters affecting performance and modeling of biofilters treating alkylbenzene-polluted air. *Appl. Microbiol. Biotechnol.*, **55**, 254–258.

Zarook S. M., Shaikh A. A. and Ansar Z. (1997). Development, experimental validation and dynamic analysis of a general transient biofilter model. *Chem. Eng. Sci.*, **52**, 759–773.

Zhou Q., Huang Y. L., Tseng D. H., Shim H. and Yang S. T. (1998). A trickling fibrous-bed bioreactor for biofiltration of benzene in air. *J. Chem. Technol. Biotechnol.*, **73**, 359–368.

Zhu X., Rihn M. J., Suidan M. T., Kim B. J. and Kim B. R. (1996). The effect of nitrate on VOC removal in trickle bed biofilters. *Water Sci. Tech.*, **34**, 573–581.

IWA Publishing's authorised EU representative for General Product Safety Regulations is Diane D'Arras, 15 rue Duret, 75116 Paris, France, e-mail: safety@iwap.co.uk.

Printed and bound by CPI Group (UK) Ltd, Croydon, CR0 4YY
12/05/2026
02108879-0016